Toxin-Induced Models of Neurological Disorders

Toxin-Induced Models of Neurological Disorders

Edited by

Michael L. Woodruff
East Tennessee State University
Johnson City, Tennessee

and

Arthur J. Nonneman
Asbury College
Wilmore, Kentucky

Springer Science+Business Media, LLC

Library of Congress Cataloging-in-Publication Data

Toxin-induced models of neurological disorders / edited by Michael L. Woodruff and Arthur J. Nonneman.
p. cm.
Includes bibliographical references and index.

1. Neurotoxicology. 2. Nervous system--Degeneration--Animal models. I. Woodruff, Michael L. (Michael Lester) II. Nonneman, Arthur J.
[DNLM: 1. Nervous System Diseases--chemically induced. 2. Neurotoxins. 3. Disease Models, Animal. WL 100 T755 1994]
RC347.5.T69 1994
616.8'0471--dc20
DNLM/DLC
for Library of Congress 94-9808
CIP

ISBN 978-1-4899-1449-1 ISBN 978-1-4899-1447-7 (eBook)
DOI 10.1007/978-1-4899-1447-7

Originally published by Plenum Press, New York in 1994.
Softcover reprint of the hardcover 1st edition 1994

Contributors

G.W. Arendash
Department of Biology and Institute for Biomolecular Science
University of South Florida
Tampa, Florida 33620

Ronald H. Baisden
Department of Anatomy and Cell Biology
J.H. Quillen College of Medicine
East Tennessee State University
Johnson City, Tennessee 37614

S. Barone, Jr.
Man Tech Environmental Technology
P.O. Box 12313
2 Triangle Drive
Research Triangle Park, North Carolina 27709

David W. Cahill
Division of Neurological Surgery
Department of Surgery
University of South Florida College of Medicine
12901 Bruce B. Downs Boulevard, MDC-16
Tampa, Florida 33612

A.J. Dunn
Department of Pharmacology and Therapeutics
Louisiana State University Medical Center
Shreveport, Louisiana 71130

Stephen B. Dunnett
Department of Experimental Psychology
University of Cambridge
Downing Street

Cambridge CB2 3EB
England

J.D. Elsworth
Neurobehavior Laboratory and Departments of Psychiatry and Pharmacology
Yale University School of Medicine
333 Cedar Street
New Haven, Connecticut 06510

Dwaine F. Emerich
CytoTherapeutics, Inc.
2 Richmond Square
Providence, Rhode Island 02906

Ralph M. Garruto
Laboratory of Central Nervous System Studies
National Institutes of Health
Bethesda, Maryland 20892

MaryEllen Groccia-Ellison
Department of Psychology and Program in Neuroscience
Williams College
Williamstown, Massachusetts 01267

Bengt G. Henriksson
Department of Geriatric Medicine
Karolinska Institutet
Huddinge University Hospital
Huddinge, Sweden

Laura M. Juárez de Ku
Department of Biological Sciences
Bowling Green State University
Bowling Green, Ohio 43402

Lee A. Meserve
Department of Biological Sciences
Bowling Green State University
Bowling Green, Ohio 43402

Abdul H. Mohammed
Department of Geriatric Medicine
Karolinska Institutet
Huddinge University Hospital
Huddinge, Sweden

Arthur J. Nonneman
Department of Psychology
Asbury College
Wilmore, Kentucky 40390

C.W. Olanow
Departments of Neurology, Pharmacology, and Psychiatry
University of South Florida College of Medicine
Tampa, Florida 33620

Karen D. Opello
Department of Psychology
Rutgers University
New Brunswick, New Jersey 08903

William W. Pendlebury
Department of Pathology
University of Vermont College of Medicine
Burlington, Vermont 05405

Paul J. Reading
Department of Experimental Psychology
University of Cambridge
Downing Street
Cambridge CB2 3EB
England

D.E. Redmond, Jr.
Neurobehavior Laboratory and Departments of Psychiatry and Pharmacology
Yale University School of Medicine
333 Cedar Street
New Haven, Connecticut 06510

R.H. Roth
Neurobehavior Laboratory and Departments of Psychiatry and Pharmacology
Yale University School of Medicine
333 Cedar Street
New Haven, Connecticut 06510

Paul R. Sanberg
Division of Neurological Surgery
Department of Surgery
University of South Florida College of Medicine
12901 Bruce B. Downs Boulevard, MDC-16
Tampa, Florida 33612

G.J. Sengstock
Department of Biology and Institute for Biomolecular Sciences
University of South Florida
Tampa, Florida 33620

J.R. Sladek, Jr.
Department of Neuroscience
University of Health Sciences
The Chicago Medical School
3333 Green Bay Road
North Chicago, Illinois 60064

Paul R. Solomon
Department of Psychology and Program in Neuroscience
Williams College
Williamstown, Massachusetts 01267

Frédéric Souverbie
Department of Geriatric Medicine
Karolinska Institutet
Huddinge University Hospital
Huddinge, Sweden

Mark E. Stanton
Health Effects Research Laboratory
Division of Neurotoxicology (MD-74B)
United States Environmental Protection Agency
Research Triangle Park, North Carolina 27711

Michael J. Strong
Department of Clinical Neurological Sciences
The University of Western Ontario
London, Ontario, Canada N6A 5A5

Erik Sundström
Department of Geriatric Medicine
Karolinska Institutet
Huddinge University Hospital
Huddinge, Sweden

J.R. Taylor
Neurobehavior Laboratory and Departments of Psychiatry and Pharmocology
Yale University School of Medicine
333 Cedar Street
New Haven, Connecticut 06510

Thomas J. Walsh
Department of Psychology
Rutgers University
New Brunswick, New Jersey 08903

Michael L. Woodruff
Department of Anatomy and Cell Biology
J.H. Quillen College of Medicine
East Tennessee State University
Johnson City, Tennessee 37614

Robert A. Yokel
Division of Pharmacology and Experimental Therapeutics
College of Pharmacy and Graduate Center for Toxicology
University of Kentucky
Lexington, Kentucky 40536

Preface

This volume discusses and evaluates the use of neurotoxins to produce animal models of human neurodegenerative diseases. The models presented use neurotoxins to induce some or most of the pathophysiological changes (including behavioral) that characterize the most studied motor neuron disorders and cognitive disorders (dementia) in humans. Within the biomedical research community, there is little doubt about the utility of such models. But with increasing frequency, individuals concerned with animal rights question the validity of such models and argue that they actually produce no useful information for understanding or developing treatments for human disorders. Each of the chapters will address this issue, considering the utility, validity, generalizability, and limitations of the models presented.

Chapter 1 begins with a brief review of the ethics of animal use in neurobiological research, including a discussion of the criteria that may be used to evaluate animal models of human disease and extrapolate from the model to appropriate questions regarding humans. The limitations of such extrapolation are also discussed, with special consideration of issues specific to the use of neurotoxins. Chapter 2 extends this theme with further consideration of issues and strategies involved in developing neurotoxin-induced models of neurodegenerative disorders and assessing risk factors associated with neurotoxin exposure. It then narrows to evaluate the use of a model systems approach to explore neurotoxin-induced learning and memory deficits in animals as related to humans with dementia. Subsequent chapters continue this more specific approach taken in the latter part of Chapter 2.

The next six chapters present and evaluate several models of toxin-induced motor neuron degeneration and the disorders resulting from it. Discussion includes various approaches to modeling the same disorder (Parkinson's disease). Some of these models have been used for years; others are new but appear to have potential for enhancing our understanding of the pathophysiology of this disorder. The last four chapters evaluate the utility of four different neurotoxin-induced models of Alzheimer's disease.

It is our hope and belief that this volume will help to foster the refinement of our models of the most important neurodegenerative disorders in humans. If fully successful, such refinement should lead to greater understanding of the etiology and pathogenesis of these disorders and the development of effective therapeutic interventions.

Michael L. Woodruff
East Tennesee State University
Johnson City, Tennessee

Arthur J. Nonneman
Asbury College
Wilmore, Kentucky

Contents

Chapter 1

Animal Models and the Implications of Their Use

Arthur J. Nonneman and Michael L. Woodruff

Chapter 2

Strategies for Developing Animal Models of Neurotoxicant-Induced Neurodegenerative Disorders: Parallel Studies of Learning and Memory in Animals and Humans

Paul R. Solomon, MaryEllen Groccia-Ellison, Mark E. Stanton, and William W. Pendlebury

Chapter 3

Experimental Paradigms of Motor Neuron Degeneration

Michael J. Strong and Ralph M. Garruto

Chapter 4

6-Hydroxydopamine Lesions of Nigrostriatal Neurons as an Animal Model of Parkinson's Disease

Paul J. Reading and Stephen B. Dunnett

Chapter 5

MPTP-Treated Mice: A Useful Model for Parkinson's Disease?

Erik Sundström, Bengt G. Henriksson, Abdul H. Mohammed, and Frédéric Souverbie

Chapter 7

Intranigral Iron Infusion as a Model for Parkinson's Disease

G.W. Arendash, G.J. Sengstock, C.W. Olanow, S. Barone, Jr., and A.J. Dunn

Chapter 8

Excitotoxic Lesions of the Neostriatum as an Animal Model of Huntington's Disease

Dwaine F. Emerich, David W. Cahill, and Paul R. Sanberg

Chapter 9

The Use of AF64A (Ethylcholine Aziridinium Ion) to Model Alzheimer's Disease

Thomas J. Walsh and Karen D. Opello

Chapter 10

Perinatal Intake of Polychlorinated Biphenyl (PCB): Alteration of Endocrine and Neurochemical Parameters Related to Alzheimer's Disease

Laura M. Juárez de Ku and Lee A. Meserve

Chapter 11

Aluminum Exposure Produces Learning and Memory Deficits: A Model of Alzheimer's Disease

Robert A. Yokel

1

Animal Models and the Implications of Their Use

ARTHUR J. NONNEMAN and MICHAEL L. WOODRUFF

1. Criticisms of the Validity of Animal Models

The introductory chapter to any book concerning animal models of human diseases should not ignore the arguments that have been made against the use of animals in biomedical research. These arguments appear to have two principal components. The first component is an ethnical/moral concern for the rights of animals, which may be viewed as the more fundamental of the two. The second component seeks to dismiss the benefit to humans that has been derived from biomedical research using animals. Generally, a concern for animals underlies both arguments.

Acceptance of one of these arguments does not entail acceptance of the other. An extreme animal-rights activist might well agree that some medical cures have been developed from animal research yet still take the position that the rights of animals supersede the right of humans to use animal experimentation. This position is discussed in Section 1.1. Moreover, a person who is concerned about excessive government spending might be against biomedical research using animal models if he or she believed that no benefit could be derived from such work, even if this same person hunted wildlife and ate meat. An example of the argument that animal models are not useful in human medicine is presented in Section 1.2.

ARTHUR J. NONNEMAN • Department of Psychology, Asbury College, Wilmore, Kentucky 40390. MICHAEL L. WOODRUFF • Department of Anatomy and Cell Biology, J.H. Quillen College of Medicine, East Tennessee State University, Johnson City, Tennessee 37614.

Toxin-Induced Models of Neurological Disorders, edited by Michael L. Woodruff and Arthur J. Nonneman. Plenum Press, New York, 1994.

1.1. Two Ethical Positions on the Use of Animals in Biomedical Research

In its basic form, the philosophical argument against the use of animals in biomedical research says that there is no difference in moral status between humans and infrahumans. The statement by animal-rights activist Ingrid Newkirk (quoted by McCabe, 1990, p. 76: "I don't believe human beings have the 'right to life.' That's a supremacist perversion. A rat is a pig is a dog is a boy.") graphically summarizes this position, and sophisticated philosophical arguments that support it have been developed by Singer (1975), Regan (1983), and Rollin (1981, 1985).

Singer's arguments may be considered seminal. Essentially, Singer takes the position that sentience, the capacity for pain and suffering, is the crucial characteristic that entails moral equality among the species. Therefore, humans are morally bound not to compromise the fundamental rights of animals to comfort, freedom, and life. Others disagree.

Cohen (1986), for example, has argued that animals lack the capacity to share in the moral community of humans and, therefore, do not possess rights. As Cohen puts it: "The holders of rights must have the capacity to comprehend rules of duty, governing all including themselves. In applying such rules, the holders of rights must recognize possible conflicts between what is in their own interest and what is just." (Cohen, 1986, p. 866). Clearly, this philosophical position does not recognize sentience as a sufficient criterion for the assignment of rights to a species.

It must be emphasized that this position does include our obligations toward animals. However, these obligations are not based on innate rights of animals that give them claim against us. Rather, our obligations toward animals arise (as they often do toward humans) from the development of a special relationship, from differences in status, or from a special circumstance. Cohen cites the example of a parent who feels obligated to pay college tuition for a child although the child may not have a right to this support.

Indeed, in the same way, scientists, as human moral agents themselves, have an ethical responsibility to safeguard "the health and well being of the animals" (and to prevent) "any unnecessary pain and suffering" (Olfert, 1992, p. 4). The compelling obligation of scientists to safeguard the health and welfare of the animals used in biomedical research includes not just the period of time during which the experiment is underway, but also the provision of sanitary quarters, sufficient food and water, adequate space, and exercise before and after the experiment. In the United States, these obligations are detailed in guidelines issued by the NIH and are enforced by USDA inspectors. Conscientious compliance by scientists with these guidelines is perhaps the best reply to the first, primarily philosophical component, of the argument against the use of animals in biomedical research. Additionally, adhering to these guidelines provides healthier, more stable animals, thereby providing better control over unwanted variables that might confound an experiment. Consideration of ethical principles and the needs of good science are entirely consistent.

1.2. Pragmatic Concerns Regarding the Use of Animals in Research

The second component of the argument against the use of animals in biomedical research generally states that the results of these experiments have not provided enough useful information to justify either the suffering inflicted on the animals or the expense of the research. The proposition that there are alternatives to animal testing is often included in this component.

Absolute adherents to this position claim that nothing of use to human medicine has been, or can be, learned from animals (e.g., Reusch, 1978), whereas moderates take the position that, while some benefit to humans has come from animal research, the benefits have not been as dramatic as the biomedical community claims. This latter position is generally based on interpretations of the history of biomedical science and on evaluation of how animal models are defined. Because consideration of this position, particularly of the way in which models are defined by those who argue against their use, provides a good framework from which to begin to evaluate the chapters that are contained in this book, a recent presentation of it by LaFollette and Shanks (1993a,b) will be discussed.

LaFollette and Shanks begin their attack on the use of animals in biomedical research by disputing two of the conclusions presented by the American Medical Association in its 1988 White Paper. One of the conclusions is that the alternatives to live-animal tests are inadequate. Another is that many of the effective treatments for disease are due to experiments done with whole animal models. Those who oppose using animals reject both of these claims by the AMA. The following sections summarize the AMA position and respond to some of the criticisms of it by LaFollette and Shanks.

1.2.1. Alternatives to Animal Models

Before continuing, it must be emphasized that the issue is not whether useful experiments can or cannot be conducted using computer simulation or tissue culture, but whether these, or other nonanimal models, can entirely substitute for animals. The principle that alternative models should be employed whenever possible to obtain needed information without using animals has long been endorsed by most biomedical scientists. The 1989 revision of the U.S. Laboratory Animal Welfare Act and similar regulations in other Western countries requires that scientists document their search for appropriate alternative models. However, nonanimal models cannot replace all animal testing.

1.2.1.a. Computer Simulation as a Substitute for Animal Models The primary pitfall for the proposition that computer simulation can replace experiments on animals becomes apparent when one considers the difficulty in writing a program to simulate an ill-defined biologic phenomenon. If adequate information on a natural function is available, then computer simulation will be accurate and useful. One example would be the ability to simulate the nerve action potential and to predict the influence that changes in intracellular and extracellular concentrations of various ions might have on magnitude, waveform, and duration. However, knowledge of this process is very extensive. The ability to use a computer simulation to predict the outcome of a proposed treatment for Alzheimer's disease is more problematic.

1.2.1.b. Tissue Culture as a Substitute for Live Animals The complete substitution of tissue culture for whole animal models is also rejected in the AMA White Paper. Although tissue culture is certainly a valuable and widely used tool in biomedical research, isolated cells do not respond as whole systems do. This is apparent for investigations of the outcome of activity within complex neural networks. For example, tissue culture techniques are unlikely to reveal the interactions among the thalamic reticular nucleus, thalamic nuclei that project to the cortex, and the cortex itself that lead to production of the rhythms of the electroencephalogram.

Additionally, tissue culture may fail to be useful even in biomedical research that is less obviously systems-oriented. For example, as Soike *et al.* (1984) and Tabor *et al.* (1983) have reviewed, many viruses do not infect *in vitro* culture systems and even when they do, *in vitro* tests for antiviral efficacy of drugs may not yield data that can be extrapolated to humans. In a review of AIDS research McClure *et al.* (1990) point out that only about 1% of compounds having antiviral activity in tissue culture also are active in living animal systems.

The observations just made should not be taken to mean that *in vitro* systems should be discarded. Rather, the evidence suggests that information from *in vivo* experiments can help refine *in vitro* procedures to the point where they can replace live animals (van der Voet *et al.*, 1991). In fact, one goal of work with live animals has been, and continues to be, the development of increasingly useful nonanimal models. For example, knowledge gained from a live-animal model permitted the development of a useful *in vitro* model of hepatitis A, which resulted in a decrease in the use of live animals (Tabor *et al.*, 1983).

1.2.2. Can Data from Animal Experiments Be Extrapolated to Humans?

In reply to the claims made in the AMA White Paper that most of the improvement in human health is due to live-animal research, LaFollette and Shanks (1993a) argue first that preventive, not interventive, medicine is principally responsible for the advances that have occurred in human health during the past century. They cite data indicating that the number of deaths from disease such as tuberculosis, pneumonia, influenza, diphtheria, and typhoid had decreased by 90% even before the development of vaccinations and pharmacological treatment of these diseases. The largest increase in the prolongation of human life has occurred, in the view of LaFollette and Shanks, from advances in preventive public health measures, not in interventive therapies.

The second argument against animal experiments is that many of the treatments attributed to research on animals are, in fact, due to clinical research on humans (LaFollette and Shanks, 1993a). The discovery of the relationship between diabetes and insulin is one example given in support of this proposition. The conclusion reached by these first two arguments is that because animal research has in fact contributed little to human welfare it should be discontinued, or at least drastically reduced. A brief reply to this position is presented in Section 1.3.

The third argument presented by LaFollette and Shanks (1993a) is that results from animal research are often misleading. The example of the birth defects produced by thalidomide is often given to support this contention. Thalidomide is a sedative/hypnotic that was widely used after its introduction in 1956 to relieve nausea and vomiting during pregnancy. Before thalidomide was introduced, preclinical tests on adult animals were conducted, and no profoundly deleterious effects were found, nor were untoward effects found in adult humans given therapeutic doses of thalidomide. The drug was widely prescribed to pregnant women in Europe, and by 1960 a dramatic increase in birth defects, primarily severe limb malformations, was reported in Germany (Manson, 1986). Before the drug was completely removed from the market, between 7,000 and 10,000 incidences of newborns with birth defects related to thalidomide had been reported.

LaFollette and Shanks (1993a) use the thalidomide disaster to argue that preclinical testing of proposed pharmacotherapeutic agents does not yield useful information. There is, of course an obvious flaw in this argument, which LaFollette and Shanks try to cover. The case of thalidomide does not invalidate animal testing; it simply indicates that more parameters must be explored (i.e., the animal testing done with thalidomide did not explore possible teratogenic effects). Indeed, one result of thalidomide was the inclusion of screens for teratogenicity in drug testing. However, even with the inclusion of testing for drug-induced birth defects, LaFollette and Shanks maintain that the potential harmful effects of thalidomide would not have been detected.

They base this argument on the well-known observation that several species of animals, including rats and mice, are resistant to the teratogenic effects of thalidomide. Further, strains within species show a variable teratogenic response to thalidomide. Thus, they argue, the ability of the drug to produce birth defects would have gone undetected even if animal screening had been conducted. Indeed, this may have been the case for thalidomide, but three things need to be noted. First, the fact that the New Zealand white rabbit (a common laboratory animal) gives a consistent teratogenic response to thalidomide (Fabro, 1981) and that five species of macaque monkeys, baboons, and marmosets react to thalidomide in a manner almost identical to humans (Wilson, 1973) is generally neglected, as it was by LaFollette and Shanks, in articles using the thalidomide story to attack animal research. Second, the thalidomide incident led to significant revision of preclinical drug testing so that now more than one species is required and teratogenicity is routinely assessed. Third, only a few cases of thalidomide-induced birth defects occurred in the United States. This was because the FDA withheld permission to release the drug for clinical use owing to what it considered inadequate preclinical screening.

Despite these facts, critics of animal research continue to use the thalidomide disaster to argue that preclinical screening using animals misses possible detrimental effects of drugs and is, therefore, invalid. The opposite argument is also made. For example, LaFollette and Shanks (1993a, p. 118) interpret the literature to indicate that "aspirin causes birth defects in rats and mice, poisons cats, and has no effects whatsoever in horses." Thus, they argue that we would be deprived of the benefits of this drug if animal screening had been used to test it. The argument is that, not only does animal testing lead to the release of dangerous drugs, but it also will prevent the release of beneficial drugs. In reply, it should be noted that contemporary preclinical testing involves the use of more than one species and that current knowledge of animal physiology and biochemistry assists in the selection of appropriate species and significantly reduces the risk of false negatives.

A related objection to live-animal research is that animals and humans differ too much for useful extrapolations to be made between species. From this point of view, only complete identity at all levels would permit extrapolation to be made from animal to human. An example of the philosophical wording of this argument is that in order for a model to be useful it and the system modeled must be causally isomorphic and "two systems are causally isomorphic if they have all and only the same causal mechanisms" (LaFollette and Shanks, 1993b). Any divergence of events in any subsystem component of the whole system produces systemic disanalogy and invalidates the model for any study of mechanism and leaves only a heuristic role for animal studies in that they may stimulate thinking about the human disorder and lead to insight in that manner, but they cannot yield direct information regarding mechanism.

1.3. Summary of the Response to Criticisms of the Use of Animal Models

In summary, those who would eliminate the use of animals from biomedical research argue first that preventive, not interventive, medicine has been primarily responsible for increased health and longevity in humans. Therefore, we should concentrate on this aspect, not development of more treatment. However, while change in lifestyle might further reduce the incidence of cancer and heart disease, the data presently available indicate that it is unlikely to have much effect on the incidence of disorders such as Alzheimer's disease.

Second, opponents argue that alternative models to animals exist that are so powerful that animals are not needed for biomedical research. The main alternatives are computer simulation and tissue culture. The weaknesses of these alternatives were mentioned above. However, if the position is accepted that the only real benefit of the use of animal models is to stimulate ideas about human diseases, then these models may be just as useful (LaFollette and Shanks, 1993a,b).

Third, opponents of animal testing argue that animal models lead to misinformation because they are inherently weak models of human systems. The real breakthroughs in interventive medicine have come as the result of clinical observations, and, as a result, we should spend money on epidemiological studies, not animal research.

Finally, many who argue this point also claim that animals enjoy the same moral status as humans because of shared attributes. This raises a troubling philosophical inconsistency. If it *is* the case that animals are moral objects because of their similarity to humans, then one cannot *at the same time* argue that the results of animal research are not valid and may *not* be extrapolated to humans because of critically important species differences. In addition to this philosophical difficulty, there are at least two other problems with the position that epidemiological studies should replace experiments using animals.

First, although an animal model may not be isomorphically analogous to the human condition, it is often sufficiently analogous to permit parametric relationships (e.g., dose-response) to be determined. A parametric relationship between variables observed under strictly controlled conditions is a primary requirement for development of knowledge concerning cause-effect relationships. Epidemiological studies, while extremely useful, provide correlational information. Most scientists recognize that correlation does not entail causation.

Second, the assumption is that to be useful a model must be completely isomorphic with that being modeled in all relevant relationships. Biomedical scientists are well aware that this is generally not the case with the models they use. Indeed, there are many types of models, and their usefulness varies with the degree to which they are isomorphic with the human system or disease being modeled. Criteria for the usefulness of models, definitions of different types of models, and some examples of their uses are presented in the next section.

2. Utility of Animal Models

What constitutes a useful animal model? What are its characteristics? And what ethical considerations do or should govern its use? The general answer for this first question is that a model is useful when it has the ability to generate information that can be extrapolated to humans. But in considering what constitutes an appropriate human question, we must return to the question of ethical principles. Rolling (1985, p. 926) suggests two guiding principles that we find useful. The first is the *utilitarian principle*. "Before embarking on a piece of research, one ought to determine, to the best of one's ability, that the potential benefit to humans (or to humans and animals) clearly outweighs the pain and suffering to be experienced by the experimental animals." The second is the *rights principle*. "In cases where research is deemed justifiable by the utilitarian principle, it should be conducted in such a way as to maximize the animal's potential for living its life according to its nature, or *telos*, and certain fundamental rights should be preserved as far as possible given the logic of the research, regardless of cost considerations." Thus, not only the economic costs but also the ethical costs of the research should be considered in choosing "appropriate" questions and the models to be used in this study (e.g., Ortenwall, 1992).

Beyond the general principle that a model's utility depends on its ability to provide information that can be extrapolated to humans, the characteristics of a useful model depend almost entirely on its purpose (Calabrese, 1984). One purpose of animal models can be to refine our clinical definitions based on upon explicitly operational criteria (Brady, 1991). Sometimes animal models help to demonstrate the existence of subtypes of a disease that otherwise might remain indistinct. Or the model may help to clarify diagnostic criteria to distinguish separate but closely related human disorders (Richardson, 1991).

In situations in which a disorder is complex and seems to have multiple causes and maintenance factors, it is likely that the disorder is not a single disease entity (Strober, 1985). In such a case, multiple models may be needed to clarify the problem because "one cannot define a model of an entity that is itself poorly defined" (Strober, 1985). Thus, while it is true that a lack of uniformity in definition or diagnosis of the human condition may complicate the development of an animal model for that condition, it is also true that the use of various animal models often can lead to improvements in diagnosis or definition (Caldecott-Hazard *et al.*, 1991). With regard to human neurodegenerative diseases, Calne (1991) argues that there is no reason to believe that any of them represent a single entity. In this case, the use of several models is required to study the various causes and mechanisms leading to similar pathology (neural and behavioral). But it is important to note that we must cross-validate these various animal models using common measures, because each model duplicates only part of the human situation (Caldecott-Hazard *et al.*, 1991).

A useful and sometimes serendipitous result of the cross-validation process with multiple animal models is the refinement of some or all of those models (Caldecott-Hazard *et al.*, 1991). Anger (1991) has made a similar point, noting that the development of new models requires comparison with the results of older, established models. In other words, the use of a model may change as research progresses. Initially it may provide data needed to better characterize the disorder it represents. But as knowledge about the disorder increases to the point where limitations in the original model become apparent, it may still be useful for the validation and refinement of the new model(s) that replace it. A specific example from the current volume can be seen by comparing the various models of Parkinson's disease, the old 6-OHDA model (Reading and Dunnett, this volume) and the newer MPTP (Sundström *et al.*, this volume; Taylor *et al.*, this volume) and intranigral iron (Arendash *et al.*, this volume) models.

Other purposes of animal models that determine the utility of a given model include tests of possible etiologic factors, screening of possible therapeutic interventions, estimating clinically effective dose-response relationships, determining mechanism(s) of the disorder or response to therapy, and contributing to the knowledge base of the scientific discipline (i.e., basic science). Many models may be useful for more than one of these roles (Crawley *et al.*, 1985; Emerich *et al.*, this volume; Willner, 1991). For example, Schwarcz *et al.* (1984) demonstrated the utility of excitotoxic models of Huntington's disease and epilepsy-induced neurodegeneration for studying etiology, mechanism, and treatment issues (see also Emerich *et al.*, this volume; Juárez de Ku and Meserve, this volume; Solomon *et al.*, this volume; Strong and Garruto, this volume, for other examples of multipurpose model use).

3. Taxonomy of Animal Models

Many ways exist to categorize or label different types of animal models, and because of the wide variety of suggested taxonomies, some have questioned whether the taxonomies serve any purpose. We contend that the wide variety of taxonomic systems simply reflects the wide range of purposes and functions of models. Moreover, thinking about the type of model one plans to use can clarify thinking about its utility, validity, and perhaps need for refinement.

The most common taxonomy is based on the degree of isomorphism between the animal model and the human disorder with regard to inducing or etiologic conditions, symptoms and pathology, mechanism, and response to treatment (e.g., McKinney and Bunney, 1969). If every aspect is fully isomorphic between animal model and human condition, including cause and mechanism, the model is *homologous*. Most would agree that such a model represents the ideal. For example, Crawly *et al.*, (1985) state, "Ideally, a laboratory-based animal model should be (the same as) the human disorder in (1) symptoms, (2) postulated etiology, (3) neurobiological mechanism, and (4) treatment response." "However, even if environmental conditions or the resulting behaviors vary for different species, the biological intermediaries may be similar." This latter case, in which there is a partial isomorphism or a parallel between human and model, but in which the cause is different, is an *analogous* model. Finally, if there is no clear isomorphism but some aspect of the disease can be predicted by the model, it is a predictive or *correlational* model (Anger, 1991; Kornetsky, 1977; Treit, 1985; Walsh and Opello, this volume; Woodruff and Baisden, this volume; Yokel, this volume).

Katz (1981) suggests a taxonomy based on the purpose or use of the model rather than the degree of isomorphism between the animal model and the human situation. They are used because they are useful—they work! He offers two model types. The *empirical/utilitarian* model (e.g., drug screening), is useful when a theoretical rationale may be offered, but is not needed. *Theoretical* models are used to test specific hypotheses about etiology, mechanism, and so forth. Most models are not pure representatives of either type but have elements of both.

Lal and Forster (1991) also offer a two-type taxonomy for development of therapeutic interventions. They are distinguished on the basis of purpose. One purpose is to retard, prevent, or reverse progress of the disease. For such models, etiologic factors and mechanism are the primary considerations, ideally homologous. The second purpose is to reduce or reverse the symptoms (not the progression) of the disease. For these models the pathology and behavior must be similar, but not necessarily homologous, and the cause need not be the same.

4. Validity of Animal Models

As exemplified by the arguments of LaFollette and Shanks outlined above, individuals opposed to animals for research argue that animal use is inappropriate because the information obtained cannot be extrapolated to humans. Usually, this argument is made on the basis of a lack of strict homology between the animal test, or model, and the human response or condition. If a model is strictly homologous (complete isomorphism in all respects), then validity is guaranteed. But the converse is not true. A model that is not homologous with the human condition it represents may still have some validity. The key to validity rests in the choice of the right model for the right reason, not necessarily in homology.

In most cases, models represent a compromise. A relatively simple experimental system is used to represent a more complex and less readily studied system. The model may represent only one aspect of a biological and/or behavioral system, but it is used because of the complexity of the natural system or because we believe we are modeling the most relevant component(s) of the problem, and the model allows a test of that belief (Kornetsky, 1977; Linssen *et al.*, 1991; Masoro, 1991). In other cases, the animal studies provide a complement for concurrent clinical or epidemiological studies. For example, Velican

(1981) argues that epidemiologic studies are made much more meaningful and powerful if they are compared to, and especially if they are reinforced by, parallel control studies in suitable animal models. Willner (1991) has made a similar point, noting that animal models often form the critical interface between the clinician and the basic researcher. "On the one hand, they [animal models] are the major channel through which developments at the basic level are brought into clinical perspective; on the other, they provide a particularly informative means of investigating the psychobiological foundations" (Willner, 1991, p. 136).

In contrast, it certainly is true that animal models can be used inappropriately, and in an effort to address the "wrong" question, the results are invalidated. Sometimes this reflects a level-of-analysis problem (Solomon *et al.*, this volume). At other times it reflects a lack of adequate isomorphism between the model and the human situation. We believe Carlton (1978) and Leibovici and Wolman (1984) are correct that there need be no logical or functional connection between the human and animal model to address some valid questions; all that is needed is a correlation, a predictive relationship. But we also agree with Treit (1985) that a problem arises when individuals try to use a correlational model inappropriately for determining mechanism. For this, some degree of isomorphism is required.

In some cases, human testing may be the only valid option. For example, Calabrese (1984) asserts that high-risk humans may differ from "normal" humans such that the "typical" animal model is invalid. In such a situation, although epidemiological studies are costly in time and money, the cost (social, ethical, medical) of *not* doing the human testing must be considered. Garruto (1991) has stressed the usefulness of human "natural experiments" for studying some toxicant-induced neurological disorders. These are most likely to occur and to be useful as alternatives to animal models when the human populations are small and genetically homogeneous, culturally and geographically isolated, and in which the incidence of the disorder is high.

When animal models are available and useful but each model only partially duplicates the human situation, cross-validation of the various models can be performed using common measures (Caldecott-Hazard *et al.*, 1991). This cross-validation process is necessary if one is to determine concurrent validity and/or construct validity of a model or models (see Russell, 1991, for a description of various types of validity related to toxicity testing).

5. Development of Animal Models

It is typically assumed that the more closely a species is related to humans the better the model it will provide. In some cases, this likely is true. For example, the effects of MPTP administration in nonhuman primates parallel much more closely the effects of MPTP in humans, and of idiopathic Parkinson's disease, than do the effects of MPTP in nonprimate species (Taylor *et al.*, this volume). But presumed similarity of relationship to humans is not the only, and not necessarily the best, consideration in developing an animal model. Sometimes species that appear very dissimilar to humans in most respects share particular characteristics with humans that suit them especially well to answer specific questions (Sundström *et al.*, this volume). A specific example is provided by Marg (1982), who argues that the cat provides a better model than the monkey of stimulus deprivation amblyopia in children. Furthermore, adequate testing with this cat model could have predicted and prevented therapeutic errors caused by premature clinical interventions in humans with developmental amblyopia (see McClure *et al.*, 1990, for a similar point). Thus, there is an interactive developmental process in which advances in animal model development and advances in clinical practice and scientific understanding are mutually interdependent (Masoro, 1991).

Carlton (1978) refers to this mutually interdependent developmental process as "correlational" model development. In using this term, Carlton is referring to the intercorrelation of data sets derived from the clinic and from the animal lab. This process serves multiple purposes: development and refinement of the animal model(s); development, refinement, or selection of clinical interventions; and cross-validation of the model(s) and the intervention(s).

In his discussion of the development of animal models in toxicology, Winneke (1992) distinguishes two stages. The first is qualitative, or endpoint-based. It focuses on comparability of function, mechanism, or development of effect in humans and animals. Work at this stage typically precedes and provides essential information needed for the development of quantitative or dose-based models. The goal of this second developmental phase is the determination of a Reference Dose (RFD) that allows accurate extrapolation from animal to human (see also Solomon *et al.*, this volume). Attaining this second stage of quantitative modeling can be very difficult. This is especially the case when the character, and perhaps even the mechanism, of the disorder changes as the disease progresses. In this case, it is often necessary to develop different models to study different developmental stages of the disease (Leibovici and Wolman, 1984).

In like manner, when different human populations exist with regard to exposure or other risk factors, it may be necessary to develop several different models to represent the different groups of humans. Calculation of a single Reference Dose, in this case, may not be possible (Calabrese, 1984).

6. Characteristics of Toxin-Induced Models

Most of this chapter has dealt with general issues in the development and use of animal models. But are there specific, perhaps even unique, features of toxin-induced models of neurodegenerative disorders? Garruto (1991) has stressed the need for chronic, rather than acute, animal models because the neurodegenerative disorders are, by nature, slow and progressive. As important as this is, an even more central issue for the development of toxin-induced animal models is the need for appropriate behavioral assessment. Wolthuis (1991) has made this point very clearly in reference to extrapolation of neurotoxicity. "Apart from controlling several physiological functions, the main function of the nervous system is . . . to regulate and steer behavior. [Therefore] measurements of behavior are indispensable if one wants to investigate the effects of chemicals on the functions of the nervous system."

It seems unlikely that anyone would argue this point. But the development of appropriate behavioral models is a tricky business. On the one hand, apparently different behaviors in different species may rely on similar biological intermediaries (Crawley *et al.*, 1985). On the other hand, seemingly identical behavior may be produced by very dissimilar processes, and failure to recognize the difference may invalidate the model. A commonly used example is the surface similarity but the neurological dissimilarity between hypothermia-induced shivering and resting tremor. Within the cognitive domain, Smith (1988) cautions us to consider noncognitive effects of experimental manipulations that may affect task performance in a way that mimics cognitive deficits such as amnesia or dementia. Indeed, a principal weakness of much biobehavioral research (not only models of neurodegenerative disorders) stems from a failure to recognize that some tasks can be solved in more than one way (Ridley and Baker, 1991). Therefore, simply looking at an animal's ability or inability to solve a task without examining *how* it solved or failed the task may invalidate the model.

In conclusion, the development of a good animal model depends on adequate characterization of the phenomena being modeled, including behavior. But when only some aspects are adequately understood or characterized, a model of these aspects may be useful for a while until newly acquired information allows replacement of the original model with a better one.

7. References

Anger, W.K., 1991, Animal test systems to study behavioral dysfunctions of neurodegenerative disorders, *Neurotoxicology* 12:403–414.

AMA White Paper, 1988, Use of animals in biomedical research: The challenge and response. American Medical Association, Washington, DC.

Brady, J.V., 1991, Animal models for assessing drug abuse, *Neurosci. Biobehav. Rev.* 15:35–43.

Calabrese, E.J., 1984, Suitability of animal models for predictive toxicology: Theoretical and practical considerations, *Drug Metab. Rev.* 15:505–523.

Caldecott-Hazard, S., Guze, B.H., Kling, M.A., Kling, A., and Baxter L.R., 1991, Clinical and biochemical aspects of depressive disorders. I: Introduction, classification, and research techniques, *Synapse* 8:185–211.

Calne, D.B., 1991, Neurotoxins and degeneration in the central nervous system, *Neurotoxicology* 12:335–340.

Carlton, P., 1978, Theories and models in psychopharmacology, *in* "Psychopharmacology. A Generation of Progress" (M.A. Lipton, A. DiMascio, and K.F. Killam, eds.), pp. 553–561, Raven Press, New York.
Cohen, C., 1986, The case for the use of animals in biomedical research, *N. Engl. J. Med.* 315:865–870.
Crawley, J.N., Sutton, M.E., and Pickar, D., 1985, Animal models of self-destructive behavior and suicide, *Psychiatr. Clin. North Am.* 8:299–310.
Fabro, S., 1981, Biochemical basis of thalidomide teratogenicity, *in* "The Biochemical Basis of Chemical Teratogenesis" (M.R. Juchau, ed.), pp. 159–178, Elsevier/North Holland, New York.
Garruto, R.M., 1991, Pacific paradigms of environmentally-induced neurological disorders: Clinical, epidemiological and molecular perspectives, *Neurotoxicology* 12:347–378.
Katz, R.J., 1981, Animal model and human depressive disorders, *Neurosci. Biobehav. Rev.* 5:231–246.
Kornetsky, C., 1977, Animal models: Promises and problems, *in* "Animal Models in Psychiatry and Neurology" (I. Hanin and E. Usdin, eds.), pp. 3–7, Pergamon Press, New York.
LaFollette, H., and Shanks, N., 1993a, Animal models in biomedical research: Some epistemological worries, *Public Aff. Q.* 7:113–130.
LaFollette, H., and Shanks, N., 1993b, The intact systems argument: Problems with the standard defense of animal experimentation, *So. J. Phil.* 21:323–333.
Lal, H., and Forster, M.J., 1991, Autoimmune mice as models for discovery of drugs against age-related dementia, *Drug. Dev. Res.* 24:1–27.
Leibovici, J., and Wolman, M., 1984, Animal models for tumor progression, *Cancer Res.* 4:165–168.
Linssen, M.J., Wilhelms, O.-H., and Timmerman, H., 1991, Animal models for testing anti-inflammatory drugs for treatment of bronchial hyperactivity in asthma, *Pharm. Weekbl. Sci.* 13:225–237.
Manson, J., 1986, Teratogens, *in* "Casarett and Doull's Toxicology: The Basic Science of Poisons" (3rd ed.; C.D. Klaassen, M.O. Amdur, and J. Doull, eds.), pp. 195–220, Macmillan, New York.
Marg, E., 1982, Is the animal model for stimulus deprivation amblyopia in children valid or useful? *Am. J. Optom. Physiol. Optics* 59:451–464.
Masoro, E.J., 1991, Use of rodents as models for the study of "normal aging": Conceptual and practical issues, *Neurobiol. Aging* 12:639–643.
McCabe, K., 1990, Beyond Cruelty, *The Washingtonian* (February 1990), pp. 72–77, 185–195.
McClure, H.M., Anderson, D.C., Ansari, A.A., Fultz, P.N., Klumpp, S.A., and Schinazi, R.F., 1990, Nonhuman primate models for evaluation of AIDS therapy, *Ann. NY Acad. Sci.* 616:287–298.
McKinney, W.T., and Bunney, W.E., 1969, Animal model of depression: Review of evidence and implications for research, *Arch. Gen. Psychiatry* 21:240–248.
Olfert, E.D., 1992, Ethics of animal models of neurological diseases, *in* "Animal Models of Neurological Disease. I: Neurodegenerative Diseases. Neuromethods 21" (A. Boulton, G. Baker, and R. Butterworth, eds.), pp. 1–22, Humana Press, Totowa, NJ.
Ortenwall, P., 1992, Endothelial cell seeding: How far can animal models fake expectations? *in* "Applied Cardiovascular Biology 1990–1991," Vol. 2. (P. Zilla, R. Fasol, and A. Callow, eds.), pp. 83–90, S. Karger, Basel.
Regan, T., 1983, "The Case for Animal Rights," University of California Press, Berkeley.
Reusch, H., 1978, "Slaughter of the Innocent," Bantam Books, New York.
Richardson, J.S., 1991, Animal models of depression reflect changing views on the essence and etiology of depressive disorders in humans, *Prog. Neuropsychopharmacol. Biol. Psychiatry* 15:199–204.

Ridley, R.M., and Baker, H.F., 1991, A critical evaluation of monkey models of amnesia and dementia, *Brain Res. Rev.* 16:5–37.

Rollin, B.E., 1981, "Animal Rights and Human Morality," Prometheus Books, New York.

Rollin, B.E., 1985, The moral status of research animals in psychology, *Am. Psychol.* 40:920–926.

Russell, R.W., 1991, Essential roles for animal models in understanding human toxicities, *Neurosci. Biobehav. Rev.* 15:7–11.

Schwarcz, R., Foster, A.C., French, E.D., Whetsell, W.O., and Köhler, C., 1984, Excitotoxic models for neurodegenerative disorders, *Life Sci.* 35:19–32.

Singer, P., 1975, "Animal Liberation: A New Ethic for Our Treatment of Animals," Avon Books, New York.

Smith, G., 1988, Animal models of Alzheimer's disease: Experimental cholinergic denervation, *Brain Res. Rev.* 13:103–118.

Soike, K.F., Rangan, S.R.S., and Gerone, P.J., 1984, Viral disease models in primates, *Adv. Vet. Sci. Comp. Med.* 28:151–199.

Strober, W., 1985, Animal models of inflammatory bowel disease: An overview, *Dig. Dis. Sci.* 30:3S–10S.

Tabor, E., Purcell, R.H., and Gerety, R.J., 1983, Primate animal models and titered inocula for the study of human hepatitis A, hepatitis B, and non-A, non-B hepatitis, *J. Med. Primatol.* 12:305–318.

Treit, D., 1985, Animal models for the study of anti-anxiety agents: A review, *Neurosci. Biobehav. Rev.* 9:203–222.

van der Voet, G.B., Marani, E., Tio, S., and de Wolff, F.A., 1991, Aluminum neurotoxicity, *in* "Progress in Histo- and Cytochemistry. Vol. 23: Histo- and Cytochemistry as a Tool in Environmental Toxicology" (W. Graumann and J. Drucker, eds.), pp. 29–61, Fischer Verlag, Stuttgart.

Velican, C., 1981, Are animal models suitable for the study of ischemic heart disease? *Rev. Roum. Méd.-Méd. Int.* 19:5–19.

Willner, P., 1991, Animal models as simulations of depression, *Trends Pharm. Sci.* 12:131–136.

Wilson, J.G., 1973, "Environment and Birth Defects," pp. 11–35, Academic Press, New York.

Winneke, G., 1992, Cross species extrapolation in neurotoxicology: Neurophysiological and neurobehavioral aspects, *Neurotoxicology* 13:15–26.

Wolthuis, O.L., 1991, Some animal models and their probability of extrapolation to man, *Neurosci. Biobehav. Rev.* 15:25–34.

2

Strategies for Developing Animal Models of Neurotoxicant-Induced Neurodegenerative Disorders

Parallel Studies of Learning and Memory in Animals and Humans

PAUL R. SOLOMON, MARYELLEN GROCCIA-ELLISON, MARK E. STANTON, and WILLIAM W. PENDLEBURY

Exposure to neurotoxic substances may play an important role in the etiology of memory impairment associated with neurodegenerative disorders. Because of this, significant effort has been devoted to developing animal models of these neurotoxin-induced disorders. In this chapter we will begin by critically reviewing some of the issues surrounding the development of animal models of neurologic conditions that may produce memory impairment. We will then argue that one important aspect of any strategy for developing neurotoxin-induced models of neurodegenerative disorders is to reduce the uncertainty involved in generalizing from animal models to human conditions. Next we will review some of our own work that has attempted to accomplish this by using a model systems approach to the study of learning and memory. The primary advantage of this approach to neurotoxicology is that it allows for the study of identically learned behaviors in animals and humans within a learning and memory paradigm whose neurobiological basis is well understood. As such, when neurotoxin-induced

---

PAUL R. SOLOMON and MARYELLEN GROCCIA-ELLISON • Department of Psychology and Program in Neuroscience, Williams College, Williamstown, Massachusetts 01267. **MARK E. STANTON** • Health Effects Research Laboratory, Division of Neurotoxicology (MD-74B), United States Environmental Protection Agency, Research Triangle Park, North Carolina 27711. **WILLIAM W. PENDLEBURY** • Department of Pathology, University of Vermont College of Medicine, Burlington, Vermont 05405.

Toxin-Induced Models of Neurological Disorders, edited by Michael L. Woodruff and Arthur J. Nonneman. Plenum Press, New York, 1994.

memory deficits are found in animal models it is possible to: (1) form testable hypotheses regarding the neurobiological basis of these deficits and (2) have a reasonable degree of certainty that similar memory deficits would occur if human exposure produced similar neurodegenerative effects. To support this position we will review studies of classical eyeblink conditioning in humans and animals that have shown parallel deficits in both species following brain lesions, aging, exposure to benchmark substances such as scopolamine, and exposure to a known neurotoxin—for example, aluminum.

1. General Strategies for Developing Animal Models

Animal models of human pathologic conditions can be developed for a variety of purposes. Three purposes that are often stated or implied, however, involve developing models to (1) study the etiology of a disorder, (2) study the mechanisms of a disorder, and (3) develop therapeutic interventions.

While it is often the case that many models address more than one of these points, it is also true that the particular bias that a model reflects will inform the research that is conducted. For example, considerable effort is being devoted toward developing pharmacological agents to enhance disordered memory in humans. To accomplish this, two basic strategies exist. The first might be called an *inductive approach* and encompasses Purposes 1 and 2 above. This approach attempts to understand the neurobiological basis of memory and then uses this information to generate possible therapeutic approaches. The key point is that when therapeutic measures are introduced, there is reason to believe that their mechanisms are understood. A second approach, a *deductive approach,* involves Purpose 3. This involves testing potentially efficacious therapeutic agents in clinical trials. The key point here is that if an agent facilitates cognition, it is not crucial to understand, at least initially, the precise mechanism by which it works. As we have pointed out elsewhere (Solomon and Pendlebury, 1993), both of these strategies ultimately may have the same outcome, but the approach (and perhaps the probability for success) is very different.

2. Levels of Analysis

A second general issue that surrounds any attempt to develop animal models is the level of analysis. In considering models of neurological disorders, many levels of analysis are possible. Because deficits in behavior and cognition are hallmarks of many neurological disorders, these are often one level of analysis. But it is also possible to become increasingly reductionistic and study neurological disorders using (1) behavioral, (2) systems, (3) cellular, and (4) molecular approaches. It is possible to argue that behavioral approaches are more likely to be cases of Purpose 3, whereas the remaining three levels of analysis are more likely to be cases of Purposes 1 and 2, but this does not necessarily have to be the case. Consider the screening of antipsychotic agents by measuring their affinity for dopamine receptors. This is an example of a cellular/molecular approach, and at face value it looks as though this is a Purpose 2 model. Yet we now realize that changes in dopamine receptors are not necessarily the "cause" of schizophrenia and may well not be the mechanism of psychotic disorders, but simply a link in the chain (e.g., Koob and Bloom, 1983). Nevertheless, it is clear that dopaminergic blockers have antischizophrenic effects. As such, examining dopamine binding may be an example of looking at a behavior, albeit in a dish. This then sounds like a Purpose 3 model—developing therapeutic interventions. Although the purpose of a particular model may not always be clear, what is clear is that eventually a successful model will need to integrate findings across all levels of analysis.

3. Determining the Validity of an Animal Model

A third issue regarding animal models asks an important question: What constitutes a valid animal model of a neurological disorder? There are certainly many approaches to this problem, but most seem to take into account the guidelines initially described by McKinney and Bunny (1969), who proposed four criteria for evaluating animal models: (1) similarity of inducing conditions, (2) similarity of behavioral states produced, (3) common underlying neurobiological mechanisms, and (4) reversal by clinically effective treatment techniques. By these criteria, some would argue that we have no adequate animal model of any neurological condition, and in the narrowest sense they are correct. But this focus may be too confining. Although we have no model that meets all these criteria, certainly approximations exist for a number of disorders. For an animal model to be useful, it does not necessarily have to reflect all aspects of the disease state. Clearly, the closer the model comes to meeting the four above-stated criteria, the better the model will be. However, a model that addresses only one aspect of the etiology or mechanisms of the disorder can be quite useful. Consider the case of Alzheimer's disease. There is certainly no adequate animal model, and some have argued that Alzheimer's disease is a uniquely human condition precluding the use of animal models. But a model that reproduces only some of the aspects of the disease may be very useful. For example, if senile plaques could be induced with no corresponding production of neurofibrillary tangles (or vice versa), most would view this as a valuable contribution.

4. Using Animal Models to Study Neurotoxin-Induced Neurological Disorders

We have reviewed several aspects of animal models and may have begun to arrive at a set of criteria for an ideal animal model that may well be unachievable. Nevertheless, one important aspect of attempting to reach the ideal is that valuable information is gleaned along the way. As the chapters in this volume suggest, one important application of an animal model is to begin to understand how neurotoxins may contribute to neurological disorders. Another important application is to assist in making judgments about the risks involved in exposure to neurotoxins. In the next sections we will briefly review the process of risk assessment and then discuss how the model system approach may contribute to this.

5. Risk Assessment in Neurotoxicology

Risk assessment is an analytical, preferably quantitative, process that can guide decisions concerning the safety of chemicals under certain exposure conditions. It is used in both a number of government agencies and in the chemical and pharmaceutical industries to manage risk associated with many types of toxicity (National Academy of Sciences, 1983), including neurotoxicity (Cranmer, 1990; U.S. Congress, Office of Technology Assessment, 1990). Human toxicity data can be used in the risk-assessment process, but it is more common to use animal data to estimate human risk. In addition, the process is an area of intensive research in toxicology. Animal models therefore play an important role in both the practice and the basic science of risk assessment.

The risk-assessment process is comprised of four steps: (1) hazard identification, (2) dose-response assessment, (3) exposure assessment, and (4) risk characterization (National Academy of Sciences, 1983). For present purposes, the second and fourth steps are most relevant. They can be illustrated by explaining the calculation of the reference dose (RfD). The RfD refers to the level of human exposure to a compound below which it is assumed that there is little or no risk of adverse health effects. For neurotoxicity, the RfD is determined through the NOAEL safety factor approach (U.S. Congress, Office of Technology Assessment, 1990; Ch. 6). The NOAEL (no-adverse-effects-level) is divided by number of safety, or uncertainty factors (UFs, Kimmel, 1990) as follows:

$$\text{RfD} = \frac{\text{NOAEL (or LOAEL)}}{\text{UF}_h \times \text{UF}_a \times \text{UF}_s \times \text{UF}_l}$$

Where: NOAEL = dose of compound having no significant effect in study
LOAEL = lowest dose of compound having a significant effect in the study
UF_h = factor for variation in human sensitivity (e.g., infants, the aged)
UF_a = factor for animal-to-human extrapolation
UF_s = factor for less-than-chronic exposure situation in study
UF_l - factor for use of LOAEL rather than NOAEL in calculation
$UF_h = UF_a = UF_s = UF_l = 10$

Thus, the RfD is derived by examining dose-response data from empirical studies, determining the dose (other than vehicle) at which toxicity is lowest, and dividing this dose by a number ranging from 10 to 10,000, depending upon uncertainties inherent in the study. For example, if the study involved animals exposed acutely to a neurotoxicant with a range of doses that yielded a LOAEL (low-adverse-effect-level) but not a NOAEL, the LOAEL would be divided by 10,000 to determine the RfD for humans (although this is possible in principle, it has never occurred in practice; Kimmel, 1990). On the other hand, if the study involved humans exposed chronically at doses including one that produced no effect (NOAEL), then the RfD would be determined by dividing this dose by 10 to allow for possible differences in sensitivity among special populations of humans (UF_h), such as infants and children or the aged.

Clearly, this is a highly conservative, simplified, and arbitrary procedure for assessing human risk on the basis of data from animal models. Not surprisingly, it is not always accurate. In some instances, it underestimates risk. For example, a comparison of RfDs calculated from human versus animal data on the developmental neurotoxicity of polychlorinated biphenyls (PCBs) showed that despite the use of uncertainty factors entailed in this method, animal studies still underestimated the human RfD by nearly one order of magnitude (Tilson *et al.*, 1990). However, in many instances the safety-factor approach probably overestimates human risk. In any case, there is widespread interest in developing alternative approaches (Cranmer, 1990).

One goal of basic research on risk assessment is to supplant contemporary approaches, such as the one just described, with approaches that employ biologically based models. These models seek to develop mechanistic accounts of toxicity that can elucidate species differences in a manner that would reduce or eliminate the need for uncertainty factors. If these models can be expressed in quantitative terms, they could serve as a basis for quantitative risk assessment. In their most abstract conception, biologically based models must consider the relation between exposure to a compound and the dose delivered to target tissues (exposure-dose), as well as the relation between this dose and its subsequent toxic effects (dose-effect). In these "exposure-dose-effect models" (Figure 1), physiologically based-pharmacokinetics (PBPK) models address the former, whereas biologically based-dose-response (BBDR) models address the latter of these relations. A quantitative model that describes the entire exposure-dose-effect relation has recently been developed for studying chemical-induced carcinogenicity. Applying appropriate values for humans to the parameters of this model yields a mechanistic basis for predicting human risk associated with exposure to a given chemical. To our knowledge, no models of this kind have yet been developed for neurotoxicity.

Neurotoxins may be involved in the etiology of age-related neurodegenerative disorders that may cause impaired memory and other aspects of dementia. The study of this phenomenon provides an ideal opportunity to address basic

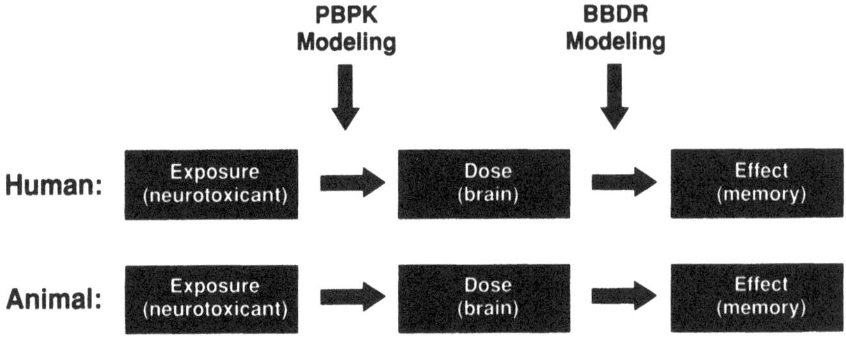

Figure 1 Biologically based models in risk assessment for neurotoxicology.

issues of risk assessment in neurotoxicology. Indeed, parallel studies of this problem in humans and animal models could provide a situation in which three major factors in risk assessment—chronic (e.g., lifetime) exposure, susceptible populations (i.e., the aged), and animal-to-human extrapolation—could be examined in an integrated fashion. Such studies would be particularly powerful if they were sufficiently mechanistic to permit the development of biologically based models.

To develop a biologically based model of neurotoxin-induced memory impairment, the effects of chemical exposure on memory must be examined in both humans and animals in terms of the intervening effects on the nervous system (Figure 1). To accomplish this, one must study a form of memory that is well understood at the neurobiological level and that lends itself readily to comparative investigation in humans and animals. Perhaps the best example of this is classical conditioning of the eyeblink reflex (see model systems approach, below). This preparation measures a discrete response that has the same topography, controlling musculature, and is governed by the same behavioral laws and parameters in humans and several other mammalian species, including monkeys, cats, rabbits, and rats. In addition, this form of learning is almost unique in the degree to which its underlying neural circuitry has been studied and characterized (Gormezano *et al.*, 1987). To develop eyeblink conditioning as a biologically based model of neurotoxin-induced memory impairment, it is necessary to determine the degree to which its neural mechanisms are similar in humans and in animal models (BBDR modeling). If this similarity can be established, then investigation of neurotoxins that produce comparable effects on eyeblink conditioning in humans and animals could provide a context in which to study species differences in pharmacokinetics and dosimetry (PBPK modeling). In the next sections we will discuss progress in pursuing these questions through the use of classical eyeblink conditioning as a model system.

6. The Model Systems Approach

The model systems approach to learning and memory advocates studying a relatively simple form of learning in a relatively simple preparation. The goal is to understand this form of learning and memory fully in terms of the neural systems and the cellular mechanisms involved. The basic strategy involves (1) characterizing the behavior, (2) identifying the neural systems, (3) identifying the synapses that show plasticity, and (4) elucidating the neurobiological mechanisms. This approach was initially used in invertebrates such as aplysia and hermissenda (see Byrne, 1987) and then adapted to mammals. The most commonly used mammalian model system is the classically conditioned eyeblink (EB) response in the rabbit. Classical conditioning of the EB response is arguably the best characterized form of learning and memory at both the behavioral and neurobiological levels. Enormous strides have been made in the past few years using this preparation, and we are now at the point of identifying the circuitry for simple forms of learning (Thompson, 1988). We have argued elsewhere (e.g., Solomon *et al.*, 1988a; Solomon and Pendlebury, 1992) that as progress continues in identifying the neurobiological mechanisms of this form of learning, it should be possible to begin to use this approach to study other aspects of learning and memory. Our research has focused on using the model systems approach to study age-related memory disorders, but the logic can be extended to all forms of memory disorders including those present in neurotoxin-induced neurological conditions (see Solomon and Pendlebury, in press). We suggest that this approach lends itself well to the criteria for evaluating animal models described earlier. The primary advantage of this approach is that *when disorders of learning and memory are detected, it is possible to form readily testable hypotheses about their neurobiological substrate.* This is possible because there is significant and growing knowledge about the neurobiology of this type of learning. With this information in hand, it may then be possible to better characterize the human risk and eventually to develop therapeutic strategies. In the framework of our earlier discussion of animal models, this is clearly an inductive (Purpose 2) approach.

7. The Model Systems Approach and Risk Assessment

As discussed earlier, a biological modeling in risk assessment would ideally involve a preparation in which (1) an identical behavior can be studied in both humans and animals, (2) a learned behavior shares similar neurobiological mechanisms in humans and animals, and (3) the behavior can be disrupted in similar ways in humans and animals by a variety of manipulations, including neurotoxins. If such a preparation can be identified, we might argue that what we learn from our animal model is applicable to the human condition. Classical EB

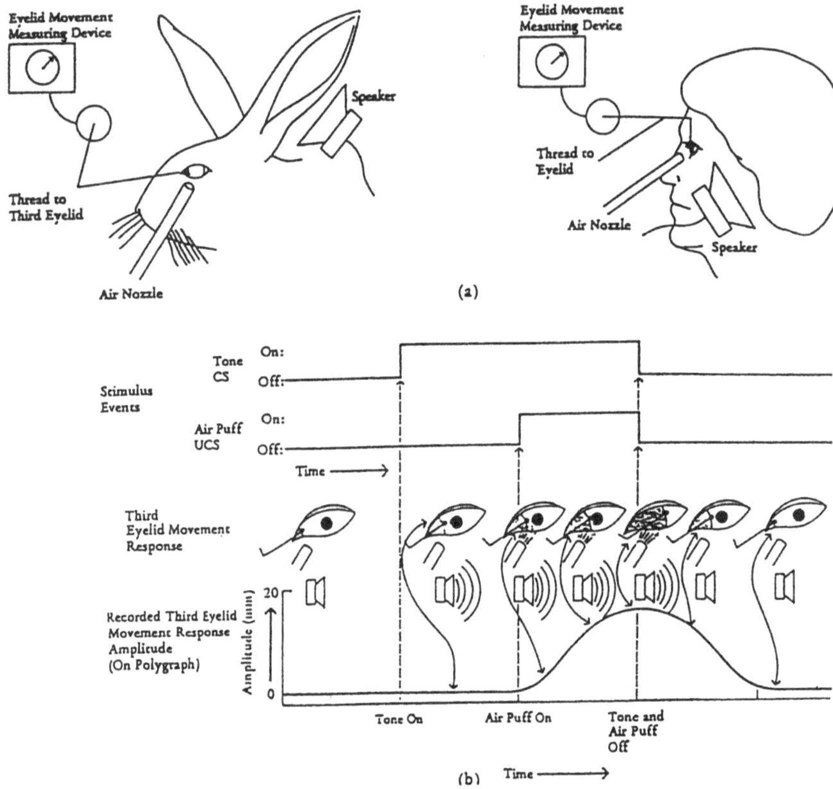

Figure 2 The classically conditioned eyeblink response in animals and humans. A typical conditioning session consists of 100 presentations of the tone and air puff. On each presentation, the tone-conditioned stimulus (CS) remains on for 500 msec. Four hundred milliseconds after the tone onset, the air puff unconditioned stimulus (UCS) remains on for 100 msec. Both stimuli terminate together. Initially, the organism responds only to the air puff, thus emitting only an unconditioned response (UCR). After pairing the tone and air puff, the organism responds during the first 400 msec of the tone and before the air puff occurs, thus emitting a conditioned response (CR). (After Lindsay, Hall, and Thompson.)

conditioning may be a candidate preparation. Figure 2 shows the preparation for both animals (rabbits) and humans. As we will discuss, this preparation allows us to study an identical behavior in humans and animals. This behavior can be disrupted in a similar manner by a variety of manipulations in both humans and rabbits, and it does appear to share common neurobiological mechanisms in both humans and animals.

8. Behavioral Aspects of Classical Eyeblink Conditioning: Similarities between Humans and Animals

One crucial step in developing a biologically based model of neurotoxin-induced memory disorders is to demonstrate that the form of learning and memory being studied follows the same rules in animals and humans. Although many researchers who use animal models work under the assumption that the forms of learning and memory studied in animals are mediated by the same processes as those seen in humans, in many instances this may assume too much. Although the behaviors studied in animals may be analogous, homologous, or similar to those seen in humans, there are relatively few instances in which the *identical* behavior is studied. To our knowledge there are only two such cases: delayed matching to sample in monkeys and humans (Zola-Morgan and Squire, 1986) and classical eyeblink conditioning. It is noteworthy that although classical conditioning is now most prominently studied in animals, it was actually first studied in humans. Even before Pavlov's studies in dogs, E.B. Twitmyer studied classical conditioning of the patella reflex in human subjects. Classical EB conditioning was also first studied in humans (Hilgard and Marquis, 1940). The advent of conditioning in rabbits came through the pioneering work of Gormezano and colleagues (Gormezano, 1966), who subsequently worked out the parameters for acquisition of this response (Gormezano *et al.*, 1983). What is crucial is that within certain boundaries, the parameters governing the acquisition of the conditioned response are the same for both rabbits and humans. (Hilgard and Marquis, 1940; Solomon *et al*, 1991a). For example, the effects of parametric manipulations such as intensity of the tone Conditioned Stimulus (CS) or air puff Unconditioned Stimulus (UCS), the time between trials (intertrial interval), or stimuli (interstimulus interval) all follow the same rules in both species (Gormezano and Moore, 1969). This type of information certainly suggests that the underlying mechanisms and neural circuitry are similar. This, coupled with the fact that we know more about the rules of EB conditioning than any other type of learning, makes it a potentially useful preparation.

9. Neurobiological Aspects of Classical Eyeblink Conditioning: Similarities between Humans and Animals

9.1. Studies of Cerebellar Damage in Animals

There is now considerable evidence that an essential part of the necessary circuitry for classical EB conditioning is in the cerebellum. Studies by Thompson and colleagues (Thompson, 1986) have demonstrated that lesions to the dentateinterpositus nucleus block acquisition of the response in naive animals and eliminate the response in trained animals. It is important to note that only the conditioned response (CR) is disrupted whereas the unconditioned response (UR) remains intact, indicating that these manipulations affect learning and not motor performance. Moreover, conditioning in the eye ipsilateral, but not contralateral, to the lesions is disrupted, indicating that the disruption is not due to other nonassociative processes. Based on these and other extensive studies, Thompson has argued that the necessary circuitry for the memory trace is in the cerebellum (Thompson, 1986).

9.2. Studies of Cerebellar Damage in Humans

If the crucial circuitry for conditioning in rabbits is in the cerebellum and there are similar neurobiological substrates for this type of learning in rabbits and humans, then we might expect disrupted conditioning in humans with cerebellar damage. Several studies suggest that this is the case. In our initial study, we had the opportunity to condition a 54-year-old woman with damage to cerebellar afferents. Neuropsychological evaluation revealed that she had normal intellec-

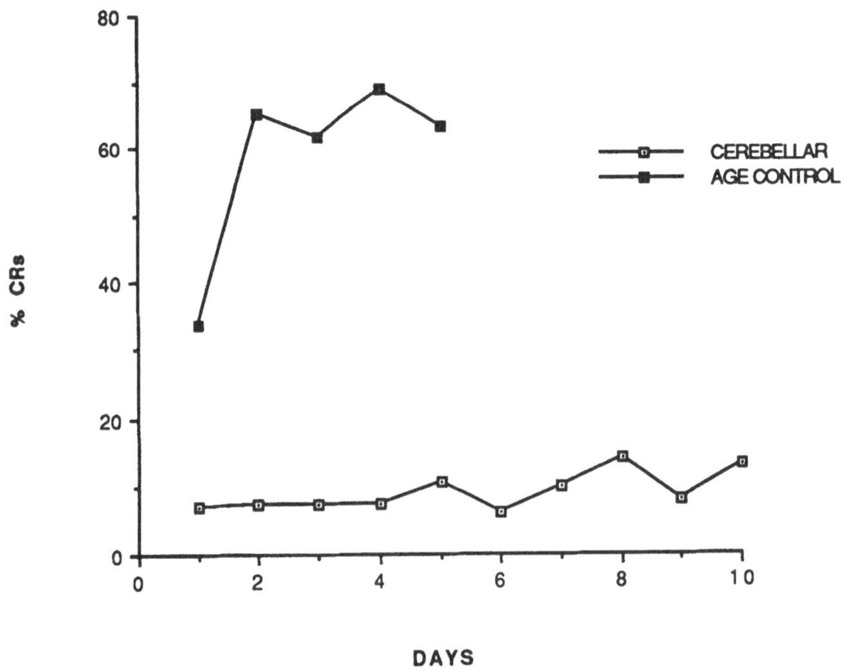

Figure 3 Mean percentage of conditioned responses (CRs) for two patients with cerebellar damage and age-matched controls.

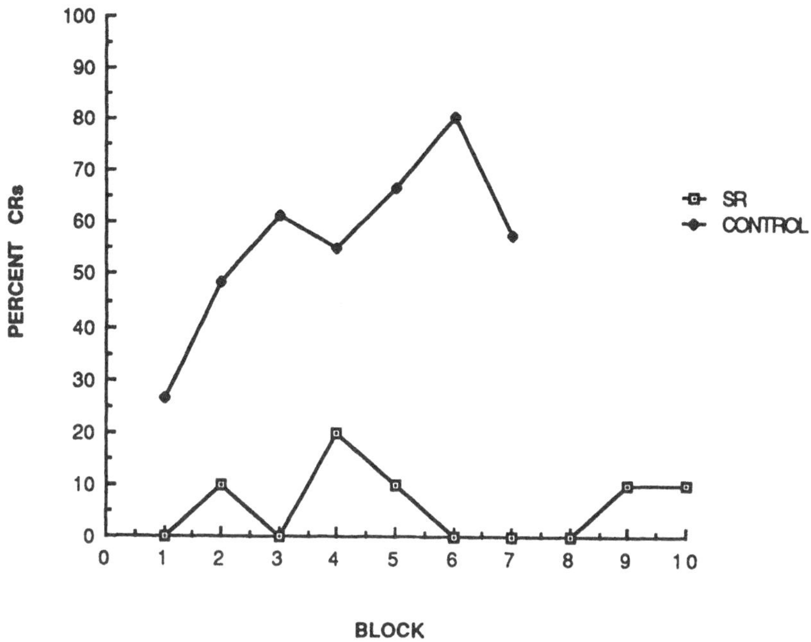

Figure 4 Mean percentage of conditioned (CRs) for age-matched control subjects over 7 blocks of conditioning and for patient SR over 10 blocks of conditioning. (From Solomon, Stowe, and Pendlebury, 1989.)

tual abilities and commensurate declarative memory. Her ability to acquire the CR, however, was disrupted. In comparison to age-matched controls who reached emitted 60% CRs after one day of conditioning, she showed no sign of acquisition over the entire day (Figure 3) (Solomon *et al.*, 1989c). We are aware of one other study that showed similar results (Lye *et al.*, 1988).

Although these studies suggest that cerebellar damage blocks acquisition of the CR, each study was conducted in only a single conditioning session. This leaves open the possibility that if sufficient conditioning trials were administered, cerebellar patients would eventually condition. To begin to address this issue, we have now tested two humans with cerebellar damage over multiple conditioning days. The first has idiopathic cerebellar degeneration and the second is a case of cerebellar damage secondary to dilantin intoxication. In both instances, and in contrast to age-matched controls who readily conditioned, there was no indication of conditioning over 10 days (Figure 4). Because most humans acquire the CR in one day, these data suggest that, as in the rabbit, cerebellar lesions permanently abolish the ability to acquire the EB response in humans.

9.3. Studies of Hippocampal Damage in Animals

Whereas it appears that the cerebellum is essential for classical EB conditioning, the hippocampus appears to play a modulatory role (Solomon, 1979, 1980; Berger *et al.*, 1986; Pennick and Solomon, 1991). Although large bilateral hippocampal lesions do not disrupt acquisition of the CR in rabbits (Solomon and Moore, 1975; Solomon, 1977), manipulations that disrupt activity in the structure, such as posttrial electrical stimulation (Salafia *et al.*, 1977), penicillin injection (Thompson *et al.*, 1980), or scopolamine-induced disruption of hippocampal EEG (Salvatierra and Berry, 1989), all disrupt conditioning. These animals do eventually acquire the response, but at a significantly slower rate than controls. These data lead to the prediction that humans with hippocampal lesions that approach the magnitude of damage seen in rabbits with bilateral aspiration lesions will acquire the CR normally, but humans with manipulations that do not completely destroy the hippocampus, but do disrupt functioning, will be retarded in acquisition. Although this is a counterintuitive position, it is well supported in the animal literature (e.g., Salvatierra and Berry, 1989; Solomon *et al.*, 1983) and is a general principle in human neuropsychology (Lezak, 1983). There is also data to support this view in the human-conditioned EB literature.

9.4. Studies of Hippocampal Damage in Humans

Weiskrantz and Warrington (1979) reported that a postencephalitic patient readily acquired the EB response, although the patient had no recollection of having been conditioned. More recently Daum, Channon, and Canavan (1989) investigated eyelid conditioning in 3 patients with suspected bilateral temporal lobe damage and found that they conditioned within the range of normal controls. Although it is difficult to find patients with damage to hippocampus that can be said to produce altered activity such as that seen in posttrial stimulation or penicillin-induced seizures in animals, one approximation may be early to mid-stage Alzheimer's patients. It is now well documented that the hippocampus is a primary site of pathology in Alzheimer's disease (AD) (e.g., Hyman *et al.*, 1984) and it is one of the earliest structures involved in the disease. Our initial work with conditioning in AD patients showed that compared to aged-matched controls, AD patients did not condition in a single conditioning session (Solomon *et al.*, 1989a, 1991b). We would note that Woodruff-Pak and colleagues (Woodruff-Pak *et al.*, 1990) have produced similar results. These studies, however, do not answer the important question of whether, like our cerebellar patients, AD patients were incapable of acquiring the conditioned response or whether given sufficient conditioning sessions, they would learn. To address this question, we are in the process of carrying out a study in which AD patients undergo multiple conditioning sessions. We have now had the opportunity to test

7 patients and 6 controls over 5 conditioning sessions. All controls showed clear signs of acquisition of the CR. As in our single-session study, AD patients did not condition in the first session, but perhaps more importantly, 6 of the 7 patients did acquire the CR over the 5 days of conditioning. These data are consistent with animal research in indicating that disease states that produce an abnormal hippocampus disrupt conditioning, but do not preclude eventual acquisition of the CR.

9.5. Studies in Aging Animals

Although there is now a growing literature on learning and memory in a variety of species across the life span, only a few published studies use classical EB conditioning (see Powell *et al.*, 1991, for a review). In one of these studies, we (Graves and Solomon, 1985) found that aged (36–60 months) animals acquired the CR in a trace-conditioning paradigm significantly more slowly than did controls (6 months). These findings have been replicated by Woodruff-Pak, Lavond, Logan, and Thompson (1987). Unfortunately, to date, none of these studies have correlated age-related behavioral changes with changes in either neuropathology or neurochemistry. To address this issue we have recently completed a study examining several types of conditioning in animals ranging in age from 6 to 48 months. The behavioral data from this study are now collected and they generally agree with the results of our earlier study in indicating that conditioning does decline with aging, with deficits occurring at about age 4. What is particularly interesting about these data is that about 25% of our aged animals conditioned nearly as well as young animals. We are currently attempting to correlate the anatomical and neurochemical data with the behavioral data in these animals in an attempt to determine what allows some animals to age more "successfully" than others. This type of information should begin to allow us to develop hypotheses regarding the underlying neurobiology of these deficits. It should then be possible to manipulate these neurobiological systems in young animals to test these hypotheses.

9.6. Studies in Aging Humans

Three early studies reported age-related deficits in EB conditioning in humans (Braun and Geiselhart, 1959; Gakkel and Zinna, see Jerome, 1959; Kimble and Pennypacker, 1963). Each of these, however, is subject to alternative explanations. For example, none of the studies attempted to determine whether nonassociative factors such as age-related changes in sensitivity to the stimuli or changes in spontaneous blink rates contributed to conditioning rate.

To address these issues we have completed several studies examining conditioning over the life span. In the first of these studies (Solomon *et al.*, 1989b), we

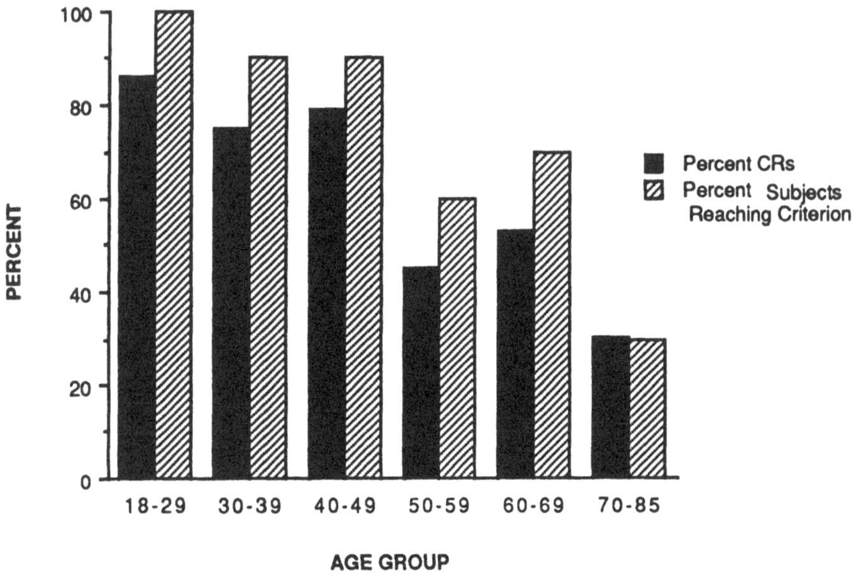

Figure 5 Mean percentage of conditioned responses (CRs) and mean percentage of subjects reaching criterion for acquisition of the CR for subjects in six age groups. (From Solomon *et al.*, 1989c.)

found an age-related decline in conditioning that became most pronounced at age 50 (Figure 5). These differences could not be attributed to nonassociative factors such as sensory or motor deficits nor to changes in spontaneous blink rates.

It is somewhat difficult to compare ages of rabbits and humans to determine whether the age-related conditioning deficits are similar in both species, but the comparisons that can be made are suggestive. Comparatively little is known about life expectancy in the rabbit. The available information suggests that rabbits have a life expectancy of about 7 to 8 years (Fox, 1980). Using this as a benchmark, it may not be unreasonable to assume that a 50-year-old human is similar chronologically to a 4- to 5-year-old rabbit. If this is the case, then it may be noteworthy that conditioning deficits in both species begin at about the same point in the life span.

9.7. Studies in Animals following Scopolamine Administration

One of the best-characterized effects in conditioning of the rabbit's EB response is disrupted acquisition of the conditioned response following central cholinergic blockade. Several studies have reported disrupted conditioning of the rabbit's classically conditioned EB response following systemic administration of atropine (Downs *et al.*, 1972) or scopolamine (Harvey *et al.*, 1983; Moore *et al.*, 1976; Powell, 1979). Our own work has shown that animals who receive injections of scopolamine hydrobromide condition significantly more slowly than do controls given a peripheral analogue (scopolamine methylbromide) or a vehicle (saline) (Moore *et al.*, 1976). Although these animals initially show retarded acquisition, they eventually acquire the CR. As such, this pattern is much like that seen in rabbits with altered activity in hippocampus and with AD patients. Interestingly, subsequent work in animals had strongly suggested that the disruptive effect of conditioning in animals is mediated through altered activity in hippocampus (Solomon and Gottfried, 1981; Salvatierra and Berry, 1989; Solomon *et al.*, 1983). Moreover, recent work carried out in collaboration with M.F. Beal has shown that there is a depletion of AchE in hippocampus of aged rabbits who showed poor conditioning.

9.8. Studies in Humans following Scopolamine Administration

We have recently had the opportunity to examine the effects of scopolamine on EB conditioning in humans (Solomon *et al.*, 1993). As in each of our other lines of research, the conditioning parameters used were identical to those used in our animal studies. In this study, 72 human volunteers received either saline,

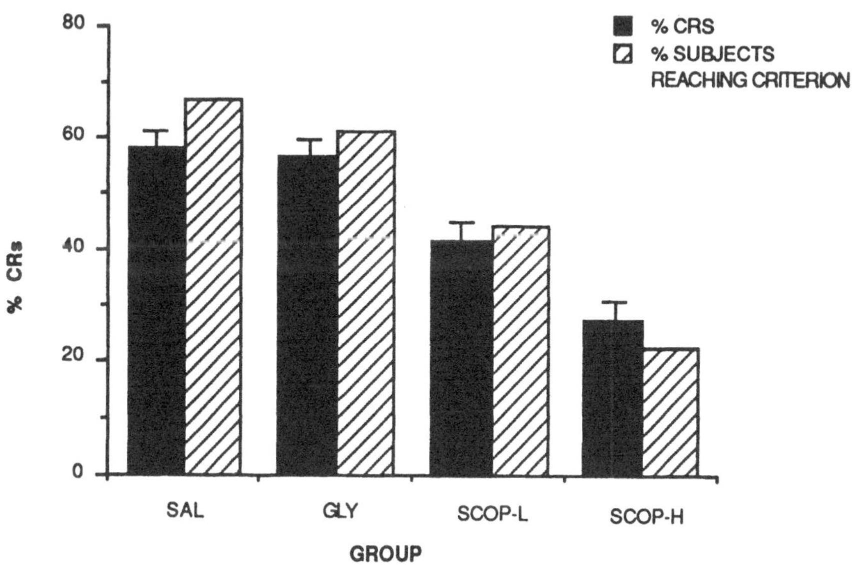

Figure 6 Mean percentage of conditioned responses (CRs) for subjects receiving a low dose of scopolamine (SCOP-L), a high dose of scopolamine (SCOP-H), glycopyrrolate (GLY), or saline (SAL).

oral scopolamine (0.6 mg or 1.2 mg), or the peripheral analogue, glycopyrrolate. Subjects who received glycopyrrolate conditioned at about the same rate as controls, but scopolamine-treated subjects showed impaired conditioning in a dose-related manner (Figure 6). As in our other studies, we were able to rule out nonassociative explanations for these effects.

Because this study was conducted within a single conditioning session, we cannot at this time address the issue of whether the rates of conditioning over multiple days would be the same in rabbits and humans, but we can suggest that this drug clearly disrupts conditioning in both species. To our knowledge, this is the first demonstration of a drug disrupting an identical learned behavior in laboratory animals and humans.

10. NEUROTOXICOLOGICAL ASPECTS OF EYEBLINK CONDITIONING: SIMILARITIES BETWEEN ANIMALS AND HUMANS

10.1. Studies in Rabbits following Aluminum Intoxication

Ultimately, any animal model of neurotoxic contributions to neurological disorders should test known neurotoxins. In our hands, this would mean testing the same neurotoxin in humans and animals in the context of a model system. One such neurotoxin is aluminum. Aluminum may be a particularly interesting substance because of its suggested relationship to a variety of degenerative neurological disorders including amytrophic lateral sclerosis and Guamanian Parkinson's disease (Garruto and Yase, 1985; Perl and Brody, 1980; Perl *et al.*, 1982). There has also been the controversial suggestion that aluminum may contribute to Alzheimer's disease.

Intrathecal administration of aluminum salts in the rabbit produces an acute encephalopathy that is characterized pathologically by the formation of neuronal cytoplasmic structures resembling neurofibrillary tangles (NFT) at the light microscopic level (Klatzko *et al.*, 1965). To characterize the aluminum-induced neurofilamentous degeneration (NFD), we have been involved in a series of studies investigating the behavioral, pathological, neurochemical, and immunocytochemical effects of aluminum intoxication in rabbits. These results have been reviewed elsewhere (Pendlebury *et al.*, 1988a; Solomon and Pendlebury, 1988), but briefly we have found:

1. Intraventicular injection of $AlCl_3$ produces NFD and disrupts both acquisition and retention of the CR. Animals that receive aluminum prior to conditioning acquire the CR significantly more slowly than do controls (Pendlebury *et al.*, 1988c), and rabbits that first acquire the CR and then receive aluminum show disrupted retention of the CR relative to controls (Solomon *et al.*, 1988b).

2. Disrupted learning and memory does not appear to stem from either sensory or motor deficits. Motor responses are intact. $AlCl_3$ rabbits are no different from controls in their ability to give the eyeblink response reflexively when presented with an air puff UCS (i.e., the response amplitudes do not differ from controls). Responses to the tone CS also appear unaltered. Threshold tests for tone are similar in control and $AlCl_3$ rabbits. We have also demonstrated that memory deficits are present when conditioned stimuli from three different modalities are used (i.e., tone, light, and vibratory). Finally, memory is also disrupted in aluminum-treated rabbits when electrical brain stimulation to the medical geniculate body is used as a CS. This result indicates that memory deficits are not due to pathology of the primary sensory pathways as they are bypassed using this technique (Solomon *et al.*, 1987).
3. There is no correlation between aluminum-induced illness and either acquisition or retention of the response (Pendlebury *et al.*, 1988c; Solomon *et al.*, 1988b).
4. Deficits in learning and memory are correlated with the degree of pathology. We have evaluated the number of neurons containing neurofilamentous accumulation in five brain areas (frontal and parietal cortex, ventral hippocampus, cerebellum, and pons) and have found the overall degree of pathology is significantly correlated with the percentage of CRs (Pendlebury *et al.*, 1988b; Solomon *et al.*, 1988). This may be particularly interesting given our data showing disrupted conditioning in rabbits and humans following scopolamine administration.
5. Using monoclonal antibodies directed against neuronal cytoskeletal components and the avidin-biotin immunocytochemical technique, we have shown that phosphorylated and nonphosphorylated neurofilament epitopes are present in $AlCl_3$-induced neurofilamentous accumulations (Munoz-Garcia *et al.*, 1986; Pendlebury *et al.*, 1988a).
6. Quantitative neurochemical analysis of $AlCl_3$-exposed rabbits revealed similarities to neurochemical deficits seen in AD patients. The most striking parallel is the significant reduction in CAT activity in the entorhinal cortex and hippocampus in aluminum-exposed rabbits (Beal *et al.*, 1989).

Most recently, we have been involved in a series of studies investigating the effects of inhaled aluminum on conditioning in rabbits. These animals were first conditioned and then surgically implanted in the nasal cavity with Gelfoam soaked in aluminum lactate. Following exposure for 6 months, they were tested for retention of the CR. Our results to date indicate that these animals show retention deficits and develop neuropathologic features characterized by perivesicular inflammation and granuloma formation in the olfactory pathways. But perhaps of greatest significance, and unlike animals in our acute studies, these animals do not become ill.

10.2. Studies in Humans following Aluminum Intoxication

We do not at this time have any data on humans, yet there are several intriguing possibilities. McGlaughlin and his colleagues have identified a group of miners who had extensive exposure to aluminum as a prophylaxis for silicosis (personal communication, 1991). These miners developed cognitive deficits beyond those of age-matched controls. It would be most interesting to evaluate these groups in the EB-conditioning situation.

11. Summary

We have attempted in this chapter to demonstrate how the model systems approach to the neurobiology of learning and memory might be a useful in understanding memory loss associated with neurotoxicant-induced neurodegenerative disorders. We have emphasized the idea that for an animal model to be useful, it is important to be able to generalize from animals to humans. One way to accomplish this is to use a model system in which the identical behavior can be studied in both humans and animals. Using this approach, if various manipulations produce similar effects in both species, the risks inherent to cross-species generalization would seem to be reduced. We have presented data from eyeblink conditioning in humans and rabbits to suggest that this is the case. We have shown that a number of manipulations and conditions produce similar effects in both species including (1) brain lesions to the cerebellum and hippocampus, (2) aging, and (3) scopolamine administration. We have also provided evidence regarding the effects of a known neurotoxicant—aluminum—on conditioning in rabbits and suggested how we might study aluminum-exposed humans. Collectively, these data suggest that using relatively simple systems in which both the behavior and neurobiology are well understood, and in which identical behaviors can be studied in humans and animals, may eventually be helpful in establishing hypotheses about the etiology and pathogenesis of neurodegenerative disorders.

ACKNOWLEDGMENTS. The manuscript was reviewed by the Health Effects Research Laboratory, U.S. Environmental Protection Agency, and approved for publication. Mention of trade names or commercial products does not constitute endorsement or recommendation for future use. The research was supported by EPA grant CR816043-01-0, NIA grant AGO5134, a grant from the Essel Foundation, and a Faculty Research Grant from Williams College. Finally, we would like to thank C.J. Gillig for his help in preparing the figures.

12. References

Beal, M.F., Mazurek, M.F., Ellison, D.W., Kowall, N.W., Solomon, P.R., and Pendlebury, W.W., 1989, Neurochemical characteristics of aluminum-induced neurofibrillary degeneration in rabbits. *Neuroscience* 29, 339–346.

Berger, T.W., Berry, S.D., and Thompson, R.F., 1986, Role of hippocampus in classical conditioning of aversive and appetitive behaviors, *in* "The Hippocampus" (R.L. Isaacson and K.L. Pribram, eds.), Plenum Press, New York, pp. 203–240.

Braun, W.H., and Geiselhart, R., 1959, Age differences in the acquisition and extinction of the conditioned eyelid response, *J. Exp. Psychol.* 57:386–388.

Byrne, J.H., 1987, Cellular analysis of associative learning, *Physiol. Rev.* 67:329–349.

Cranmer, J.M., 1990, Neurotoxicity and risk assessment, *Neurotoxicology* 11:177–390.

Daum, I., Channon, S., and Canavan, A.G.M., 1989, Classical conditioning in patients with severe memory problems, *J. Neurol. Neurosurg. Psychiatry* 52:47–51.

Downs, D., Cardoza, C., Schneiderman, N., Yehle, A.L., Van Decar, D.H., and Zwilling, G., 1972, Central effects of atropine upon aversive classical conditioning in rabbits, *Psychopharmacologia* 23:319–333.

Fox, R.R., 1980, The rabbit (*Oryctolagus cuniculus*) and research on aging, *Exp. Aging Res.* 6(3):235–248.

Garruto, R.M., and Yase, Y., 1985, Neurodegenerative disorders of the Western pacific: The search for pathogenesis, *Trends Neurosci.* 9:368–373.

Gormezano, I., 1966, Classical conditioning, *in* "Experimental Methods and Instrumentation in Psychology" (J.B. Sidowski, ed.), Mc-Graw Hill, New York, pp. 385–420.

Gormezano, I., Kehoe, E.J., and Marshall, B.S., 1983, Twenty years of classical conditioning with the rabbit, *Prog. Psychobiol. Physiol. Psychol.* 10:197–275.

Gormezano, I., and Moore, J.W., 1969, Classical Conditioning, *in* "Learning Processes" (M.H. Marx, ed.), Macmillan, New York, pp 129–206.

Gormezano, I., Prokasy, W.F., and Thompson, R.F. (eds.), 1987, "Classical conditioning III: Behavioral, Physiological, and Neurochemical Studies in the Rabbit, Erlbaum, Hillsdale, NJ.

Graves, C.A., and Solomon, P.R., 1985, Age-related disruption of trace but not delay classical conditioning of the rabbit's nictitating membrane response, *Behav. Neurosci.* 99:88–96.

Harvey, J.A., Gormezano, I., and Cool-Hauser, V.A., 1983, Effects of scopolamine and methylscopolamine on classical conditioning of the rabbit's nictitating membrane response, *J. Pharmacol. Exp. Ther.* 225:42–49.

Hilgard, E.R., and Marquis, D.G., 1940, "Conditioning and Learning," Appleton-Century-Crofts, New York.

Hyman, B.T., Van Hoesin, G.W., Damasio, A.R., and Barnes, C.L., 1984, Alzheimer's disease: Cell-specific pathology isolates the hippocampal formation, *Science* 225:1168–1170.

Jerome, E.A., 1959, Age and learning: Experimental studies, *in* "Handbook of Aging and the Individual" (J.E. Birren, ed.), University of Chicago Press, Chicago. pp. 655–699.

Kimble, B.A., and Pennypacker, H.S., 1963, Eyelid conditioning in young and aged subjects, *J. Genet. Psychol.* 103:283–289.

Kimmel, C.A., 1990, Quantitative approaches to human risk assessment for non-cancer health effects, *Neurotoxicology* 11:189–198.

Klatzko, I., Wisniewski, H., and Streicher, E., 1965, Experimental production of neurofibrillary degeneration. I: Light microscopic observations, *J. Neuropathol. Exp. Neurol.* 24:187–199.

Koob, G.F., and Bloom, F.E., 1983, Cellular and molecular mechanisms of drug dependence, *Science* 242:715–723.

Lezak, M., 1983, "Neuropsychological Assessment" (2nd ed.), Oxford University Press, New York.

Lindsay, G., Hall, C.S., & Thompson, R.F., 1978, Psychology (Second Edition). New York: Worth Publishers, New York, pp. 208.

Lye, R.H., O'Boyle, D.J., Ramsden, R.T., and Schady, W., 1988, Effects of unilateral lesion on the acquisition of eyeblink conditioning in man, *J. Physiol.* 403:58.

McKinney, W.T., and Bunny, W.E., 1969, Animal models of depression. I: Review of the evidence, implications for research, *Arch. Gen. Psychiatry* 21:240–248.

Moore, J.W., Goodell, N.A., and Solomon, P.R., 1976, Central cholinergic blockade by scopolamine and habituation, classical conditioning, and latent inhibition of the rabbit's nictitating membrane response, *Physiol. Psychol.* 4(3):395–399.

Munoz-Garcia, D., Pendlebury, W.W., Kessler, J.B., and Perl, D.P., 1986, An immunocytochemical comparison of cytoskeletal protein in aluminum-induced and Alzheimer-type neurofibrillary tangles, *Acta Neuropathol.* 70:243–248.

National Academy of Sciences, 1983, "Risk Assessment in Federal Government: Managing the Process," National Academy Press, Washington, DC.

Pendlebury, W.W., Beal, M.F., Kowall, N.W., and Solomon, P.R., 1987, Immunocytochemical, neurochemical, and behavioral studies in aluminum-induced neurofilamentous degeneration. *J. Neur. Trans.* (supplement 24) 214–217.

Pendlebury, W.W., Beal, M.F., Kowall, N.W., and Solomon, P.R., 1988a, Neuropathologic, neurochemical, and immunocytochemical characteristics of aluminum-induced neurofilmanetous degeneration, *Neurotoxicology* 9(3):503–510.

Pendlebury, W.W., Perl, D.P., Schwentker, A., Pingree, T., and Solomon, P.R., 1988b, Aluminum-induced neurofibrillary degeneration disrupts acquisition of the rabbit's classically conditioned nictitating membrane response, *Behav. Neurosci.* 102(5):615–620.

Penick, S., and Solomon, P.R., 1991, Hippocampus, context, and conditioning. *Behav. Neurosci.* 105:609–613.

Perl, D.P., and Brody, X., 1980, Alzheimer's disease: X-ray spectrometric evidence of aluminum accumulation in neurofibrillary tangle-bearing neurons, *Science* 208:297–299.

Perl, D.P., Gajdusek, D.C., Garruto, R.M., Yanagihara, R.T., and Gibbs, C.G., 1982, Intraneuronal aluminum accumulation in amyolateral sclerosis and Parkinsonism-dementia of Guam, *Science* 217:1053–1055.

Powell, D.A., 1979, Peripheral and central muscarinic cholinergic blockade: Effects on Pavlovian conditioning, *Bull. Psychonomic Soc.* 14:161–164.

Powell, D.A., Buchanan, S.L., and Henrnandez, L.L., 1991, Classical (Pavlovian) conditioning models of age-related changes in associative learning and their neurobiological substrates, *Prog. Neurobiol.* 11:201–228.

Salafia, W.R., Romano, A.G., Tynan, T., and Host, K.C., 1977, Disruption of rabbit (*Oryctalogus cuniculus*) nictitating membrane conditioning by post-trial electrical stimulation of hippocampus, *Physiol. Behav.* 18:207–212.

Salvatierra, A.T., and Berry, S.D., 1989, Scopolamine disruption of septo-hippocampal activity and classical conditioning, *Behav. Neurosci.* 103(4):715–721.

Solomon, P.R., 1977, Role of hippocampus in blocking and conditioned inhibition of the rabbit nictitating membrane response, *J. Comp. Physiol. Psychol.* 91(2):407–417.

Solomon, P.R., 1979, Temporal versus spatial information processing theories of hippocampal function, *Psychol. Bull.* 86(6):1272–1279.

Solomon, P.R., 1980, A time and a place for everything? Temporal processing views of hippocampal function with special reference to attention, *Physiol. Psychol.* 8(2):254–261.

Solomon, P.R., and Gottfried, K.E., 1981, Disruption of acquisition of the classically conditioned rabbit nictitating membrane response following injection of scopolamine into the medial septal nucleus, *J. Comp. Physiol. Psychol.* 95(2):322–330.

Solomon, P.R., and Moore, J.W., 1975, Latent inhibition and stimulus generalization of the classically conditioned nictitating membrane response in rabbits (*Oryctolagus cuniculus*) following dorsal hippocampal ablation, *J. Comp. Physiol. Psychol.* 89(10):1192–1203.

Solomon, P.R., and Pendlebury, W.W., 1988, A model systems approach to age-related memory disorders, *Neurotoxicology* 9:443–462.

Solomon, P.R., and Pendlebury, W.W., 1992, Aging and Memory: A model systems approach, *in:* N. Butters and L. R. Squire (Eds.), *Neuropsychology of Memory*, Guilford Press, New York, pp. 262–276.

Solomon, P.R., and Pendlebury, W.W., 1993, Alzheimer's disease: Therapeutic strategies for the 1990s, *Neurobiol. Aging,* in press.

Solomon, P.R., Solomon, S.D., Vander Schaaf, E.R., and Perry, H.E., 1983, Altered activity in the hippocampus is more detrimental to conditioning than removal of the structure, *Science* 220:329–331.

Solomon, P.R., Koota, D., Kessler, J.B., and Pendlebury, W.W., 1987, Disrupted retention of the classically conditioned eyeblink response in the aluminum-intoxicated rabbit using brain stimulation as a conditioned stimulus, *Soc. Neurosci. Abstr.* 13(part 1):642.

Solomon, P.R., Beal, M.F., and Pendlebury, W.W., 1988a, Age-related disruption of classical conditioning: A model systems approach to age-related memory disorders, *Neurobiol. Aging* (Special Issue) 9:935–946.

Solomon, P.R., Pingree, T., Baldwin, D., Koota, D., Perl, D.P., and Pendlebury, W.W., 1988b, Disrupted retention of the classically conditioned nictitating membrane response in rabbits with aluminum-induced neurofibrillary degeneration, *Neurotoxicology* 9:209–222.

Solomon, P.R., Levine, E., Bein, T., and Pendlebury, W.W., 1989a, Disrupted eyelid conditioning in patients with Alzheimer's disease, *Soc. Neurosci. Abstr.* 15:888.

Solomon, P.R., Pomerleau, D., Bennett, L., James, J., and Morse, D.L., 1989b, Acquisition of the classically conditioned eyeblink response in humans over the lifespan, *Psychol. Aging* 4:34–41.

Solomon, P.R., Stowe, G.T., and Pendlebury, W.W., 1989c, Disrupted eyelid conditioning in a patient with damage to the cerebellar inputs, *Behav. Neurosci.* 103:898–902.

Solomon, P.R., Groccia-Ellison, M., Levine, E., and Blanchard, S., 1991a, Do temporal relationships in conditioning change across the lifespan? Perspectives from eyeblink conditioning in humans and rabbits, *in* "The Development and Neural Bases of Higher Cognitive Functions" (A. Diamond, ed.), MIT Press, Cambridge, MA., pp. 212–238.

Solomon, P.R., Levine, E., Bein, T., and Pendlebury, W.W., 1991b, Disruption of classical conditioning in patients with Alzheimer's Disease, *Neurobiol. Aging* 12:283–287.

Solomon, P.R., Groccia-Ellison, M., Flynn, D., Mirak, J., Edwards, K.E., and Stanton, M.E., 1993, Disruption of the human eyeblink response following central cholinergic blockade with scopolamine, *Behav. Neurosci.* 107:271–279.

Thompson, R.F., 1986, The neurobiology of learning and memory, *Science* 233:941–947.

Thompson, R.F., 1988, A model systems approach to age and the neuronal bases of learning and memory, *in* "Perspectives on Behavioral Medicine" (M.W. Riley, J.D. Matarazzo, and A. Baum, eds.), Academic Press, New York, pp. 118–131.

Thompson, R.F., Berger, T.W., Berry, S.D., Hoehler, F.K., Kettner, R.D., and Weisz, D.J., 1980, Hippocampla substrate of classical conditioning. *Physiological Psychology,* 8:269–279.

Tilson, H.A., Jacobson, J.L., and Rogan, W.J., 1990, Polychlorinated biphenyls and the developing nervous system: Cross-species comparisons, *Neurotoxicol. Teratol.* 12:239–248.

U.S. Congress, Office of Technology Assessment, 1990, *Neurotoxicity: Identifying and Controlling Poisons of the Nervous System,* G.P.O., Washington, D.C.

Weiskrantz, L., and Warrington, E.K., 1979, Conditioning in amnesic patients, *Neuropsychologia* 17:187–194.

Woodruff-Pak, D.S., Finkbinder, R.G., and Sasse, D.K., 1990, Eyeblink conditioning discriminates Alzheimer's patients from non-demented controls, *Neuroreport* 1:45–59.
Woodruff-Pak, D.S., Lavond, D.G., Logan, C.G., and Thompson, R.F., 1987, Classical conditioning in 3-, 30-, and 45-month-old rabbits: Behavioral learning and hippocampal unit activity, *Neurobiol. Aging* 8:101–108.
Zola-Morgan, S., and Squire, L., 1986, Memory impairment in monkeys following lesion of the hippocampus, *Behav. Neurosci.* 100:155–160.

3

Experimental Paradigms of Motor Neuron Degeneration

MICHAEL J. STRONG and RALPH M. GARRUTO

1. Introduction

A motor neuron, by strict definition, is any neuron that functionally innervates a muscle fiber. In common usage, however, this term refers to both the supraspinal descending neurons of the motor pathways (upper motor neurons, UMN) and lower motor neurons (LMN). By analogy, the term *motor neuron degeneration* does not imply a disease state or process restricted to a single motor pathway. Because the manifestations of motor neuron degeneration, whether UMN or LMN, are ultimately governed by the motor unit consisting of the alpha motor neuron, its axon, and muscle fibers, only a limited repertoire of clinical phenomenology is possible for a relatively broad group of degenerative or toxic states. Therefore, in designing paradigms of motor neuron degeneration, it is important to differentiate among those degenerative processes that can affect upper or lower motor neurons selectively or in a combination, and yet still yield relatively similar disease states.

To be useful in this context, an experimental paradigm of motor neuron degeneration should be evaluated by a series of questions: Does the natural or experimental model at hand recapitulate the unique topographic specificity of human motor neuron degeneration; does it recapitulate the morphological and ultrastructural characteristics of the disease; and finally, are the biological characteristics, to the extent known, replicated? Clearly, no experimental paradigm of motor neuron degeneration meets each of the above criteria and, for amyotrophic lateral sclerosis, the most common of these disease entities, no

MICHAEL J. STRONG • Department of Clinical Neurological Sciences, The University of Western Ontario, London, Ontario, Canada N6A 5A5. RALPH M. GARRUTO • Laboratory of Central Nervous System Studies, National Institutes of Health, Bethesda, Maryland 20892.
Toxin-Induced Models of Neurological Disorders, edited by Michael L. Woodruff and Arthur J. Nonneman. Plenum Press, New York, 1994.

definitive model has yet been produced. However, by addressing specific characteristics of the disease process, each model contributes to our understanding of the human degenerative process. In this review, we begin by addressing the functional neuroanatomy of the motor system, its developmental biological properties, and the relevant human disease states. Thereafter, experimental paradigms are described along with their limitations, relevance, and relative contribution to our understanding of the degenerative process of motor neurons.

2. Functional Anatomy of the Motor System

The motor system is traditionally described as possessing both UMN and LMN components. The former, by convention, has implied those neurons and their projections coursing longitudinally within the pyramids of the medulla (pyramidal tract). However, given that descending pathways, distinct from those in the pyramids, also mediate motor neuron activity, UMN fibers are more correctly considered as including all descending supraspinal tracts that directly or indirectly influence the activity of motor neuron pools within the ventral spinal cord (for a detailed review, see Brodal, 1981).

2.1. Corticofugal Motor Pathway

Cortically originating (corticofugal) motor fibers that directly, or via interneurons, mediate the activity of the peripheral motor neuron originate from an extensive region of the cerebral hemispheres, including not only the motor cortex proper (precentral gyrus or area 4 of Brodmann) and the more rostrally located area 6, but additionally the somatic sensory cortex (areas 3, 2, and 1 of Brodmann) (Figure 1). Efferent fibers arise from large Betz cells in cortical layer V and converge to form, with the thalamocortical fibers, the corona radiata. These fibers course through the posterior limb of the internal capsule, ventrally within the cerebral peduncles, and within the brainstem through the mesencephalon to the basis pontis where the medial and lateral divisions of the corticopontine fibers terminate in the pontine nuclei (Figure 2).

Large bundles of fibers additionally continue through the basis pontis, including fibers (e.g., corticobulbar tracts) supplying the somatic efferent nuclei of the cranial nerves (e.g., hypoglossal, accessory nerve, facial nucleus). The remaining fibers cross caudally within the brainstem at the level of the pyramidal decussation and are thereafter termed *corticospinal fibers*. The majority of the corticospinal fibers descend contralateral to their site of origin in the lateral corticospinal tract of the spinal cord (Figure 3). However, a minority descend ipsilaterally in the ventral corticospinal pathways.

In the spinal cord, the corticospinal tracts make both monosynaptic and polysynaptic connections with the alpha-motor neurons. The lateral corticospinal

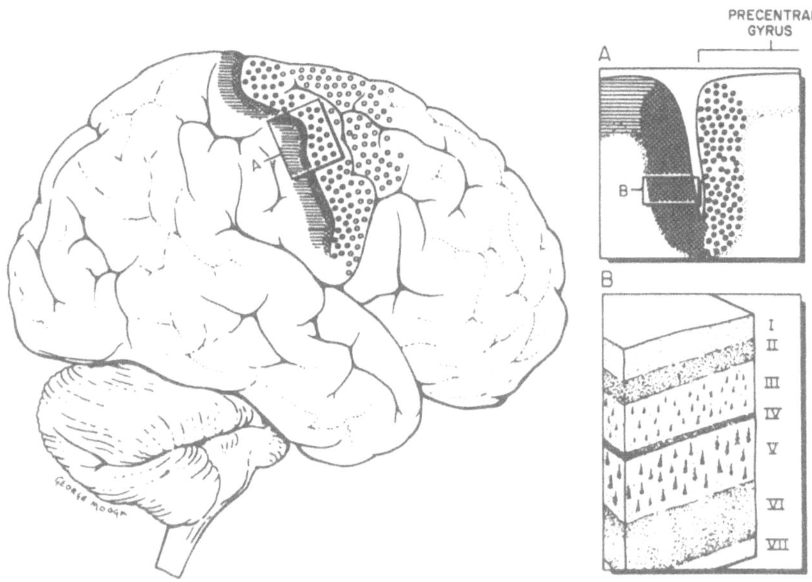

Figure 1 Lateral view of the cerebral hemisphere illustrating the precentral gyrus (closed dots: Brodmann area 4), supplementary motor cortex (open dots: Brodmann area 6), and postcentral gyrus (crosshatch: Brodmann area 3; horizontal lines: Brodmann area 1). (**A**) Idealized section taken at the level of the central sulcus perpendicular to the cortical surface. (**B**) Drawing of a Nissl-stained section taken through the postcentral gyrus perpendicular to the cortical surface demonstrating the cortical layers I through VI (VII represents subcortical white matter). Efferent motor fibers arise from large pyramidal neurons in layer V.

tract projects to lamina IV and V (of Rexed), to the interneurons in the intermediate zones, and to the dorsolateral ventral horn motor neurons subserving distal muscles. The ventral corticospinal tracts ultimately project bilaterally to the medial ventral horn motor neurons subserving axial and proximal muscles. The presence of monosynaptic interactions between corticospinal efferents and the motor neuron is phylogenetically determined, with lower vertebrates having all corticospinal fibers terminating in lamina IV through VII with interneurons required to innervate alpha motor neurons. In humans, although the majority of alpha motor neuron dendrites ascend and descend several segments within the spinal cord, dorsal dendritic projections into layer VI allow monosynaptic interactions with descending supraspinal tracts. Functionally, the corticospinal tract is facilitatory to alpha motor neurons supplying flexor muscles and inhibitory to those supplying extensor muscles.

A phylogenetically more primitive series of descending supraspinal path-

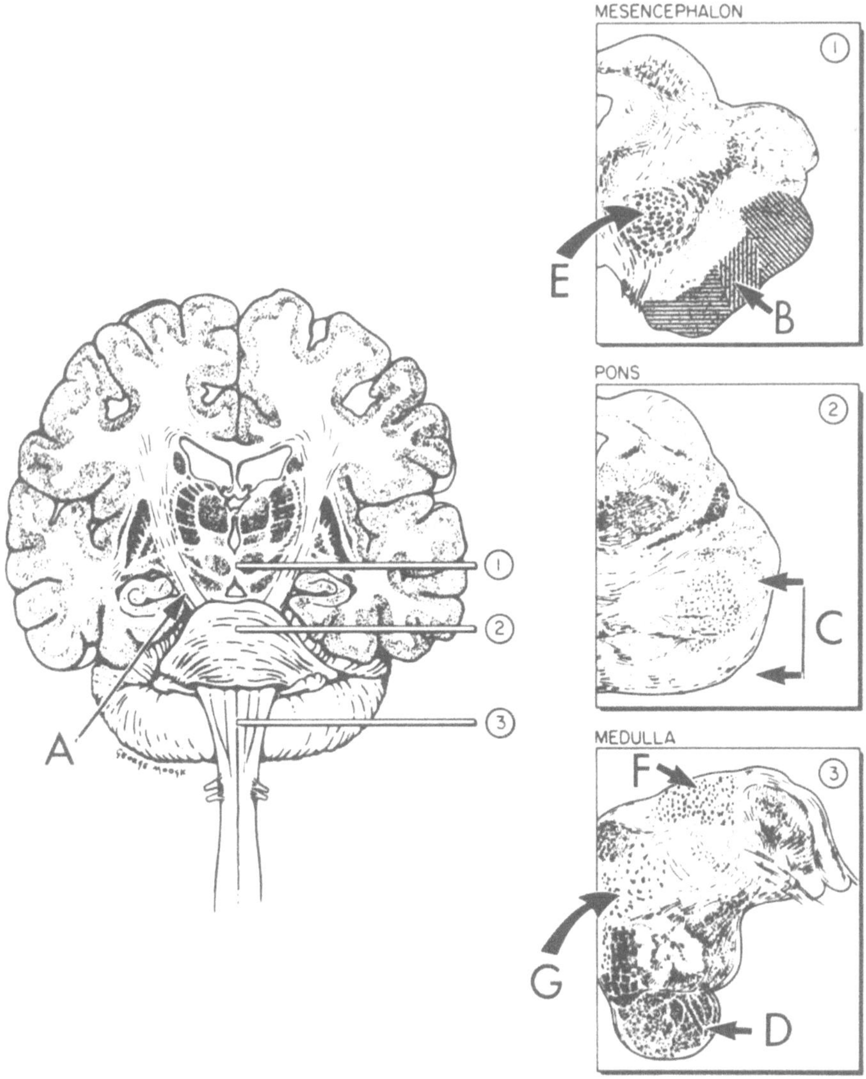

Figure 2 Corticofugal motor pathways in a frontal view, coronal section. Cortically originating fibers course through the internal capsule (**A**) to run ventrally within the cerebral peduncle (**B**). Fibers thereafter descend through the pons in the corticopontine tracts (**C**) to the medulla where the majority cross in the pyramidal decussation (**D**) to run contralateral to their side of origin. More primitive descending supraspinal fibers arise from the nucleus ruber in the mesencephalon (**E**), the vestibular complex in the medulla (**F**), and the reticularis gigantocellularis (**G**).

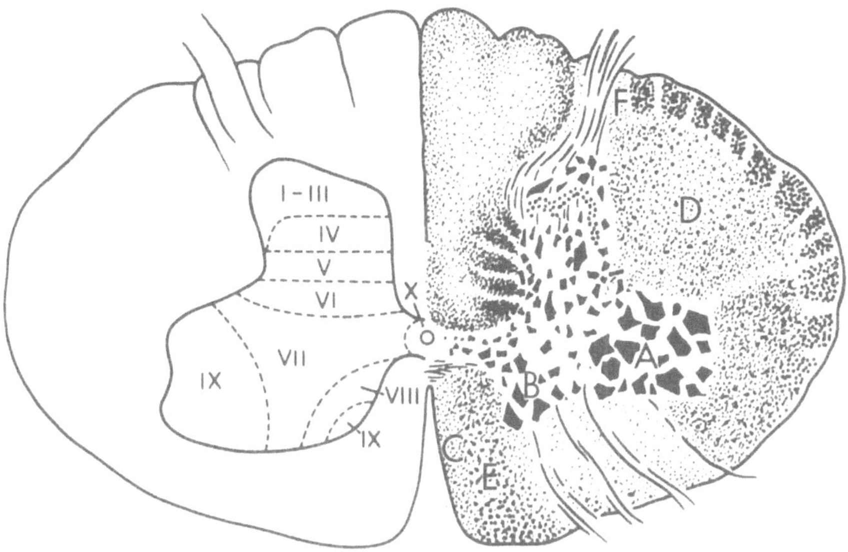

Figure 3 Diagram of a transverse section of the spinal cord at the midcervical level. The left side depicts the lamina of Rexed. (**A**) lateral motor nucleus; (**B**) medial motor nucleus; (**C**) ventral corticospinal tracts; (**D**) lateral corticospinal tracts; (**E**) medial longitudinal fasciculus (vestibulospinal and reticulospinal tracts); (**F**) dorsolateral fasciculus (rubrospinal tract).

ways also exists, providing an additional control over the activity of the alpha motor neuron. The most important of these includes the rubrospinal, vestibulospinal, and reticulospinal tracts.

The rubrospinal tract arises from the magnocellular portion of the nucleus ruber (red nucleus, a named derived from the red color conferred upon the nucleus by high iron content) at the level of mesencephalon (Figure 2). These efferent fibers are almost entirely crossed and course in the dorsolateral fasciculus with fibers terminating in lamina V, VI, and components of VII. The nucleus ruber also sends fibers to the lateral reticular nucleus, inferior olive, and external cuneate nucleus—from whence fibers to the cerebellum arise. Somatotopically organized fibers of the rubrospinal tract are involved in excitation of the alpha-motor neuron and static gamma motor neurons of flexor muscles via interneurons. The nucleus ruber receives monosynaptic excitatory input from the cortical fibers.

Vestibulospinal fibers arise from a cluster of nuclei that form the vestibular complex at the pontomedullary junction. The lateral vestibular nucleus gives rise to the lateral vestibulospinal tract of the spinal cord, which terminates in Rexed lamina VII, with many of the lamina VII interneurons projecting contralaterally. The medial vestibulospinal fibers correspond to the descending fibers arising from the vestibular nucleus remaining in the medial longitudinal fasciculus. These fiber tracts mediate increased extensor tone in ipsilateral limbs.

Reticulospinal fibers arise from the pontine and medullary reticular formation, the latter predominantly from the nucleus reticularis gigantocellularis. Pontine reticulospinal fibers terminate in lamina VIII and to a lesser degree in lamina VII, whereas medullary reticulospinal fibers terminate in lamina VII, with some in lamina VIII and IX. There is no somatotopic organization, and the functional significance of these pathways is not clear.

Lesser descending supraspinal tracts (e.g., tectospinal and interstitiospinal tracts) also contribute to the UMN component of the motor system. However, their participation in the motor neuron degenerations is not entirely clear, and thus further discussion with regards to these pathways will not be given (the reader is referred to Brodal, 1981, for further discussion).

2.2. Motor Unit

The motor unit consists of the alpha motor neuron, its peripheral axon, and the innervated skeletal muscle fibers (the motor unit territory). Within the ventral horn, these neurons are arranged in longitudinal columns with their dendrites oriented in a rostrocaudal fashion over several segments. Alpha motor neurons projecting to the proximal muscles are medially placed, with those projecting to distal muscles dorsolaterally placed. The size of the motor unit (the number of skeletal muscle fibers innervated by a single alpha motor neuron) is highly variable, but inversely proportional to the degree of fine motor control possessed by the motor unit. Hence, a motor unit supplying the orbicularis oculi may contain only two to three fibers per unit, whereas the quadriceps may possess a thousand fibers.

Gamma motor neurons are small motor neuron fibers that innervate the interfusal fibers of the muscle spindles, a highly specialized muscle fiber that regulates the excitability of muscle spindle afferents. While important in conveying information about muscle tension and velocity of stretch, it is not known if they are involved in the degenerative processes of the motor neuron. Hence, for the purposes of this review, LMN will mean the alpha motor neuron as being the integral participant in the degenerative process.

3. Developmental Biology

In virtually all species, there is a developmentally regulated proliferation and regression of LMN populations during neurogenesis. This has been extensively documented in studies of the chick embryo lateral motor column, but additionally in other developing neuronal populations (Hamburger and Levi-Montalcini, 1949; Lumsden, 1990). By embryonic day 5 to 6.5 in the chick embryo, the maximal proliferation of neurons within the developing lateral motor column has occurred (Hamburger, 1975; Oppenheim *et al.*, 1989). Following this proliferation, a significant reduction occurs in the number of motor neurons in a process that appears to be related to specific interactions among the neurons, their target tissue (the skeletal muscle), and primary sensory efferents (Okado and Oppenheim, 1984) with as many as 50% to 53% of all motor neurons being lost between days 5.5 and 9 of incubation, with an accompanying loss of peripheral axons (Chu-Wang and Oppenheim, 1978a,b). Little overlap occurs between the period of neuronal proliferation and regression. The signal for regression is related to target-derived trophic factors (Hughes, 1968) inasmuch as the process can be inhibited in the presence of homogenates of embryonic hind limb tissue (Oppenheim *et al.*, 1988) or by the transplantation of a supernumerary limb bud (Hamburger, 1939). Moreover, neuronal regression can be induced by the removal of a limb bud prior to the onset of the naturally occurring regressive process (Chu-Wang and Oppenheim, 1978a). Similar observations have been made for the induction of ventral horn motor neurons in the fetal mouse (Harris-Flanagan, 1969), the South African clawed toad (Prestige, 1967; Lamb, 1974, 1977), and rat (Tada *et al.*, 1979).

In humans, the process of alpha motor neuron regression is not restricted to the embryonic or developmental period, but has been demonstrated to occur with increasing age in the human. Based on morphometric studies, the alpha motor neuron pool appears to be stable until age 60, thereafter demonstrating significant reductions in numbers with a mean loss of 24% by age 95 (Tomlinson and Irving, 1977). In absolute numbers, a linear decline of 170–260 neurons per decade per lumbar segment (L3, L4, and L5) has been observed by Kawamura and colleagues (1977) with similar anatomic correlations being observed for the cervical spinal cord (Tsukagoshi *et al.*, 1979, 1980). Using electrophysiologic techniques, motor unit estimates of the biceps brachii and brachialis muscles have demonstrated a similar reduction in motor unit numbers with aging (Brown *et al.*, 1988).

Thus, in most animals studied, there is a clear proliferation of motor neurons during the developmental period, followed thereafter by neuronal regression. In humans, an absolute reduction in motor neuron numbers following the sixth or seventh decade of life occurs, and, clearly, any discussion of degenerative processes of the motor system must take into account target-derived or trophic factor-dependent processes that may give rise to naturally occurring degenerative states.

4. Motor Neuron Diseases

Based on the preceding neuroanatomic discussion, the degenerative diseases of motor neurons can be considered as being restricted to the descending supraspinal pathways, the LMN, or combinations thereof. While these degenerative processes are often considered as a continuum of disease, for reasons that will soon become clear, it is likely that these disorders are distinct clinical/pathological entities that are manifestations of unique biological processes. Several examples typifying this are given below.

4.1. Primary Lateral Sclerosis

The supraspinal descending motor system is selectively involved in a rare neurodegenerative condition—primary lateral sclerosis. Until recently, the entity of primary lateral sclerosis as a distinct neurodegenerative process had been questioned (Ungar-Sargon *et al.*, 1980; Sotaniemi and Myllyla, 1985; Younger *et al.*, 1988; Russo, 1982). Clinically, the disease manifests as an adult-onset progressive spastic quadraparesis, a spastic dysarthria, and emotional lability (pseudobulbar affect) (Gastaut *et al.*, 1988). Neuropathological changes are restricted to the descending supraspinal pathways with little, if any, involvement of the LMN (Beal and Richardson, 1981; Fisher, 1977; Pringle *et al.*, 1992; Younger *et al.*, 1988). Atrophy of the pre- and postcentral gyrus is accompanied by a complete absence of Betz cells, a reduction in the pyramidal neurons of layer III and V of the precentral gyrus, and degeneration of the descending corticospinal tracts throughout their course in the hemispheres, brainstem, and spinal cord (Figure 4). The anterior horn cells appear normal in number and appearance, although rare cells may demonstrate chromatolysis. Two of the three pathologically described cases demonstrate pallor in the fasciculus gracilis in the posterior columns. No ultrastructural studies are as yet available.

4.2. Neurolathyrism

Closely related to primary lateral sclerosis is neurolathyrism, a neurotoxic degenerative disorder of the UMN recognized for centuries as the consequence of ingestion of the chickling pea (*Lathyrus sativus*) during famines (Selye, 1957). Although described in a number of different geographic regions and cultures, neurolathyrism continues to be most prominent and endemic in Bangladesh, India, and Ethiopia where prevalence rates range from 300 to 2,500/100,000 population (Gardner and Sakiewicz, 1963; Spencer *et al.*, 1986; Ludolph *et al.*, 1987). Afflicted individuals develop a symmetric spastic paraplegia, rarely with UMN signs in the upper extremities. Early in the illness, muscle cramping and stiffness, myoclonus, tremulousness, myokymia, urgency and frequency of micturition, nocturnal erections and ejaculation occur to varying degrees. Acutely, sensory symptoms may be present. Fasciculations are not observed. While symptoms may not resolve with cessation of chickling pea ingestion, leaving the individual with varying degrees of a spastic paraplegia, the illness is nonprogres-

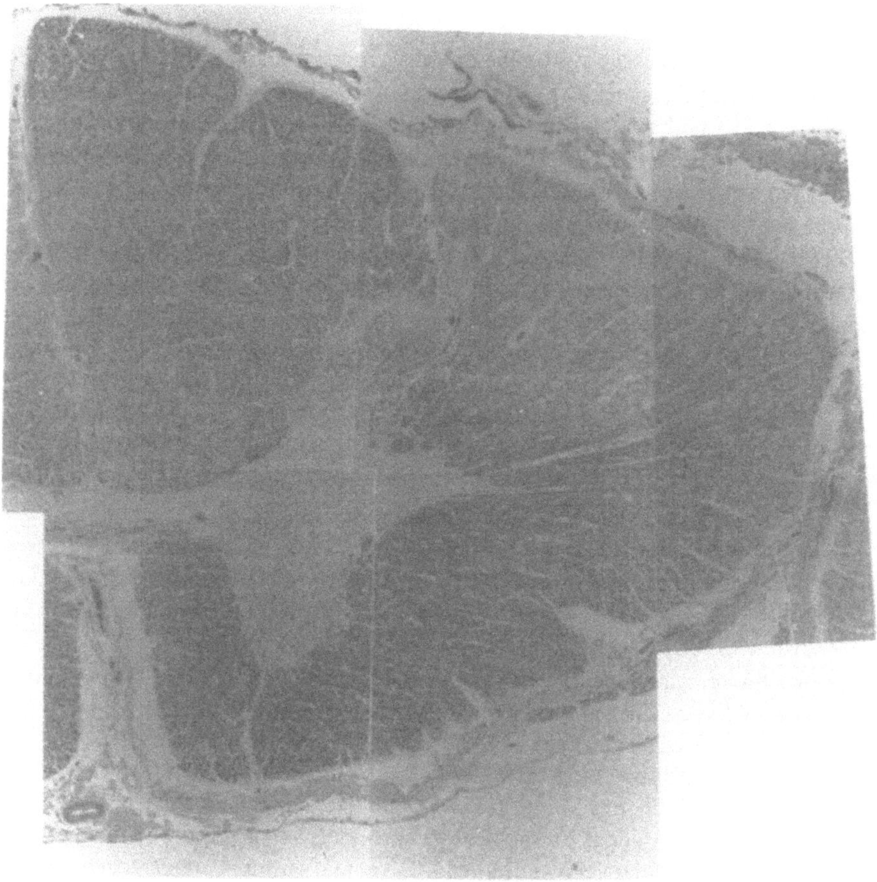

Figure 4 Light micrograph transverse section of the cervical spinal cord taken from a patient with primary lateral sclerosis In contrast to the prominent staining for myelin in the ventral fasciculus, a discrete pallor can be seen in the region of the lateral corticospinal tract, and to a lesser degree in the posterior columns, where tract degeneration has occurred (×10; Nissl-stained slide kindly provided by Professor J. Kiernan).

sive and self-limiting. The few neuropathological studies available report extensive corticospinal tract degeneration within the spinal cord, and to a lesser degree of the spinocerebellar tract and dorsal column. Neuroaxonal spheroids have been described in a single case (Striefler *et al.*, 1977). Anterior horn cell populations are maintained, but degeneration of Betz cells may be extensive. While neither vacuolar nor neurofilamentous pathology has been described, small eosinophilic inclusions have been observed (Hirano *et al.*, 1976).

4.3. Spinal Muscular Atrophies

The spinal muscular atrophies (SMA) represent a spectrum of predominantly inherited degenerative disorders of spinal motor neurons, often with involvement of bulbar motor nuclei, but without supraspinal descending pathway involvement (Emery, 1971). The acute infantile form (SMA type 1; Werdnig-Hoffmann disease) is inherited in an autosomal recessive pattern and manifests as a prenatal onset of generalized muscle weakness with death by 3 years of age (Zellweger *et al.*, 1969; Pearn, 1978; Hausmanowa-Petrusewicz, 1991). It is the most common neurological cause of infantile death in the Western hemisphere, with prevalence rates of 1/25,000 (Pearn, 1978). By virtue of its rapid clinical progression, it is appropriately considered a distinct variant among the spinal muscular atrophies.

Distinct from SMA type 1, a broad spectrum of SMA exists, including an early postnatal onsetting variant with either a rapid progression (SMA type 2; Kugelberg-Welander variant) or a slowly progressive, chronic variant with good life expectancy (Zellweger *et al.*, 1969; Munsat *et al.*, 1969). An adult-onsetting variant with a chronic benign course and bulbar involvement is also recognized (Zellweger *et al.*, 1969; Emery, 1971; Meadows *et al.*, 1969a,b). Several pedigrees of SMA type 2 possess linkage to markers on chromosome 5q11 (Brzustowicz *et al.*, 1991), although no clinical phenomenon clearly distinguishes the remaining pedigrees without linkage (Munsat *et al.*, 1990). As many as 50% of the chronic progressive infantile variant are nongenetic in etiology (Bundey and Lovelace, 1975).

Detailed neuropathological studies of SMA are limited (Namba *et al.*, 1970; Brandt, 1950; Kohn, 1968), but the available data demonstrate extensive spinal motor neuron loss, gliosis of the ventral horn, neuronophagia, and a loss of the motor nuclei of cranial nerves. Thalamic and spinal sensory ganglia degeneration with wallerian degeneration of sural nerves has also been observed (Carpenter *et al.*, 1978; Shishikura *et al.*, 1983; Towfighi *et al.*, 1985). Ballooned neurons and the remaining shrunken neurons in the spinal motor neurons demonstrate increased argentophilia by silver staining techniques, corresponding to regions of increased immunoreactivity against antibodies recognizing phosphorylated epitopes of the neurofilament subunit proteins (Lippa and Smith, 1988; Sobue *et al.*, 1990). Ubiquitin immunoreactivity has been found in a diffuse perikaryal pattern in ballooned neurons and in central perikaryal regions of surviving neurons (Lee *et al.*, 1989). Ultrastructural studies demonstrate fragmentation of the endoplasmic reticulum and an increase in neurofilamentous material in perikarya and proximal neurites. In the most severely chromatolytic neurons of type 1 SMA, neurofilaments are displaced to the periphery by massive increases in mitochondria (Chou and Fakadej, 1971). Murayama *et al.* (1991) have suggested that these features indicate a progression of the disease process from a diffuse increase in neurofilament subunit proteins to an accumulation of mitochondria and other organelles centrally within the perikaryon.

4.4. Amyotrophic Lateral Sclerosis

In contrast to primary lateral sclerosis and the spinal muscular atrophies, amyotrophic lateral sclerosis (ALS) combines the features of both UMN and LMN loss. This topographically specific motor neuron degeneration represents the prototypic motor neuron degeneration, the etiology of which remains a mystery.

Three variants of ALS are recognized—classical sporadic, familial, and western Pacific. With the exception of minor variations, these entities are clinically indistinguishable. Classical sporadic ALS is an age-dependent neurodegenerative process that varies little worldwide and has an incidence rate of 1.0–1.8/100,000 population (Juergens *et al.*, 1980; Hudson *et al.*, 1986; Gunnarsson *et al.*, 1990). The mean duration to death is 22.5 months although long-term (> 1 decade) survival is recognized in 10% to 12% of patients. Recent studies indicate increasing incidence rates, disproportionate to secular increases in longevity of the population (Lilienfeld *et al.*, 1989; Durrleman and Alperovitch, 1989). Clinically, the disease manifests with progressive wasting and weakness, involving both bulbar and spinal motor units, with signs of supraspinal descending pathway involvement, while sparing intellect, oculomotor, and sensory modalities (Mitsumoto *et al.*, 1988). In contrast to the age-dependent sporadic variant, the familial variant of ALS has a peak age at onset of 48 years, a median survival of 24 months, and an autosomal dominant pattern of inheritance—although autosomal recessive forms have also been reported (Strong *et al.*, 1991b). Linkage to the Cu/Zn superoxide dismutase gene on chromosome 21 has been reported (Siddique *et al.*, 1991; Rosen *et al.*, 1993).

The western Pacific variant of ALS is epidemiologically unique. Three foci have been recognized among geographically and genetically distinct Pacific populations: one among the isolated and remote Auyu and Jakai people of southern West New Guinea, the second among the Japanese from the Kii Peninsula, and the last, and most systematically studied focus, among the Chamorro people of Guam and the Northern Mariana Islands. In all three foci, ALS has declined, and on Guam dramatic declines in the incidence of ALS from 50/100,000 population to levels that approximate those of classical ALS have been documented (Garruto *et al.*, 1985; Rodgers-Johnson *et al.*, 1986). Coupled with the observation of a 20-year latency in disease onset migrants (Garruto *et al.*, 1980, 1981a) and the failure to find evidence for a genetic or infectious etiology (Garruto, 1991; Gibbs and Gajdusek, 1982; Reed *et al.*, 1975), this has suggested that the western Pacific variant of ALS is an environmentally-induced, place-specific disease phenomenon (Garruto and Yase, 1986). The documentation of abnormally low levels of calcium and magnesium with high levels of aluminum in garden soil and drinking water within each hyperendemic region (Gajdusek and Salazar, 1982; Garruto *et al.*, 1984; Iwata *et al.*, 1978; Shiraki and Yase, 1975; Tracey *et al.*, 1955), coupled with abnormalities in vitamin D and calcium metabolism (Yanagihara *et al.*, 1984; Yase, 1978), radiographic and densitometric decreases in bone mass (Plato *et al.*, 1982), and increased levels of bone aluminum in afflicted individuals, suggest that trace-metal toxicity, specifically aluminum, might play an etiological role in the western Pacific focus of ALS (Garruto, 1991). Several investigators, using widely divergent techniques, have observed the co-localization of calcium, aluminum, and silicon within degenerating neurons of Guamanian ALS and parkinsonism-dementia patients (Garruto *et al.*, 1984; Linton *et al.*, 1987; Perl *et al.*, 1982; Yoshida, 1977). An alternative environmental hypothesis is that an ingested toxin, derived from the use of the cycad nut as a dietary supplement or poultice, is pathologically linked to the occurrence of ALS in the western Pacific (Spencer *et al.*, 1987). However, restricted use of the material to individuals afflicted with the disease has not been demonstrated (Reed *et al.*, 1975); neither cycad nor any one of its primary constituents recapitulates the neuropathological features typical of Guamanian or classical ALS, and the traditional purification of cycad by Chamorro women effectively detoxifies the flour (Duncan *et al.*, 1988; Garruto *et al.*, 1988; Hirano, 1973). However, the proposal that excitatory amino acid toxicity might produce a motor neuron degeneration is not without merit and will be reviewed later.

No single neuropathological feature is pathognomic for ALS. Rather, it is the combination of widespread alpha motor neuron loss in a topographically specific pattern with chromatolytic anterior horn cells, atrophic proximal axonal processes, atrophy of ventral roots, and demyelination of the corticospinal tracts that establishes the diagnosis (Hirano, 1991). Sphincteric motor neurons of Onufrowicz nucleus are characteristically spared (Pullen *et al.*, 1992). Although not an invariable feature of all clinically confirmed cases of ALS (Brownell *et al.*, 1970), corticospinal tract degeneration, when present, is extensive, involving the entirety of the tract with a concomitant loss of Betz cells (Lawyer and Netsky, 1963). Additionally, pallor of the spinocerebellar tracts and posterior columns with degeneration of Clarke's column occurs in the familial variant (Hirano *et al.*, 1967), while, in the western Pacific variant, widespread neurofibrillary tangles are present (Rodgers-Johnson *et al.*, 1986). Neuronophagia or inflammatory changes are rarely encountered (Lawyer and Netsky, 1953).

The earliest neuropathological changes consist of chromatolysis with a loss of Nissl substance and an accumulation of neurofibrillary material within the perikarya of the anterior horn cells and axonal processes (Hirano and Inoue, 1980; Sasaki and Maruyama, 1992) (Figure 5). These latter structures, termed *axonal spheroids,* are present in numbers much greater than expected during

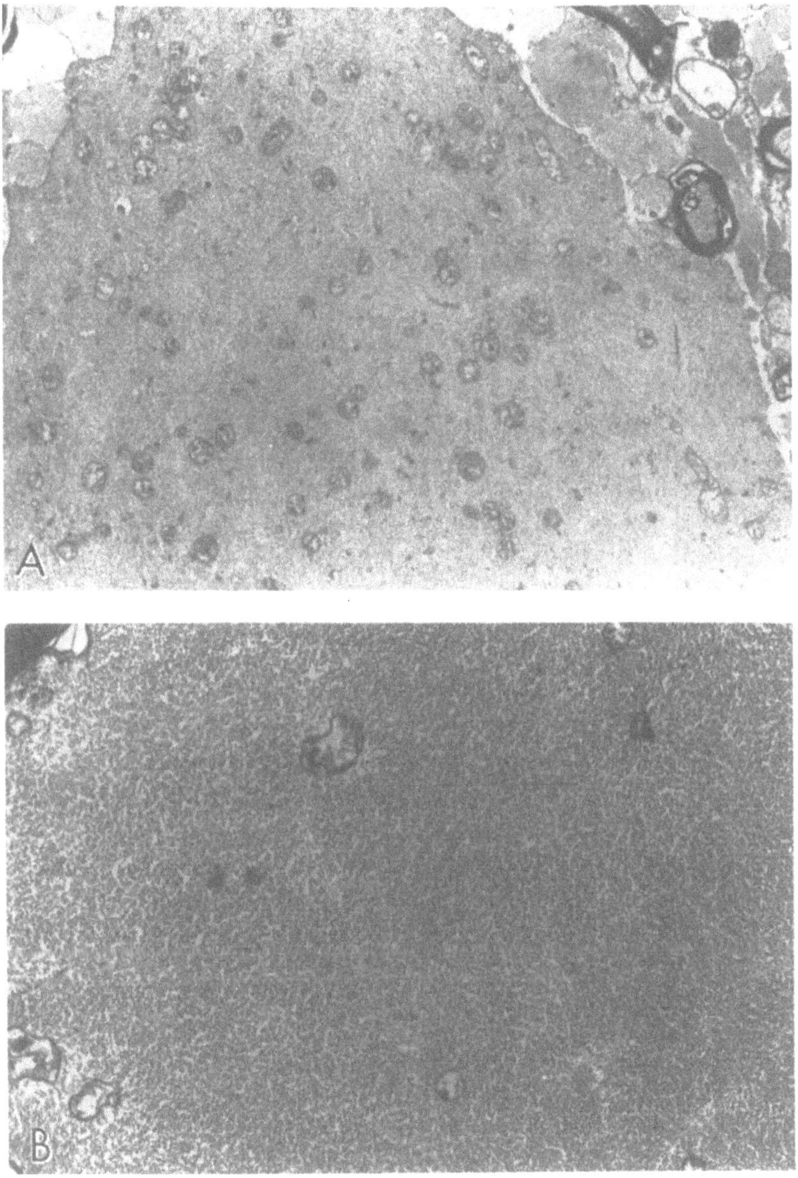

Figure 5 Neurofilamentous accumulations in amyotrophic lateral sclerosis. (**A**) Electron micrograph of a part of a spheroid in the anterior horn region of the spinal cord in sporadic ALS (×10,000). The 10-nm neurofilamentous bundles show a characteristic interwoven pattern. (**B**) Cross section of a distended myelinated axon containing an enormous increase in 10-nm neurofilaments (×13,000). (Both photographs kindly provided by Dr. Asao Hirano (**A**). Reproduced with permission from Hirano, A., in *Research Progress in Motor Neurone Disease* (F. Clifford Rose, ed.), Pitman Books, London, 1984.)

"normal" aging (Carpenter, 1968; Averback, 1981; Delisle and Carpenter, 1984). The perikaryal inclusions are morphologically varied and can be described in terms of tinctorial, ultrastructural, and immunohistochemical characteristics (Chou, 1978). Many consist of interwoven skeins or parallel arrays of 10 nm neurofilament (NF) subunit proteins.

Neurofilaments are cytoskeletal phosphoproteins thought to be integral to the structural integrity of the neuron, with molecular weights (by SDS-PAGE estimation) of 200 kDa (NF-H), 160 kDa (NF-M), and 68 kDa (NF-L). The individual NF subunit proteins self-assemble in the perikaryon into a single NF triplet protein and are transported into the axon. Phosphorylation of the subunit protein occurs coincident with this assembly, although the majority of the NF phosphorylation occurs within the axon such that little phosphorylated NF is normally detectable within the perikaryon. Phosphorylation and dephosphorylation of the triplet protein occurs along the entire length of the axon and likely serves to modulate not only the interaction between the NF triplet and other cytoskeletal elements, but also NF catabolism (Greenwood *et al.*, 1991; Clark and Lee, 1991).

In contrast to this normal somatofugal localization of phosphorylated NF, perikaryal and axonal accumulations of phosphorylated NF are invariably observed early during the course of ALS, regardless of the disease variant (Manetto *et al.*, 1988; Munoz *et al.*, 1988; Leigh *et al.*, 1989a,b). In the absence of biochemical proof to the contrary, this immunohistochemical observation has been taken to suggest a fundamental abnormality in the process of NF phosphorylation in the induction of the disease state.

More recently, it has been observed that the majority of the intracytoplasmic inclusions are ubiquitin-conjugated (Chou, 1988; Aldridge *et al.*, 1989) and that either diffuse ubiquitin immunoreactivity (Murayama *et al.*, 1990a) or skeins of ubiquitin-immunoreactive material can be observed within motor neurons (Leigh and Swash, 1991). Ubiquitin-immunoreactive filamentous inclusions have also been observed in anterior horn cells of Guamanian ALS patients, and are distinct from the microtubule-associated protein tau-reactive neurofibrillary tangles, suggesting that the ubiquitin-conjugated fibril accumulation may be an early feature of ALS (Matsumoto *et al.*, 1990). Although this phenomenon is not unique to ALS, it does suggest either an ineffective ATP-dependent nonlysosomal proteolytic degradation or a resistance of the cytoskeletal protein to this pathway.

Beyond these observations, however, are a wide array of intracellular inclusions, including hyaline inclusions consisting of 5–15 nm fibrils in association with a variety of organelles (Hirano, 1991), lipofuscin, Lewy body-like inclusions (Lowe *et al.*, 1989; Murayama *et al.*, 1989, 1990b; Wakayama, 1992) and Bunina bodies (Figure 6). These latter 2- to 5-μm-diameter conglomerates, consisting of homogeneous electron-dense material with a rim of ribosomal particles (Tomonaga *et al.*, 1978), may reflect a condensation of the larger hyaline inclusions (Chou, 1978).

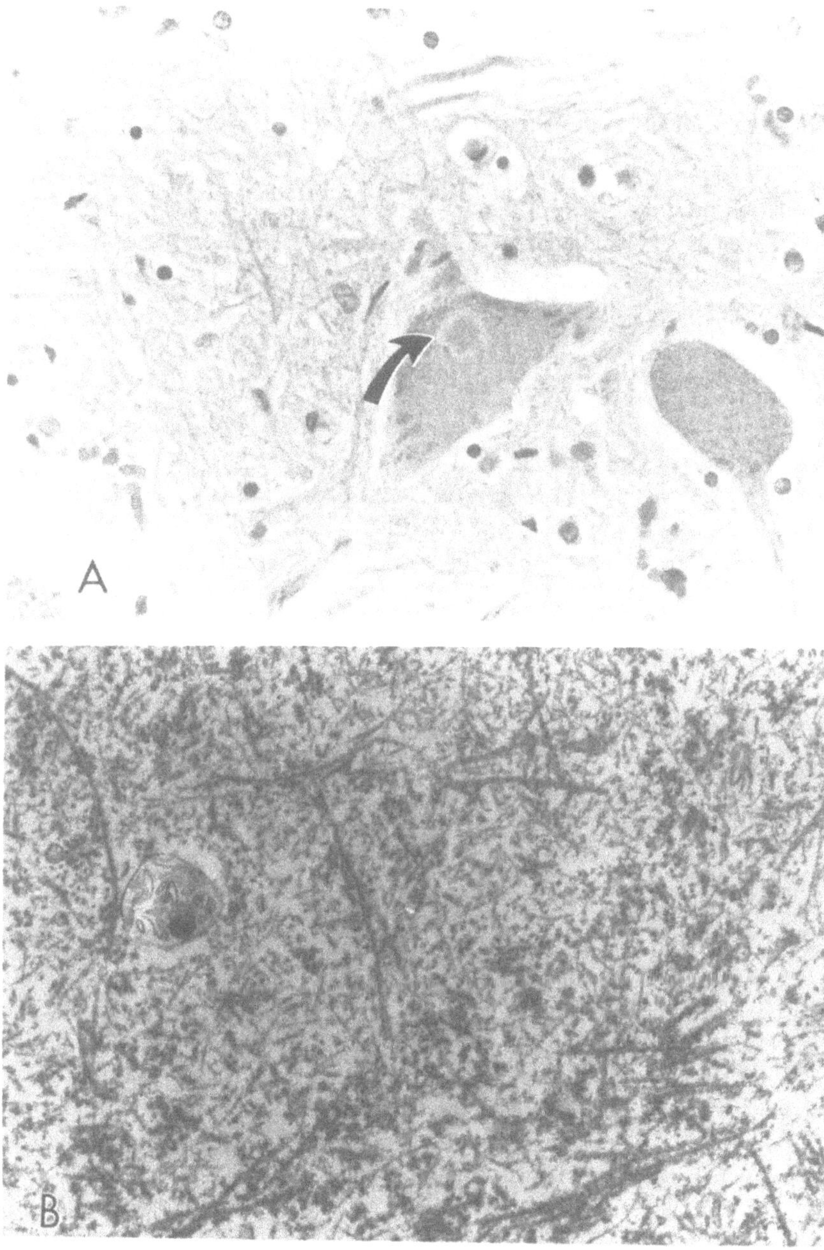

Figure 6 Light micrograh of a Lewy body-like inclusion in an anterior horn cell neuron in a patient with amyotrophic lateral sclerosis. (**A**) Discrete intracellular inclusion (arrow) surrounded by a peripheral halo in chromatolytic-appearing neurons (×40). (**B**) An electron micrograph of a portion of a Lewy body-like inclusion showing scattered linear densities and granules within a focal accumulation of randomly distributed 10-nm filaments in a chromatolytic anterior horn cell (×34,000; kindly provided by Dr. A. Hirano).

5. Experimental Paradigms of Motor Neuron Degeneration

5.1. Spontaneous and Genetic Disorders

Although a number of spontaneously occurring motor neuron degenerations have been described, only a few have proven to be reproducible or have been established as reliable models (Table I). Each of these will be discussed in some detail. For the majority of models, the mechanism(s) of disease induction is unknown.

5.1.1. Canine Models

Among the first-described spontaneous canine motor degenerations was that of a progressive thigh and hind limb paralysis onsetting by 11 to 14 weeks of age in offspring of matings between Great Danes and St. Bernards (Stockard, 1936). Although the paralysis was not lethal, these animals developed a moderately severe affliction restricted to the lumbar region, sparing the trunk, forelimbs, and pelvic floor. At autopsy, a pronounced microglial infiltration of the anterior horn and ventral roots was present with degenerating and chromatolytic anterior horn cells. Hyalinized perikarya and axonal processes were rarely noted. The disease was attributed to an interaction of three autosomal dominant genes.

Hartley (1963) described a spontaneous motor neuron degeneration in New Zealand sheep dogs (primarily of the collie-type, although also in a terrier, pug, and dachshund) that manifested as a rapidly progressive flaccid paralysis, initially with posterior limb incoordination, and ultimately with extensive hind limb wasting. At autopsy, an extensive loss of anterior horn cells, diffuse and mild ventral horn gliosis, and destruction of peripheral motor neuron fibers was documented in the absence of supraspinal descending pathway involvement.

The more traditional canine model, however, is that described by Cork *et al.* (1979, 1982) in which an autosomal dominant progressive spinal muscular atrophy (HCSMA) occurs in Brittany spaniels. Three phenotypes are recognized: an accelerated form with quadriplegia onsetting at less than 3 months of age; an intermediate form with weakness by 6 months and quadriparesis by 6 years; and a chronic variant with weakness and atrophy onsetting at more than 1 year of age. The former represent pups homozygous for the trait, whereas the latter are heterozygous, with a sex-linked modifier gene proposed in the chronic variant (Sack *et al.*, 1984). The disease manifests as a progressive weakness of the limb girdle and trunk musculature, with mild facial and tongue involvement, hyporeflexia and intact extraocular muscle function. The pathology consists predominantly of axonal spheroids in the spinal cord white matter, a loss of motor neurons in the spinal cord and brainstem (trigeminal and hypoglossal nucleus) and enlarged dendritic structures. The latter demonstrate accumulations of NF, mitochondria, and tubulovesicular structures. In the accelerated form, chroma-

Table I. Genetic Models of Motor Neuron Degeneration

Model	Inheritance pattern	Age at onset	Fatal	UMN	LMN	NF aggregates	Ubiquitinated inclusions
Canine							
Brittany	Autosomal dominant	i. Accelerated <3 months	Yes	No	Yes	Yes	N/A[f]
		ii. Intermediate[g] 1.5 years	No	No	Yes	Yes	N/A
		iii. Chronic[g]	No	N/A	N/A	N/A	N/A
Boxer	Autosomal dominant	Onset 1–2 months; hind limb ataxia		Yes[d]	Yes[e]	Yes	N/A
Pointer	Autosomal dominant	5 months	Yes 9 months	No	Yes	No	No
Murine							
Mnd	Autosomal dominant	Hind limb paralysis 5–11 months	9–14 months	Yes[a]	Yes	No[b]	Yes
Wobbler (*wr*)	Autosomal recessive	Forelimb paralysis 3 weeks	3 months–>1 year	Yes[c]	Yes	No	N/A
Wasted (*wst*)	Autosomal recessive	Hind limb paralysis 14–18 days	12–14 days	No	Yes	Yes	N/A
pmn[h]	Autosomal dominant	Hind limb paralysis 3 weeks	Yes, 3–4 weeks	No	Yes	No	N/A

[a]Rare involvement of nucleus ruber, reticular nuclei; no long tract degeneration documented.
[b]Peripheral displacement/margination of NF in anterior horn cells by cytoplasmic clearings.
[c]Single report only (Campbell, 1972).
[d]Extensive neuroaxonal spheroid formation throughout brainstem nuclei, autonomic pathways and cerebellar nuclei.
[e]Dorsal and ventral spinal root involvement.
[f]N/A = immunoreactivity not assessed.
[g]Heterozygous for trait.
[h]Progressive motor neuronopathy.

tolytic neurons and axonal swellings of argentophilic material are prominent, whereas neuronal loss is most prominent in the intermediate form. The motor cortex, descending supraspinal pathways, thalamus, brainstem centers of extraocular movement, and medial sacral nucleus (comparable to the Onufrowicz nucleus in humans) are spared. Axonal caliber is significantly reduced, particularly large ventral motor neuron axons, a finding interpreted as a developmental delay in axonal growth (Cork *et al.*, 1989). Impaired slow axonal transport also has been demonstrated (Griffin *et al.*, 1982).

Progressive axonopathy in boxer dogs is a more widespread degenerative process in which motor neuron degeneration is a significant component (Griffiths *et al.*, 1980). By 2 months of age, hind limb ataxia, hypotonia, loss of reflexes, and proprioceptive deficits occur and progress over the ensuing 12–18 months, with an apparently static course thereafter (Griffiths, 1985). The neuropathological hallmark of this autosomal dominant disease is the extensive number of axonal spheroids in spinal cord white matter, brainstem nuclei, and dorsal and ventral spinal nerve roots (Griffiths *et al.*, 1985, 1986, 1989). Ultrastructurally and immunohistochemically, the spheroids are composed of arrays of aggregated NF, vesicles, and tubulovesicular profiles. The remaining neurons demonstrate perikaryal immunoreactivity to phosphorylated NF. Although axonal spheroids are prominent within the lateral and ventral columns of the spinal cord, the tract degeneration is nonspecific and includes autonomic involvement (Griffiths *et al.*, 1985). A defect in axonal transport has been proposed.

A hereditary spinal muscular atrophy also has been described in pointer dogs (Inada *et al.*, 1978; Igata *et al.*, 1979; Izumo *et al.*, 1983). By 5 months of age, afflicted dogs develop extremity weakness, suppressed reflexes, and muscle wasting. No sensory disturbances are present. Death occurs within 9 months. Neuropathologically, there is evidence of a chronic axonal degeneration with neurogenic muscle atrophy but normal spinal motor neuron numbers and no evidence of tract degeneration. The major ultrastructural findings within the motor neurons are membranous cytoplasmic bodies and Zebra bodies with tinctorial characteristics of lipid accumulation.

5.1.2. Murine Models

Originally described by Falconer (1956) as a spontaneous mutation in C57B1/Fa mice, and subsequently by Duchen and Strich (1968), the Wobbler mouse has become a standard model of motor neuron degeneration. This autosomal recessive trait (*wr*) has been mapped to the proximal region of chromosome 11 (Kaupmann *et al.*, 1992). Offspring homozygous for the trait are developmentally normal until the fourth or fifth week postpartum, whereupon a progressive fore limb paralysis and gonadal dysfunction ensues. Afflicted mice display a high-stepping gait, propensity to walk on the dorsum of their paws, and a reduced climbing ability (Figure 7). Facial muscles are atrophied. Survival of

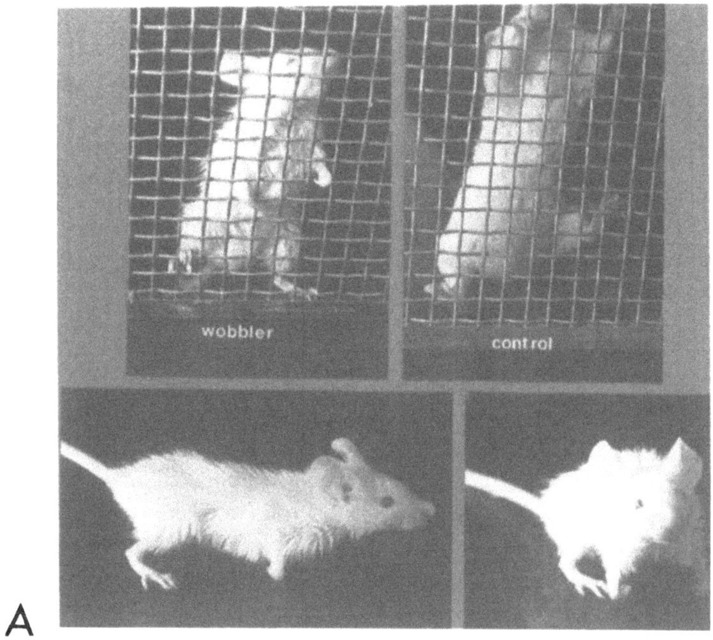

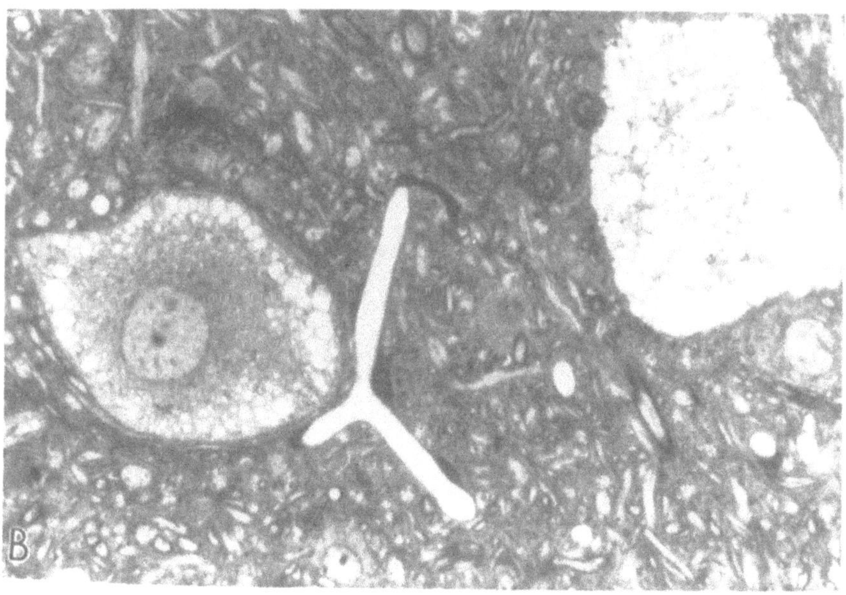

Figure 7 Wobbler mouse. (**A**) Mice affected by an autosomal recessive trait (*wr*) develop an inability to climb using their forelimb paws (upper photograph) and a propensity to walk on the dorsum of their paws (lower photographs). (**B**) Photomicrograph demonstrating pronounced vacuolar degeneration of spinal motor neurons (×40). Photographs kindly provided by Dr. H. Mitsumoto.

more than 1 year is observed in a presumably "burnt out" variant, but the majority of mice die by 6 months of age. Progeny are not produced.

The vacuolar degeneration that is the neuropathological hallmark of this disease is restricted to spinal motor neurons and brainstem nuclei (magnocellular reticular neurons, motor nuclei of V and VII, rarely in the nucleus ruber, substantia nigra, anterior and posterior colliculi, and deep cerebellar nuclei) (Figure 7B). Anterior horn cell numbers are depleted (Papapetropoulos and Bradley, 1972), with the remaining showing perikaryal and axonal vacuolar degeneration and a wide array of cytoplasmic pathology, including proliferation of microtubules and NF, smooth ER, lipid accumulation, and increased lysosomal numbers (Andrews, 1975; Mitsumoto and Bradley, 1982). A single report exists of ascending and descending tract degeneration within the ventral columns, medial lemniscus, and inferior cerebellar peduncles (Campbell, 1972). No microglial or glial reaction is present. Axonal regeneration is, however, a prominent feature (Mitsumoto and Bradley, 1982). Corresponding to the initial forelimb manifestation, the cervical regions are more extensively involved than are the lumbar regions, with a concomitant loss of large myelinated fibers (Mitsumoto *et al.*, 1990). Affected neurons have marked dendritic changes with a loss of spinous processes (Ma and Vacca-Galloway, 1991). Impaired slow axonal transport has been postulated as the underlying pathogenic mechanism (Bird *et al.*, 1971; Mitsumoto and Gambetti, 1986) although reduced protein synthesis and RNA levels are also evident prior to the onset of neuropathological changes (Murakami *et al.*, 1980, 1981).

A second murine model, the *Mnd* mouse (Messer and Flaherty, 1986), is an autosomal dominant motor neuron degeneration localized to chromosome 8 (Messer *et al.*, 1992) that manifests as hind limb weakness and ataxia beginning at 5–11 months of age, followed by a premature death by 9–14 months of age. No animals survive more than 1–2 months following disease onset, and ultimately all develop severe spasticity. The extent and topographic distribution of pathological lesions parallels the clinical features and severity of disease, with the most severely afflicted mice having few spinal motor neurons. The remaining motor neurons are chromatolytic with vacuolar pathology. There is also extensive brainstem involvement, including the hypoglossal, vagal, facial, motor trigeminal nuclei, the nucleus ruber, and reticular formation of the pons and medulla (Messer, 1992; Messer *et al.*, 1987).

The overwhelming pathological feature of this latter model is that of cytoplasmic vacuolization with a redistribution of NF to the periphery of the neuronal perikarya, leaving large cleared areas devoid of neurofilamentous immunoreactivity, but containing 25-nm tubular deposits associated with lysosomes (Callahan *et al.*, 1991). With disease progression, these regions increase in size. Rare paracrystalline and microtubular structures are also present (Mazurkiewicz *et al.*, 1986). Axonal spheroids are not a prominent feature of the pathology (Messer *et al.*, 1987). At all stages of the disease, an abnormal pattern of ubiquitin immunoreactivity can be observed (Mazurkiewicz, 1991). While co-localization of ubiquitin with peripherally marginated NF occurs, the entirety of the cytoplasm is ubiquitin-immunoreactive, including the clear vacuolar regions. This immunoreactivity can be demonstrated prior to the induction of the disease, suggesting a role for aberrant ubiquitin-mediated nonlysosomal degradation in the disease process. Recent studies have shown that both neuronal and visceral inclusions are composed of sudanophilic lysosomal material immunoreactive with antibodies to subunit c of mitochondrial ATP synthase, indicating that this model best reflects the human neuronal ceroid lipofuscinosis (Bronson *et al.*, 1993).

Vacuolar degeneration of spinal motor neurons is also a prominent feature of a third autosomal recessive trait termed *wasted.* Mice homozygous for the trait (*wst/wst*) develop tremulousness, hind limb weakness, and lymphoid hypoplasia by day 14 to 18 postnatally and die by day 28 to 31 (Shultz *et al.*, 1982; Lutsep and Rodriguez, 1989). Extensive vacuolar degeneration is observed throughout the anterior horn region and brainstem motor nuclei, with proximal axonal swellings composed of phosphorylated high molecular weight NF. The earliest ultrastructural change is that of intracytoplasmic and intra-axonal accumulations of electron-dense material and aggregates of NF. Subsequent to this, the vacuolar degeneration becomes the prominent feature.

In contrast to the vacuolar degeneration of the previous murine models, spinal motor neuron perikarya are relatively spared in the autosomal recessive progressive motor neuronopathy (*pmn*) murine model (Schmalbruch *et al.*, 1991). Afflicted mice develop hind limb paralysis and pelvic girdle atrophy by the third week of life, followed thereafter by forelimb paralysis and death by 6 to 7 weeks of age. Neurogenic muscle atrophy is prominent and axonal sprouting is observed. Anterior horn cells show slight chromatolysis. Corticospinal tracts are normal, although terminally degeneration is present in the fasciculus gracilus, rubrospinal tract, and to a lesser degree, the reticulospinal tract. Treatment of affected mice with ciliary neurotrophic factor (CNTF) ameliorates the rate of disease progression and the extent of morphological change (Sendtner *et al.*, 1992).

5.1.3. Syrian Hamster

A progressive sex-linked hind limb paralysis has been described in male Syrian hamsters (Nixon and Connelly, 1968; Homburger and Bajusz, 1970). By 6–10 months of age, afflicted hamsters develop a progressive hind limb paralysis with contractures. Hirano (1977, 1978) has defined this entity as an inherited peripheral neuropathy affecting the LMN, histologically characterized as myelin abnormalities confined to the large myelinated fibers. Characteristically, filamentous accumulations are present within the inner loops of the myelin sheathes, frequently forming a crystalloid pattern. The anterior horn cells are preserved and central neuroaxis pathology is absent.

5.1.4. Equine Motor Neuron Disease (EMND)

A spontaneous fatal motor neuron degeneration has been described recently in quarter horses and thoroughbred horses (Cummings *et al.*, 1990; Mohammed *et al.*, 1993). The disease starts with a short strided gait, excessive recumbency with rapid fatigue, sweating, fasciculations, and trembling. The peak age of onset is sixteen years and the duration one to several months. Neuropathologically, degeneration and loss of motor neurons is observed within the spinal cord and brainstem with secondary axonal degeneration and muscle denervation. Remaining motor neurons are shrunken and necrotic with some undergoing neuronophagia. Neurofilamentous accumulations occur in affected motor neurons as do eosinophilic cytoplasmic inclusions reminiscent of Bunina bodies observed in amyotrophic lateral sclerosis. In contrast to the latter, however, the eosinophilic inclusions of EMND consist of aggregates of membranous degradation material (Cummings *et al.*, 1993). Descending supraspinal tract degeneration is not observed, although the corticospinal tract is rudimentary in the horse.

5.1.5. Miscellaneous

Several spontaneously occurring motor neuron degenerations have been described in species apart from those listed above, including a degeneration in Yorkshire pigs in which 5-week-old animals develop a hind limb ataxia followed by forelimb paralysis (Higgins *et al.*, 1983). Pathological analysis demonstrated distended neurons, chromatolysis, and a mild reactive gliosis affecting brachial and lumbar motor neurons, reticular brainstem neurons, nucleus ruber, hypoglossal, facial and trigeminal nuclei. Oculomotor nuclei were extensively involved. Ultrastructurally, neurons contained an increased quantity of 10-nm filamentous material.

Shields and Vandevelde (1978) described a spontaneous LMN degeneration in 6- to 8-week-old rabbits manifesting as a progressive tetraplegia. Neuronal loss and an astrocytic reaction affecting the lateral ventral horns and both sensory and motor brainstem nuclei were observed. Ultrastructurally, neurons were filled with 10-nm fibrillar material.

Neurofilamentous accumulations have also been described in two consecutive Zebra offspring (Higgins *et al.*, 1977) in which ataxia and tetraparesis developed at 2 weeks of age. Widespread neuronal degeneration was observed with prominent chromatolytic changes, dorsolateral and ventromedial spinal cord demyelination, and neuronal argentophilia.

Swayback in lambs (Cancilla and Barlow, 1966) represents a progressive motor system degeneration in which brainstem and anterior horn neurons develop central chromatolysis, peripheral hyalinization, and accumulations of bundles or fascicles of 6- to 7.5-nm fibrils and 15-nm tubular structures.

Although not a motor system degeneration per se, the recently described hereditary hypotrophic axonopathy in Japanese quail is also of importance in that a deficiency of neurofilament subunit proteins yields thin, atrophic axons and dendrites composed primarily of microtubules (Yamasaki *et al.*, 1991).

5.1.6. Summary

The spontaneous and genetic motor neuron disorders are of value in understanding human motor neuron degenerations in that they mirror, to a large degree, the spinal muscular atrophies. With the exception of the boxer dog and the rare observation in the Wobbler mouse of UMN involvement, pathological changes are restricted to lower motor neurons, and in the case of the *pmn* murine model, restricted to peripheral motor neurons. Perikaryal neurofilamentous inclusion formation is not a major pathological feature and, if present, corresponds to a displacement of apparently normal NF by other intracellular aggregates. Ubiquitinated inclusions are observed only in the *Mnd* murine model, although it is not clear that immunoreactivity to ubiquitin protein has been looked for in all models. The majority of this group of motor neuron degenerations are rapidly progressive and fatal, although chronic, slowly progressive variants are observed. Finally, the offsprings of animals in many of these models are infertile, rendering them unsuitable for long-term or repetitive studies. Nonetheless, the experimental models described above are suitable for understanding the molecular and biological characteristics of a distinct form of motor neuron degeneration affecting almost exclusively the LMN.

5.2. Neurotoxic Models

5.2.1. Trace Metals: Aluminum

The hypothesis that chronic dietary deficiencies associated with the western Pacific environment might induce a secondary form of hyperparathyroidism accompanied by enhanced gastrointestinal absorption of aluminum has been extensively studied in nonhuman primates. Yase and colleagues (Yase, 1988; Yano *et al.*, 1987) observed anterior horn cell pathology in macaques (*Macaca fuscata*) fed a low calcium and magnesium diet for 7 months with aluminum lactate supplementation of the animals' drinking water. The neuropathology consisted of argentophilic bodies, chromatolysis, axonal and dendritic swellings—the latter due to an accumulation of interwoven masses of neurofilament. Subsequently, the effects of a chronic dietary intake of excessive aluminum and manganese in the presence of restricted calcium intake were investigated by Garruto *et al.* (1989) in six juvenile cynomolgus monkeys (*Macaca fascicularis*). Two monkeys were additionally supplemented with unprocessed cycad flour (2 g/day). During the 41–46 months of the study, no abnormal clinical features were observed. The calcium-deprived monkeys developed chromatolytic, swollen neurons, with neurofibrillary thickening and perikaryal argentophilic masses occurring in some neurons in the anterior horns of the spinal cord and in the trigeminal and hypoglossal nuclei, reticular formation, and substantia nigra. This pathology was most prominent in the aluminum- and manganese-supplemented monkeys. Intraneuronal inclusions were either small, basophilic, electron-dense masses surrounded by a clear halo (reminiscent of Bunina bodies), or larger, peripherally located hyaline bodies composed of interwoven masses of NF. Neuroaxonal spheroids were observed in the anterior and lateral columns and spinocerebellar tracts of the spinal cord. Ultrastructurally and immunohistochemically, these spheroids consisted of phosphorylated neurofilamentous material.

While providing an excellent model of the neuropathological features of amyotrophic lateral sclerosis, chronic oral models of aluminum neurotoxicity in Old World nonhuman primates are expensive, labor-intensive, and cannot be readily assessed clinically. Unfortunately, attempts at replicating the model by direct intracisternal inoculums in less expensive New World monkeys yielded a fulminant encephalomyelopathy with disseminated, nonspecific pathology (Garruto *et al.*, 1990). To circumvent these difficulties, a progressive spastic myelopathy model was successfully developed using New Zealand white rabbits (Strong *et al.*, 1991c).

Conventional acute models of aluminum neurotoxicity have proven to be inappropriate for the study of motor neuron degeneration (Strong *et al.*, 1990). Traditionally, acute intoxication studies used inoculations of 1000 μg or more of either organic or inorganic aluminum compounds administered by a variety of routes that induced aggregates of phosphorylated NF with ultrastructural characteristics reminiscent of ALS, but accompanied by an acute fulminant encephalopathy marked by seizuring, quadraparesis, and death 10 to 14 days postinoculation. Additionally a diffuse, nonspecific neuronal degeneration accompanied by suppression of NF gene transcription occurs in the acute model but not in human ALS (Clark *et al.*, 1989; Krekoski *et al.*, 1988; Muma *et al.*, 1988; Parhad *et al.*, 1989).

The notion that chronic, sublethal exposure of New Zealand white rabbits to aluminum compounds might induce a motor system degeneration had been initially addressed by Bugiani and Ghetti (1982) and Wisniewski and colleagues (1982), although DeBoni *et al.* (1976) had observed a dosage-dependent induction of neurofilamentous lesions with prominent motor phenomenology following repeated subcutaneous inoculums of organic aluminum compounds. In the former studies, insoluble aluminum compounds were placed within the cisterna magna in mature New Zealand white rabbits in an attempt to produce chronically elevated aluminum levels. Although survival to 90 days was noted in some animals, all died from fulminant encephalomyelopathies indistinguishable from the acute aluminum inoculums. Further studies using either a single intracisternal inoculum of $AlCl_3$ in developmentally immature rabbits (Wisniewski *et al.*, 1980) or repeated subcutaneous aluminum tartrate inoculums in young adult rabbits (Uemura, 1984; Uemura and Ireland, 1984) provided evidence that a chronic myelopathy could be induced. The latter also demonstrated a significant potential for a dynamic remodeling of the neurofilamentous inclusions. However, clinical/pathological correlations were never performed.

In contrast to previous models, repeated intracisternal inoculations of low dose (100 μg) $AlCl_3$ at 4 weekly intervals in young adult New Zealand white rabbits (age 8–9 weeks) induce a topographically specific motor system degeneration (Strong *et al.*, 1991c). Neurofilamentous inclusions developed in spinal motor neurons and select brainstem nuclei (thought to subserve upper motor neuron functions) and demonstrated a spectrum of immunoreactivity to antibodies against phosphorylated and nonphosphorylated epitopes of NF subunit proteins (Figure 8). Although many inclusions were intensely immunoreactive with antibodies recognizing phosphorylated epitopes of NF, inclusions were also present that demonstrate no immunoreactivity or are only faintly immunoreactive. Some inclusions react only to antibodies recognizing nonphosphorylated epitopes. Axonal spheroids and a variable number of intracytoplasmic inclusions were also uniquely immunoreactive to SMI 34, an antibody that recognizes an "age-related" phosphorylation epitope of NF (Blanchard and Ingram, 1989) but does not recognize acute aluminum-induced neurofilamentous inclusions (Sternberger *et al.*, 1985). Regardless of the phosphorylation state, all inclusions consisted of interwoven skeins of morphologically identical 10-nm intermediate filaments. Ubiquitin immunoreactivity has been observed following the formation of inclusions. *In vitro* studies using dissociated fetal rabbit motor neurons growth in monolayer cultures suggest that the topographic specificity of chronic aluminum neurotoxicity relates, in part, to unique motor neuron thresholds to aluminum-induced neurofilamentous degeneration (Strong and Garruto, 1991c).

5.2.2. Excitatory Amino Acids

Neurotoxicity induced by excitatory amino acids has been implicated in a number of human and experimental neurodegenerative disease states, both acute and chronic (Greenamyre, 1986). The pharmacological effects of these excitatory neurotransmitters, of which L-glutamate and L-aspartate are primary, are mediated by distinct receptor types identified by either the most specific agonist for the receptor (e.g., N-methyl-D-aspartate (NMDA), quisqualate, kainate) or its antagonist (e.g., L-2-amino-4-phosphonobutyrate (AP4)) (Watkins and Olverman, 1987; Cotman and Iversen, 1987). Because these receptors are predominantly dendritic, very specific lesions are induced by inoculations of excitatory amino acids, resulting in a loss of neurons at the site of inoculation while sparing the transgressing axonal processes (Pisharodi and Nauta, 1985). In *in vitro* stud-

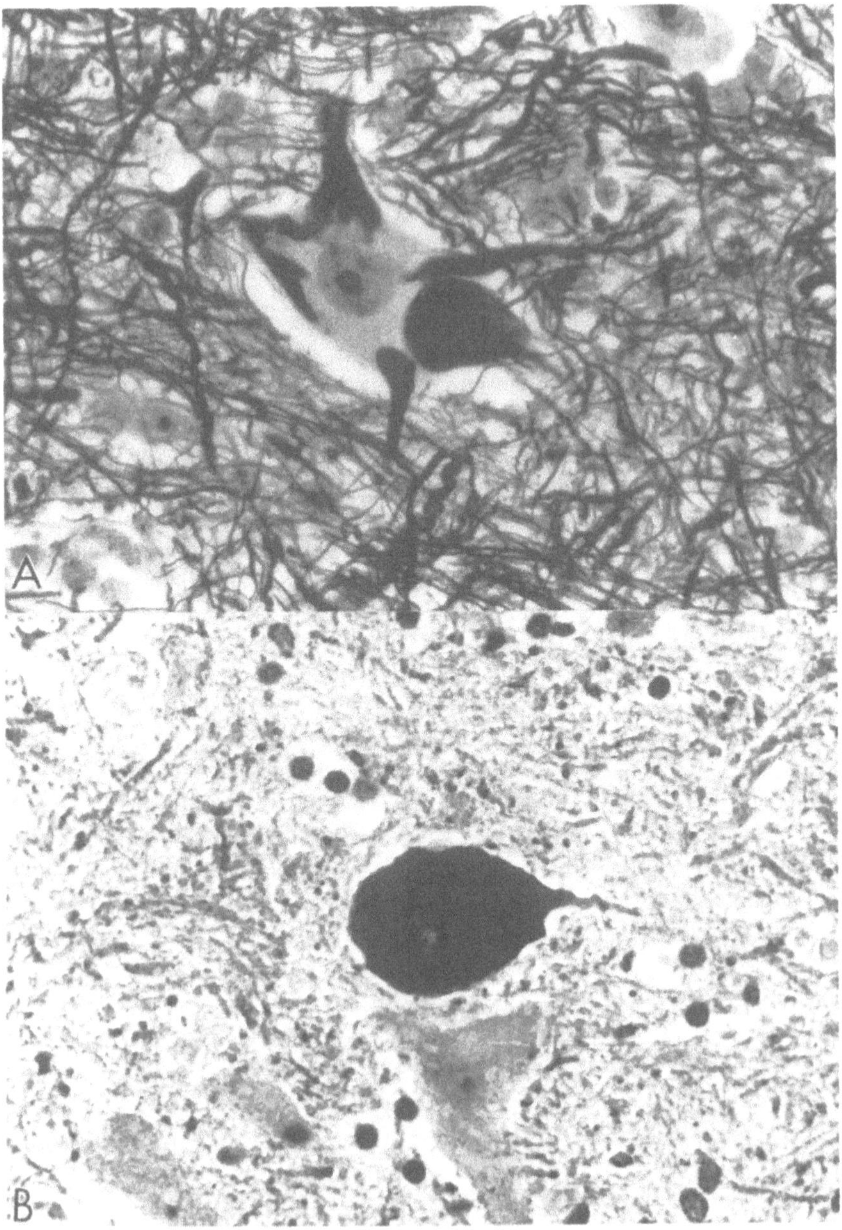

Figure 8 Light micrograph of aluminum-induced neurofilamentous inclusions. (**A**) Filamentous perikaryal inclusion in a spinal motor neuron from a New Zealand white rabbit inoculated at monthly intervals with 100 μg of $AlCl_3$ for 8 months. (Bielchowsky silver stain; ×40). (**B**) Neuroaxonal spheroid immunostained with a monoclonal antibody directed against "age-dependent" phosphorylation sites of the high molecular weight neurofilament subunit protein (SMI 34, dilution 1:1,000; ×40).

ies, acute excitatory amino acid toxicity induces two forms of neuronal degeneration (Meldrum and Garthwaite, 1990). NMDA-mediated toxicity acutely induces swelling, vacuolization, and mitochondrial expansion, while non-NMDA-mediated toxicity induces darkly stained neurons with nuclear condensation, microvacuolization, and intense ballooning of dendritic structures. However, *in vivo* studies generally demonstrate a mixture of cytopathology and, thus, this dichotomization may not be entirely valid or useful.

While the mechanism(s) by which excitatory amino acid neurotoxicity is mediated are unknown, dendritic depolarization likely results in altered membrane permeability and chloride conductance with increased intracellular calcium. With repeated exposures, the threshold for subsequent toxicity is lowered (Rothman and Olney, 1987; Meldrum and Garthwaite, 1990). Using local lumbar inoculations of kainic acid and NMDA, Bornstein *et al.* (1980) demonstrated that spinal motor neurons are acutely sensitive to excitatory amino acid neurotoxicity. Significant reductions in motor neuron numbers were noted, with subsequent studies demonstrating the existence of specific subgroups of spinal motor neurons that might be selectively sensitive to kainic acid, raising the possibility of selective neuronal damage (Curtis and Malik, 1985).

Little is known about chronic excitatory amino acid-mediated neurotoxicity, particularly as it relates to spinal motor neurons (Rothman *et al.*, 1987). While *in vivo* paradigms of chronic excitatory amino acid-induced motor neuron degeneration are rare, several naturally occurring animal states do exist and have provided models for this paradigm. As will become evident, the direct application of excitatory amino acids, either intrathecally or intraparenchymally, tends to induce the full spectrum of cytopathology described above, while chronic oral exposure induces a more selective and subtle neuropathology.

5.2.2.a. β-N-oxalylamino-L-alanine (BOAA) This direct-acting glutamate agonist is the principal neurotoxic agent of *Lathyrus sativus* (Rao *et al.*, 1964), and the likely causative agent of neurolathyrism. Repeated dietary administration of subconvulsive doses of BOAA, with or without *Lathyrus sativus* supplementation, induces a stereotyped clinical syndrome in *Macaca fascicularis* (Spencer *et al.*, 1986). Early in their clinical course, monkeys developed tremor, myoclonic jerks, increased lower extremity tone, and hind limb extensor posturing—features that were reversible following discontinuation of BOAA. Neuropathologically, mild to moderately severe chromatolytic changes were observed within motor neurons. In mice, intracerebroventricular inoculation of BOAA (ED_{100} = 50 μg) acutely induces an encephalopathy marked by arm and leg rigidity, convulsions, hypersensitivity to touch, and resting tremor—often accompanied by convulsive head tilts (Ross and Spencer, 1987), features that can be blocked by specific glutamate receptor antagonists. In spinal cord explants, BOAA rapidly induces a vacuolar degeneration in a time- and dose-dependent manner (Nunn *et al.*, 1987).

5.2.2.b. β-N-methylamino-L-alanine (BMAA) This neurotoxic glutamate antagonist, significantly less potent than BOAA (Ross and Spencer, 1987; Weis *et al.*, 1989), is a primary constituent of the leaves of the cycad plant (including the genera *Cycas, Bowenia, Lepidozamia,* and *Macrozamia*). Cattle and sheep grazing on cycad develop an acute gait disorder marked by swaying of the hind limbs, a failure to fix joints properly, and "goose-stepping" of the hind limbs (Spencer, 1987). The primary pathological features of BMAA toxicity have been well-characterized and include hepatotoxicity, enterotoxicity, and cerebellar disorganization (Sanger *et al.*, 1972). However, the use of the false sago palm (*Cycas circinalis*) as a dietary staple in the western Pacific (Whiting, 1964) prompted a re-evaluation of the potential for BMAA to induce a motor system degeneration (Spencer, 1987).

Spencer *et al.* (1987) orally supplemented cynomolgus monkeys by daily gavage with large subconvulsive doses of BMAA. Within 2–12 weeks, wrist drop, upper extremity clumsiness, and a decreased ability to pick up small objects was noted, followed thereafter by muscle weakness and reduced bulk. Monkeys progressed to develop extensor hind limb posture, stooped posture, unkempt coats and reduced vigilance, accompanied by both a resting and action tremor. With more prolonged exposure ($>$ 13 weeks), the monkeys became expressionless, motionless, and maintained a crouched, bradykinetic posturing with prominent staring. The predominant neuropathological feature was that of chromatolysis of Betz cells accompanied by neurofilament accumulation in the basilar portion of dendrites. Similar findings were observed in lamina VI–IX of Rexed with anterior horn cells less severely affected. Abnormal neuritic distention was observed in the pars compacta of the substantia nigra in a single animal while sparing the remainder of the basal ganglia, hippocampus, and cerebellum. Contrary to original thinking, this is not a model for human ALS, and BMAA is no longer suspected to be the causative agent in the western Pacific ALS foci.

5.2.2.c. N-Methyl-D-aspartate (NMDA) Recently Nag and Riopelle (1990) demonstrated that continuous lumbar intrathecal infusions of NMDA in rats over the course of two weeks will induce a dosage-dependent neuronal degeneration that can be blocked by specific NMDA antagonists or enhanced by the co-administration of glycine. The earliest pathological changes consist of an increased number of fibrillary astrocytes followed by neuronal loss with features suggestive of intracellular fluid accumulation. With increasing exposure, the remaining neurons became pyknotic and shrunken, with multifocal necrosis at the highest dosages. However, neuropathological features were not restricted to the anterior horns, and astrocytic proliferation was a hallmark.

5.2.2.d. Kainic Acid In Wistar rats, a single intrathecal inoculum of kainic acid induces tail and hind limb spasms followed by forelimb signs and paraplegia with hind limb spastic extension (Hugon *et al.*, 1989). Animals sacrificed at 48 hours demonstrated shrunken, hyperchromatic neurons with condensed and irregular nuclei. By 2 weeks, anterior horn cells developed a condensed and vacuolated cytoplasm, enlarged proximal axons, and dendrites with a modest astrocytic reaction. Ultrastructurally, intracellular aggregates of lipid-like material, mitochondria, dense bodies, and neurofilament were observed. Immunohistochemically, perikaryal phosphorylated high molecular weight neurofilament was observed, with occasional "spheroid-like" proximal axonal inclusions (Hugon and Vallat, 1990).

5.2.3. Retrogradely Transported Toxins

Two models of retrogradely induced motor neuron degeneration are of importance to our understanding of human motor neuron degeneration.

5.2.3.a. Doxorubicin Doxorubicin, an anthracycline antineoplastic antibiotic that inhibits RNA synthesis, induces somatic motor neuron and dorsal root ganglion degeneration within 6–8 days of an intraneural inoculation of the sciatic nerve or following an intralingual inoculation (Bigotte and Olsson, 1982, 1983; Yamamoto *et al.*, 1984). In the former, L_4–L_6 spinal motor neurons of Wistar rats developed extensive chromatolysis and an accumulation of bundles of neurofilament with additional features of white matter vacuolization, wallerian degeneration of the posterior columns, and a mild astrocytic and oligodendroglial proliferation. Corticospinal tracts were spared and no inflammatory response was evident.

5.2.3.b. Ricinis Communis Agglutinin The potent protein synthesis inhibitor ricinus communis agglutinin (RCA) induces a rapid and profound degeneration of dorsal root ganglia and large motor neurons ipsilateral to a sciatic nerve injection (Yamamoto *et al.*, 1985). Histologically, chromatolysis and neuronal degeneration were present, with the primary ultrastructural features being mitochondrial swelling and loss of cristae. No corticospinal involvement was present. The value of these models lies in their documentation that a retrogradely transported neurotoxin will induce motor neuronal degeneration, and, in the case of doxorubicin, may be accompanied by neurofilamentous masses.

5.2.4. Miscellaneous

Various neurotoxins are known to induce a clinical state that can be appropriately characterized as a motor system degeneration, while the neuropathological analysis clearly demonstrates the presence of either an axonal or distal dying-back neuropathy affecting systems outside of the motor system proper. However, the five paradigms that follow are of theoretical importance in the study of motor neuron degeneration, predominantly in their illustration of several principles, including dose-dependency, latency of disease induction, and the limited phenotypic reservoir of the motor system.

5.2.4.a. Cycloleucine Following a single intraperitoneal inoculation of the methionine adenosyltransferase inhibitor cycloleucine, mice develop a progressive weakness, ataxia, and hind limb paresis (Lee *et al.*, 1992). Although long utilized as a model of subacute combined degeneration (B_{12} deficiency state), mice aged 10 weeks did not develop the diffuse vacuolar degeneration that was pronounced in mice aged 21 days or 6 weeks, but rather developed a prominent distal motor axonopathy.

5.2.4.b. ρ-Bromophenylacetylurea Within 5 days of a single intraperitoneal inoculation of ρ-bromophenylacetylurea, Sprague-Dawley rats develop a progressive hind limb splaying, paddling movements, and eventually a total hind limb paralysis—the ultrastructural hallmark of which is a large accumulation of tubulomembranous material in both central and peripheral preterminal components of axons (Diezel and Quadbeck, 1960; Blakemore and Cavanaugh, 1969; Ohnishi and Ikeda, 1980). Although the predominant clinical neurotoxic manifestation reflects peripheral lower motor neuron involvement, the dying-back neuropathy affects both motor and sensory peripheral nerves with prominent posterior column degeneration. The site of action of this agent is unknown, although it has been hypothesized to induce, in a specific manner not shared by other acetylurea-based compounds, an abnormality in the transport of membranous organelles—predominantly at the preterminal site of turnaround from anterograde to retrograde transport (Brimijoin *et al.*, 1982; Nagata and Brimijoin, 1986b; Nagata *et al.*, 1987; Jakobsen and Brimijoin, 1981).

5.2.4.c. β,β′-Iminodipropionitrile (IDPN) The characteristic syndrome of intraperitoneal IDPN inoculation consists of motor hyperactivity, retropulsion, backwards somersaults, circling, head twitching, and over-alertness ("waltzing syndrome"), accompanied by giant axonal swellings (Hartmann *et al.*, 1958; Chou and Hartmann, 1964, 1965). Initially, rats develop neuroaxonal thickening affecting both spinal motor and brainstem reticular neurons, followed thereafter by giant axonal balloons. The axon distal to the neurofilamentous inclusion atrophies, a phenomenon particularly prominent following chronic low-dose IDPN intoxication (Clark *et al.*, 1980). A similar syndrome can be induced in cats, although hind limb weakness would appear to be more pronounced (Griffin *et al.*, 1982). The motor neural perikaryon remains unaffected, with increased neurofilament immunoreactivity apparent only in the neuronal perikaryon of dorsal root ganglia neurons (Gold and Austin, 1991). Of note, when combined with acrylamide, another less potent neurotoxic agent, the toxicity of the acrylamide is significantly potentiated (Gold and Halleck, 1989).

5.2.4.d. 2,4-Dithiobiuret 2,4-Dithiobiuret, a thiourea derivative utilized in the preparation of resins and thermoplastics, as a plant root growth promoter or a rodenticide and as a polymer stabilizer, induces a reversible hind limb weakness when inoculated intraperitoneally in rats (Astwood *et al.*, 1945). With daily administration, this reducing agent produces a delayed flaccid skeletal muscle weakness involving the limbs, head, neck, and trunk musculature, accompanied by hypothermia and ataxia. Death occurs in 50% of all animals within 4–5 days, although more prolonged administration at a lower dosage will induce a similar syndrome with a latency to onset characteristic of a cumulative dose threshold (Atchison *et al.*, 1981b). The illness is characterized by an alteration in presynaptic motor nerve terminal function with reduced acetylcholine release (Atchison *et al.*, 1981a). Ultrastructural features consist of a distal terminal axonopathy with motor end-plate distention combined with tubulovesicular profiles and loss of synaptic vesicles (Jones, 1989; Sahenk, 1990).

5.2.4.e. N-butyl Benzenesulfonamide The repeated intracisternal or intraperitoneal inoculation of N-butyl benzenesulfonamide, a plasticizer used in the polymerization of polyamide compounds and as a starting agent in the synthesis of agricultural herbicides, induces a dose-dependent chronic progressive spastic myelopathy in New Zealand white rabbits (Strong *et al.*, 1991a). The clinical state is marked by hind limb splaying, hyperactive reflexes, hypertonia, focal paralysis, and impaired backpedaling. This newly discovered organotoxin seems to typify our concept of a degenerative process with long latency and slow progression. The intramedullary portions of spinal motor neuron axons developed fusiform argentophilic thickening immunoreactive to monoclonal antibodies against phosphorylated neurofilaments and associated with occasional axonal spheroids. The most striking pathological feature consisted of swollen spinal motor neuron dendrites, intensely immunoreactive to monoclonal antibodies against microtubule-associated protein-2, an immunoreactivity that was lost with more prolonged exposures. Ultrastructurally, postsynaptic zones demonstrated an accumulation of vacuoles and multilamellar bodies. In a separate series of experiments, it was shown that the concomitant intracisternal administration of N-butyl benzenesulfonamide and aluminum chloride resulted in a potentiation of both the clinical and neuropathological observations (Strong and Garruto, 1991a).

5.2.5. Summary

Neurotoxic models of motor neuron degeneration provide valuable information regarding the putative mechanisms of human motor neuron disease, and also provide a variety of avenues of inducing either selective UMN, LMN, or combined degenerations. The UMN-specificity, slow clinical progression, and relatively subtle neuropathology of the motor neuron degeneration induced by BOAA correlates best with the human disease processes of neurolathyrism or primary lateral sclerosis. Its analogue, BMAA, while producing a clinical state of motor dysfunction, is more appropriately considered a model for environmentally induced extrapyramidal disease, although the doses of BMAA required to induce a disease state are excessive. The induction of a purely LMN degeneration can be achieved by a variety of chemical neurotoxins, the primary value of which rests in the demonstration of long latency to disease onset and slow progression. In contrast to these models, trace metal neurotoxicity (e.g., aluminum salts) induces a chronic, progressive motor neuron degeneration when administered in sublethal doses by intracisternal inoculation. The topographic specificity, neurofilamentous aggregation, and immunohistochemical profile of the resulting inclusions render this the most appropriate model currently available for the study of ALS.

5.3. Viral-Mediated Polioencephalopathies

Although transmissibility of either a conventional or unconventional virus has never been demonstrated in either the sporadic or western Pacific variant of ALS (Gibbs and Gajdusek, 1982), various viral-mediated motor neuron diseases have been recognized, including those induced by picornavirus (poliomyelitis), human immunodeficiency virus type 1 (HIV 1), and the human T-cell leukemia virus type 1 (HTLV-1) (reviewed by Jubelt, 1991). For the most part, experimental paradigms of viral-mediated motor neuron degeneration produce an acute noninflammatory spongiform polioencephalopathy bearing little relationship to the morphological characteristics of human motor neuron disease. However, the implication that a motor neuron degeneration can be induced by retroviral infection (RNA containing viruses that are reverse transcribed into genomic DNA), and that may require a second permissive virus infection or the presence of a recessive trait allowing viral replication, has theoretical implications for our understanding of motor neuron degenerations.

5.3.1. Murine Leukemia Virus (MuLV)

A spontaneously occurring, lethal paralysis due to an infectious, ecotropic murine leukemia retrovirus has been described in wild mice recessive for the dominant ecotropic virus restriction gene (FV-4^R or Akvr-1) from the Lake Casitas region of Southern California (Officer *et al.*, 1973; Gardner *et al.*, 1973, 1991; Gardner, 1978, 1985). Infection induces a hind limb flaccid paralysis with fasciculations and amyotrophy of lumbar paraspinal and lower extremity musculature. Both the incubation period and the severity of spongiform degeneration present are inversely proportional to the virus concentration in the inoculum when administered intracerebrally or intraperitoneally in Swiss mice (Brooks *et al.*, 1980). Pathologically, the predominant feature is that of spongiform change within spinal gray matter that precedes the degeneration of spinal motor neurons and the accompanying reactive gliosis (Andrews and Gardner, 1974; Swarz *et al.*, 1981). The cerebral cortex is spared. Neonatal mice are the most susceptible.

A similar age-dependent hind limb paralysis can be induced in susceptible neonatal laboratory mice recessive for the Fv-$1^{n/n}$ ecotropic virus restriction gene

(Oldstone *et al.*, 1980) inoculated with temperature-sensitive mutants of the Moloney strain of MuLV (Zachary *et al.*, 1986). The pathology involves primarily the anterior horns of the lumbosacral spinal cord, accompanied by spongiform changes in the spinal gray matter and lateral and ventral funiculi, with cytoplasmic vacuolar changes in neurons and glial cells, and gliosis as a late feature of the disease. Inflammatory pathology is not present. Depending on the mutant, disseminated neuroaxis involvement will occur, with neuropathological changes in the cortex and cerebellum (ts-1 Moloney MuLV), cerebellar gray matter, brainstem and upper cervical spinal cord (ts MoBA-1 MuLV), or dentate nucleus, brainstem, and basal ganglia (Cas-Br-E MuLV) (reviewed in Hoffmann *et al.*, 1988; Jolicoeur, 1990, 1991).

5.3.2. Attenuated Human Poliovirus

In an attempt to induce a persistent viral infection, Jubelt and Meagher (1984) inoculated cyclophosphamide-treated mice with attenuated human poliovirus. Affected mice developed a fatal chronic poliomyelitis manifested as a flaccid paralysis of the limbs, with a disease latency dependent on the initial dose of viral inoculum. Paralyzed mice demonstrated mild to moderate brainstem inflammation involving both sensory and motor nuclei, with inflammatory pathology in the anterior horns of the spinal cord. The latter observation correlated with the clinical observation of paralysis. Poliovirus antigen was detected in involved anterior horn cells and in cerebral neurons.

5.3.3. Lactic Dehydrogenase-Elevating Virus (LDV)

LDV infection causes an age-dependent progressive hind limb paralysis in certain mouse strains homozygous at the allele of the locus permitting N-tropic, ecotropic MuLV replication (Fv-$1^{n/n}$). Following cyclophosphamide treatment, mice become susceptible to the generally nonpathogenic togavirus LDV, and concomitant with the induction of paralysis, they develop increased levels of endogenous MuLV expression. A destructive inflammatory pathology occurs within the spinal cord and brainstem, accompanied by anterior horn cell chromatolysis, neuronophagia, perivascular cuffing, and spongiform changes (Contag and Plagemann, 1989; Contag *et al.*, 1989).

5.4. Immune-Mediated Motor Neuron Degeneration

Two models of immune-mediated motor neuron degeneration have been described in guinea pigs, and they include an experimental autoimmune motor neuron degeneration and an experimental autoimmune gray matter degeneration (Englehardt *et al.*, 1991).

5.4.1. Experimental Autoimmune Motor Neuron Degeneration

This disorder can be induced with monthly immunizations of purified bovine spinal motor neurons. Following the fourth inoculation, guinea pigs slowly develop an increase in muscle tone, weight loss, sluggish movements, hind limb footdrop when held in an upright position, and forelimb weakness (Englehardt *et al.*, 1989). By both electrophysiological and morphological measures, denervation of affected muscles occurs. Neuropathological changes are restricted to LMN and consist of neuronal degeneration, satellitosis, neuronophagia, and a loss of spinal motor neurons. No cervical or corticospinal tract involvement was observed. IgG immunoreactivity was present in the perikarya of motor neurons and at the neuromuscular junction. Both human and guinea pig pyramidal and spinal motor neurons, when incubated with experimental autoimmune motor neuron degeneration guinea pig serum, demonstrate IgG immunoreactivity, suggesting that the immunoreactivity observed *in vivo* is a result of the retrograde transport of an IgG antibody specific to motor neuron epitopes (Englehardt and Appel, 1990).

5.4.2. Experimental Autoimmune Gray Matter Degeneration

This is a related model induced by monthly immunizations of homogenates of bovine spinal cord ventral horns. The disease manifests as an abrupt UMN and LMN degeneration marked by a loss of abdominal muscle tone, paraparesis or tetraparesis, and hind limb greater than forelimb involvement. The pathology is similar to experimental autoimmune motor neuron degeneration with the additional features of perivascular inflammation, degeneration of large cortical neurons, and IgG deposition in motor neurons, within foci of inflammation, around the external membrane of the LMN and at the neuromuscular junction. In both experimental autoimmune motor neuron degeneration and experimental autoimmune gray matter degeneration, mice injected with immunoglobulins derived from moderately or severely afflicted guinea pigs developed an acute hind limb weakness accompanied by IgG deposition in the motor neurons and at the neuromuscular junction and an increased rate of miniature end-plate potentials (Appel *et al.*, 1991), confirming the ability to transfer passively the motor neuron dysfunction. However, no neuropathological lesions were induced.

5.5. Motor Neurons *in Vitro*

Although the focus of our approach to the paradigms of motor neuron degeneration has centered on *in vivo* models and the induction of clinicopathologic entities that bear a relationship to distinct human disease states or processes, the analysis of motor neurons *in vitro* plays a crucial role in our understanding of mechanisms of motor neuron survival and their dependence on

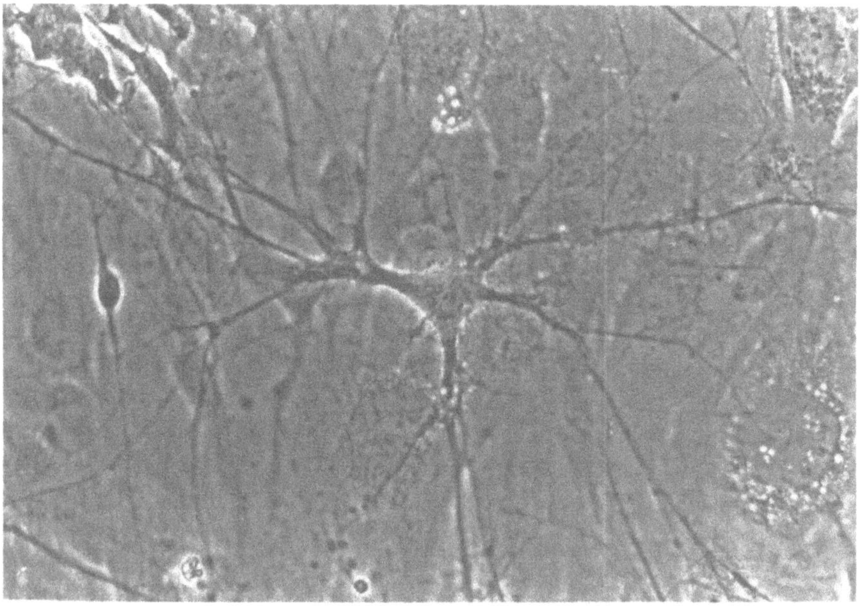

Figure 9 Dissociated fetal rabbit motor neuron 21 days postplating (×40).

various trophic factors. As alluded to at the onset of this chapter, motor neuron loss occurs with increasing age and as a natural process in neuronal development. Understanding these processes will be crucial to our understanding of naturally occurring degenerative states.

Various techniques exist for the isolation and long-term maintenance of dissociated motor neurons from a variety of species. Many utilize neuronal characteristics, including velocity separation techniques based on size differences (Capps-Covey and McIlwain, 1975; Berg and Fischbach, 1978), position in the cell cycle (Masuko *et al.*, 1979), isopycnic separation based on buoyant density (Schnarr and Schaffner, 1981; Strong and Garruto, 1989), and fluorescence-activated cell sorting of retrogradely labeled motor neurons (O'Brien and Fischbach, 1986; Schaffner *et al.*, 1987) (Figure 9). The survival *in vitro* of dissociated motor neurons is dependent on a number of muscle-derived trophic factors (Longo *et al.*, 1982; Godfrey *et al.*, 1980; Tanaka and Obata, 1982; Dohman *et al.*, 1986, 1987). McManaman *et al.* (1988, 1989, 1990) have demonstrated that rat skeletal muscle contains at least two such factors, one of which is a 22–24 kDa polypeptide capable of inducing choline acetyltransferase development that they have designated CDF (choline acetyltransferase development factor). Administration of CDF to developing chick embryos *in vivo* can prevent the naturally occurring death of motor neurons (McManaman *et al.*, 1990).

6. Conclusions

This review of experimental paradigms of motor neuron degeneration has focused primarily on understanding the correlation between the neuroanatomical and ultrastructural characteristics of human motor neuron degeneration and their experimental counterparts. The principle of limited phenotypic reserve for the expression of motor dysfunction, particularly within the experimental paradigms, is repeatedly illustrated, and the necessity for careful clinicopathological correlations emphasized and highlighted. The phenomena of hind limb paralysis, gait disturbance, and amyotrophy are clinical phenomenology that correlate with a variety of LMN lesions, whether genetic or acquired. Where spasticity and hyperreflexia would seem to correlate with the presence of UMN lesions, the majority of animals possess only a primitive corticospinal tract innervation of motor neurons, with remaining supraspinal innervation arising from primitive brainstem nuclei. Hence, the investigator is left to assume that the degeneration of large cortical pyramidal neurons and brainstem nuclei (e.g., gigantocellularis, reticular nucleus, raphe nucleus, and vestibular nucleus) is the indicator of UMN pathology in the absence of tract degeneration. The presence of additional neuropathological features such as satellitosis, neuronophagia, and perivascular cuffing finds little correlation with the human degenerative state.

Ultrastructural and, to the extent known, biochemical and immunohistochemical correlates are also crucial. Many models demonstrate the perikaryal accumulation of neurofilamentous aggregates found in amyotrophic lateral sclerosis, leading Cork *et al.* (1988) to suggest that this lesion is a common response to neuronal injury. However, when examined at the ultrastructural level, it is apparent that the neurofilamentous aggregates observed in many experimental models are the consequence of a margination or displacement of neurofilament rather than specific intraneuronal inclusions. The presence of axonal spheroids would appear to represent a more specific pathological feature. Ubiquitin immunoreactivity, perhaps as a marker of abortive nonlysosomal degradation, appears to be a common feature of all experimental paradigms in which its presence or absence has been sought, including its recent observation in a chronic aluminum neurotoxicity model (Takeda *et al.*, 1991).

When experimental models are evaluated utilizing clinical, ultrastructural, biochemical, and immunohistochemical characterizations, it is clear that the majority of *in vivo* models available today are more appropriately considered models of either spinal muscular atrophies (including virtually all spontaneous and genetic models) or toxic motor neuronopathies. Primary lateral sclerosis and neurolathyrism can be paralleled closely by experimental paradigms utilizing excitatory amino acid toxicity while amyotrophic lateral sclerosis finds its closest parallel in chronic aluminum neurotoxicity. The challenge in future years will be to manipulate these models in a systematic way to study further the etiology and therapeutics of the motor neuron diseases.

ACKNOWLEDGMENTS. Dr. Strong is supported by a Medical Research Council of Canada Scholarship and the Amyotrophic Lateral Sclerosis Society of Canada. The authors are grateful to Ms. R. Casciano for valuable assistance in preparing the manuscript.

7. References

Aldridge, J.L.F., Lennox, G., Doherty, F., Jefferson, D., Landon, M., and Mayer, R.J., 1989, Inclusion bodies in motor cortex and brainstem of patients with motor neuron disease are detected by immunocytochemical localization of ubiquitin, *Neurosci. Lett.* 105:7–13.

Andrews, J. M., 1975, The fine structure of the cervical spinal cord, ventral root and brachial nerves in the Wobbler (wr) mouse, *J. Neuropathol. Exp. Neurol.* 34:12–27.

Andrews, J. M., and Gardner, M.B., 1974, Lower motor neuron degeneration associated with type C RNA virus infection in mice: Neuropathological features, *J. Neuropathol. Exp. Neurol.* 33:285–307.

Appel, S.H., Engelhardt, J.I., Garcia, J., and Stefani, E., 1991, Immunoglobulins from animal models of motor neuron disease and from human amyotrophic lateral sclerosis patients passively transfer physiological abnormalities to the neuromuscular junction, *Proc. Natl. Acad. Sci.* 88:647–651.

Astwood, E.B., Hughes, A.M., Lubin, M., Van-DerLaan, W.P., and Adams, R.D., 1945, Reversible paralysis of motor function in rats from the chronic administration of dithiobiuret, *Science* 102:196–197.

Atchison, W.D., Lalley, P.M., Cassens, R.G., and Peterson, R.E., 1981a, Depression of neuromuscular function in the rat by chronic 2,4-dithiobiuret treatment, *Neurotoxicology* 2:329–346.

Atchison, W.D., Yang, K.H., and Peterson, R.E., 1981b, Dithiobiuret toxicity in the rat: Evidence for latency and cumulative dose thresholds, *Toxicol. Appl. Pharmacol.* 61:166–171.

Averback, P., 1981, Unusual particles in motor neuron disease, *Arch. Pathol. Lab. Med.* 105:490–493.

Beal, M.F., and Richardson, E.P., 1981, Primary lateral sclerosis: A case report, *Arch. Neurol.* 38:630–633.

Berg, D.K., and Fischbach, G.D., 1978, Enrichment of spinal cord cell cultures with motoneurons, *J. Cell Biol.* 77:83–98.

Bigotte, L., and Olsson, Y., 1982, Retrograde transport of doxorubicin (adriamycin) in peripheral nerves of mice, *Neurosci. Lett.* 32:217–221.

Bigotte, L., and Olsson, Y., 1983, Cytotoxic effects of adriamycin on mouse hypoglossal neurons following retrograde axonal transport from the tongue, *Acta Neuropathol.* (Berl.) 61:161–168.

Bird, M.T., Shuttleworth, E. Jr., Doestner, A., and Reinglass, J., 1971, The Wobbler mouse mutant: An animal model of hereditary motor system disease, *Acta Neuropathol.* (Berl.) 19:39–50.

Blakemore, W.F., and Cavanaugh, J.B., 1969, "Neuroaxonal dystrophy" occurring in an experimental "dying back" process in the rat, *Brain* 92:789–804.

Blanchard, B.J., and Ingram, V.M., 1989, Age-related neurofilament phosphorylation in Alzheimer disease, *Neurobiol. Aging* 10:253–259.

Bornstein, J.C., Lodge, D., and Curtis, D.R., 1980, Destruction of spinal motoneurones by excitant amino acids, *Proc. Aust. Physiol. Pharm. Soc.* 11:52p.

Brandt, S., 1950, Course and symptoms of progressive infantile muscular atrophy, *Arch. Neurol. Psychiatry* 63:218–228.

Brimijoin, S., Carlson, G., Nguyen, N.N., and Jakobsen, J., 1982, Analogs of ρ-bromophenylacetylurea: A structure activity study of their neurotoxicity in rats, *Muscle Nerve* 5:S122–S126.

Brodal, A., 1981, "Neurological Anatomy: In relation to Clinical Medicine" (3rd ed.), Oxford University Press, New York.

Bronson, R.T., Lake, B.D., Cook, S., Taylor, S., and Davisson, M.T., 1993, Motor neuron degeneration of mice is a model of neuronal ceroid lipofuscinosis (Batten's Disease), *Ann. Neurol.* 33:381–385.

Brooks, B.R., Swarz, J.R., and Johnson, R.T., 1980, Spongiform polioencephalomyelopathy caused by a murine retrovirus, *Lab. Invest.* 43:480–486.

Brown, W.F., Strong, M.J., and Snow, R., 1988, Methods for estimating numbers of motor units in biceps-brachialis muscles and losses of motor units with aging, *Muscle Nerve* 11:423–432.

Brownell, B., Oppenheimer, D.R., and Hughes, J.T., 1970, The central nervous system in motor neurone disease, *J. Neurol. Neurosurg. Psychiatry* 33:338–357.

Brzustowicz, L.M., Wilhelmsen, K.C., and Gilliam, T.C., 1991, Genetic analysis of childhood-onset spinal muscular atrophy, *in* "Amyotrophic lateral sclerosis and other motor neuron diseases," *Advances in Neurology* (L.P. Rowland, ed.), Raven Press, pp. 181–187.

Bugiani, O., and Ghetti, B., 1982, Progressing encephalomyelopathy with muscular atrophy, induced by aluminum powder, *Neurobiol. Aging* 3:209–222.

Bundey, S., and Lovelace, R.E., 1975, A clinical and genetic study of chronic proximal spinal muscular atrophy, *Brain* 98:455–472.

Callahan, L.M., Wylen, E.L., Messer, A., and Mazurkiewicz, J.E., 1991, Neurofilament distribution is altered in the *Mnd* (Motor neuron degeneration) mouse, *J. Neuropathol. Exp. Neurol.* 50:491–504.

Campbell, M.J., 1972, Ultrastructural observations on the Wobbler mouse, *J. Neuropathol. Exp. Neurol.* 31:190.

Cancilla, P.A., and Barlow, R.M., 1966, Structural changes of the central nervous system in swayback (Enzootic Ataxia) of lambs. II. Electron microscopy of the lower motor neuron, *Acta Neuropathol.* 6:251–259.

Capps-Covey, P., and McIlwain, D.L., 1975, Bulk isolation of large ventral spinal neurons, *J. Neurochem.* 25:517–521.

Carpenter, S., Karpati, S., Rothman, S., Watters, G., and Andermann, F., 1978, Pathological involvement of primary sensory neurons in Werdnig-Hoffmann disease, *Acta Neuropathol.* 42:91–97.

Chou, S.M., 1978, Pathognomy of intraneuronal inclusions in ALS, *in* "Amyotrophic Lateral Sclerosis" (T. Tsubaki and Y. Toyakura, eds.), pp. 135–176, University of Tokyo Press, Tokyo.

Chou, S.M., 1988, Motor neuron inclusion in ALS are heavily ubiquitinated, *J. Neuropathol. Exp. Neurol.* 47:334.

Chou, S.M., and Fakadej, A.V., 1971, Ultrastructure of chromatolytic motoneurons and anterior spinal roots in a case of Werdnig-Hoffmann disease, *J. Neuropath. Exp. Neurol.* 30:368–379.

Chou, S.M., and Hartman, H.A., 1964, Axonal lesions and waltzing syndrome after IDPN administration in rats. With a concept—"Axostasis," *Acta Neuropathol.* 3:428–450.

Chou, S.M., and Hartmann, H.A., 1965, Electron microscopy of focal neuroaxonal lesions produced by β-β-iminodipropionitrile (IDPN) in rats. 1: The advanced lesions, *Acta Neuropathol.* 4:590–603.

Chu-Wang, I.-W., and Oppenheim, R.W., 1978a, Cell death of motoneurons in the chick embryo spinal cord. I: A light and electron microscopic study of naturally occurring and induced cell loss during development, *J. Comp. Neurol.* 177:33–58.

Chu-Wang, I.-W., and Oppenheim, R.W., 1978b, Cell death of motoneurons in the chick embryo spinal cord. II: A quantitative and qualitative analysis of degeneration in the ventral root, including evidence for axon outgrowth and limb innervation prior to cell death, *J. Comp. Neurol.* 177:59–86.

Clark, A.W., Tran, P.M., Parhad, I.M., and Krekoski, C.A., 1989, Neuronal gene expression in amyotrophic lateral sclerosis, *J. Neuropathol. Exp. Neurol.* 48:381.

Clark, E.A., and Lee, V.M.Y., 1991, Dynamics of mammalian high-molecular-weight neurofilament subunit phosphorylation in cultured rat sympathetic neurons, *J. Neurosci. Res.* 30:116–123.

Clark, A.W., Griffin, J.W., and Price, D.L., 1980, The axonal pathology in chronic IDPN intoxication, *J. Neuropathol. Exp. Neurol.* 39:42–55.

Contag, C.H., and Plagemann, P.G.W., 1989, Age-dependent poliomyelitis of mice: Expression of endogenous retrovirus correlates with cytocidal replication of lactate dehydrogenase-elevating virus in motor neurons, *J. Virol.* 63:4362–4369.

Contag, C.H., Harty, J.T., and Plagemann, P.G.W., 1989, Dual virus etiology of age-dependent poliomyelitis of mice: A potential model for human motor neuron diseases, *Microb. Pathog.* 6:391–401.

Cork, L.C., Griffin, J.W., Munnell, J.F., Lorenz, M.D., Adams, R.J., and Price, D.L., 1979, Hereditary canine spinal muscular atrophy, *J. Neuropathol. Exp. Neurol.* 38:209–221.

Cork, L.C., Griffin, J.W., Choy, C., Padula, C.A., and Price, D.L., 1982, Pathology of motor neurons in accelerated hereditary canine spinal muscular atrophy, *Lab. Invest.* 46:89–99.

Cork, L.C., Struble, R.G., Gold, B.G., DiCarlo, C., Fahnestock, K.E., Griffin, J.W., and Price, D.L., 1989, Changes in size of motor axons in hereditary canine spinal muscular atrophy, *Lab. Invest.* 61:333–342.

Cork, L.C., Troncoso, J.C., Klavano, G.G., Johnson, E.S., Sternberger, L.A., Sternberger, N.H., and Price, D.L., 1988, *J. Neuropathol. Exp. Neurol.* 47:420–431.

Cotman, C.W., and Iversen, L.L., 1987, Excitatory amino acids in the brain—focus on NMDA receptors, *TINS* 10:263–265.

Cummings, J.F., DeLahunta, A., George, C., Fuhrer, L., Valentine, B.A., Cooper, B.J., Summers, B.A., Huxtable, C.R., and Mohammed, H.O., 1990, Equine motor neuron disease; a preliminary report. *Cornell Vet.* 80:357–379.

Cummings, J.F., DeLahunta, A., Summers, B.A., Mohammed, H.O., Divers, T.J., Valentine, B.A., and Trembicki-Graves, K., 1993, Eosinophilic cytoplasmic inclusions in sporadic equine motor neuron disease: an electron microscopic study, *Acta Neuropathol.* 85:291–297.

Curtis, D.R., and Malik, R., 1985, A neuropsychological analysis of the effect of kainic acid on nerve fibres and terminals in the cat spinal cord, *J. Physiol.* 368:99–108.

DeBoni, U., Otvos, A., Scott, J.W., and Crapper, D.R., 1976, Neurofibrillary degeneration induced by systemic aluminum, *Acta Neuropathol.* (Berl.) 35:285–294.

Diezel, P.B., and Quadbeck, G., 1960, Nervenschadigung durch ρ-bromphenylacetyl-Harnstoff, *Arch. Exp. Pathol.* 238:534–541.

Dohrmann, U., Edgar, D., and Sendtner, M., 1986, Muscle-derived factors that support survival and promote fiber outgrowth from embryonic motor neurons in culture, *Dev. Biol.* 118:209–221.

Dohrmann, U., Edgar, D., and Thoenen, H., 1987, Distinct neurotrophic factors from skeletal muscle and the central nervous system interact synergistically to support the survival of cultured embryonic spinal motor neurons, *Dev. Biol.* 124:145–152.

Duchen, L.W., and Strich, S.J., 1968, An hereditary motor neurome disease with progressive denervatin of muscle in the mouse: the mutant 'wobbler', *J. Neurol. Neurosurg. Psychiat.* 31:535–542.

Duncan, M.W., and Kopin, I.J., 1989, Quantification of the putative neurotoxin 2-amino--3-(methylamino) propanoic acid (BMAA) in cycadales: Analysis of the seeds of some members of the family cycadaceae, *J. Anal. Toxicol.* 13:169–175.

Duncan, M.W., Kopin, I.J., Garruto, R.M., Lavine, L., and Markey, S.P., 1988, 2-amino-3 (methylamino)-propionic acid in cycad-derived foods in an unlikely cause of amyotrophic lateral sclerosis/parkinsonism, *Lancet* (September) 10:631–632.

Durrleman, S., and Alperovitch, A., 1989, Increasing trend of ALS in France and elsewhere: Are the changes real?, *Neurology* 39:768–773.

Emery, A.E., 1971, The nosology of the spinal muscular atrophies, *J. Med. Genet.* 8:481–495.

Engelhardt, J., and Appel, S.H., 1990, Motor neuron reactivity of sera from animals with autoimmune models of motor neuron destruction, *J. Neurol. Sci.* 96:333–352.

Engelhardt, J.I., Appel, S.H., Jakab, K., Garcia, J., and Stefani, E., 1991, Immune-mediated models of motor neuron destruction in the guinea pig, *in* "Advances in Neurology, Amyotrophic Lateral Sclerosis and Other Motor Neuron Diseases," Vol 56, (L.P. Rowland, ed.), pp. 369–380, Raven Press, New York.

Engelhart, J., Appel, S.H., and Killian, J.M., 1989, Experimental autoimmune motoneuron disease, *Ann. Neurol.* 26:368–376.

Falconer, D.S., 1956, Wobbler (wr), *Mouse Newsletter* 15:23.

Fisher, C.M., 1977, Pure spastic paralysis of corticospinal origin, *Can. J. Neurol. Sci.* 4:251–258.

Gajdusek, D.C., and Salazar, A.M., 1982, Amyotrophic lateral sclerosis and Parkinsonian syndrome in high incidence among the Auyu and Jakai people in West New Guinea, *Neurology* 32:107–126.

Gardner, M.B., 1978, Type C viruses of wild mice: Characterization and natural history of amphotrophic, ecotropic, and xenotropic MuLV, *in* "Current Topics in Microbiology and Immunology" Vol 79, pp. 215–259, Springer-Verlag, New York.

Gardner, M.B., 1985, Retroviral spongiform polioencephalomyelopathy, *Rev. Infect. Dis.* 7:99–110.

Gardner, M.B., 1991, Retroviral leukemia and lower motor neuron disease in wild mice: Natural history, pathogenesis, and genetic resistance, *in* "Advances in Neurology, Amyotrophic Lateral Sclerosis and Other Motor Neuron Diseases" (L.P. Rowland, ed.), pp. 473–480, Raven Press, New York.

Gardner, M.B., Henderson, B.E., Officer, J.E., Rongey, R.W., Parker, J.C., Oliver, C., Estes, J.D., and Huebner, R.J., 1973, A spontaneous lower motor neuron disease apparently caused by indigenous type-C RNA virus in wild mice, *J. Natl. Cancer Inst.* 51:1243–154.

Gardner, A.F., Sakiewicz, N., 1963, A review of neurolathyrism including the Russian and Polish literature, *Exp. Med. Surg.* 21:164–191.

Garruto, R.M., 1991, Pacific paradigms of environmentally induced neurological disease: Clinical, epidemiological and molecular perspectives, *Neurotoxicology* 12:347–378.

Garruto, R.M., and Yase, Y., 1986, Neurodegenerative disorders of the western Pacific: The search for mechanisms of pathogenesis, *Trends Neurosci.* 9:368–374.

Garruto, R.M., Gajdusek, D.C., and Chen, K-M., 1980, Amyotrophic lateral sclerosis among Chamorro migrants from Guam, *Ann. Neurol.* 8:612–619.

Garruto, R.M., Fukatsu, R., Yanagihara, R., Gajdusek, D.C., Hook, G., and Fiori, C.E., 1984, Imaging of calcium and aluminum in neurofibrillary tangle-bearing neurons in parkinsonism-dementia of Guam, *Proc. Natl. Acad. Sci. USA* 81:1875–1879.

Garruto, R.M., Gajdusek, D.C., and Chen, K-M., 1981a, Amyotrophic lateral sclerosis and parkinsonism-dementia among Filipino migrants to Guam, *Ann. Neurol.* 10:341–350.

Garruto, R.M., Yanagihara, R., and Gajdusek, D.C., 1985, Disappearance of high-incidence amyotrophic lateral sclerosis and parkinsonism-dementia on Guam, *Neurology* 35:193–198.

Garruto, R.M., Yanagihara, R., and Gajdusek, D.C., 1988, Cycads and amyotrophic lateral sclerosis/parkinsonism-dementia, *Lancet* 2:1079.

Garruto, R.M., Yanagihara, R., Gajdusek, D.C., and Arion, D., 1989, Concentrations of heavy metals and essential minerals in garden soil and drinking water in the Western Pacific, *in* "Amyotrophic Lateral Sclerosis in Asia and Oceania" (K.M. Chen and Y. Yase, eds.), pp. 265–329, National Taiwan University, Taipei.

Garruto, R.M., Shankar, S.K., Yanagihara, R., Salazar, A.M., Amyx, H.L., and Gajdusek, D.C., 1989b, Low-calcium, high-aluminum diet-induced motor neuron pathology in cynomolgus monkeys, *Acta Neuropathol.* (Berl.) 78:210–219.

Garruto, R.M., Yanagihara, R., and Strong, M.J., 1990, Insights from Pacific ALS: development of models of chronic metal-induced motor neuron degeneration in nonhuman primates, *in* "Amyotrophic Lateral Sclerosis: New Advances in Toxicology and Epidemiology," (F.B. Norris and F.C. Rose, eds.), pp. 143–156, Smith-Gordon Publishers, London.

Gastaut, J.L., Michel, B., Figarella-Branger, D., and Somma-Mauvais, H., 1988, Chronic progressive spinobulbar spasticity, *Arch. Neurol.* 45:509–513.

Gibbs, C.J., Jr., and Gajdusek, D.C., 1982, An update on long-term *in vivo* and *in vitro* studies designed to identify a virus as the cause of amyotrophic lateral sclerosis, parkinsonism, dementia and Parkinson's disease, *in* "Human Motor Neuron Diseases" (L.P. Rowland, ed.), pp. 343–351, Raven Press, New York.

Godfrey, E.W., Schrier, B.K., and Nelson, P.G., 1980, Source and target cell specificities of a conditioned medium factor that increases choline acetyltransferase activity in cultured spinalcord cells, *Dev. Biol.* 77:403–418.

Gold, B., and Hallbeck, M.M., 1989, Axonal degeneration and axonal caliber alterations following combined β,β^1-iminodipropionitrile (IDPN) and acrylamide administration, *J. Neuropathol. Exper. Neurol.* 48:653–668.

Gold, B.G., and Austin, D.R., 1991, Regulation of aberrant neurofilament phosphorylation in neuronal perikarya. III. Alterations following single and continuous β,β′-iminodipropionitrile administrations, *Brain Res.* 563:151–162.

Greenamyre, J.T., 1986, The role of glutamate in neurotransmission and in neurologic disease, *Arch. Neurol.* 43:1058–396.

Greenwood, J.C., Troncoso, A.C., and Johnson, G.V.W., 1991, Phosphorylation, calmodulin binding and calpain-mediated degradation of individual neurofilament proteins, *Soc. Neurosci. Abstr.* 17:378.

Griffin, J.W., Cork, L.C., Adams, R.J., and Price, D.L., 1982, Axonal transport in hereditary canine spinal muscular atrophy (HCSMA), *J. Neuropathol. Exp. Neurol.* 41:370.

Griffiths, I.R., 1985, Progressive axonopathy: An inherited neuropathy of boxer dogs. 1: Further studies of the clincal and electrophysiological features, *J. Small Anim. Prac.* 26:381–392.

Griffiths, I.R., Duncan, I.D., and Barker, J., 1980, A progressive axonopathy of boxer dogs affecting the central and peripheral nervous system, *J. Small Anim. Prac.* 21:29–43.

Griffiths, I.R., Kyriakides, E., and Barrie, J., 1989, Progressive axonopathy: An inherited neuropathy of boxer dogs: An immunocytochemical study of the axonal cytoskeleton, *Neuropathol. Appl. Neurobiol.* 15:63–75.

Griffiths, I.R., Kyriakides, E., and Scott, J., 1986, Progressive axonopathy: An inherited neuropathy of boxer dogs: Quantitative and morphometric analysis of the peripheral nerve lesion, *J. Neurol. Sci.* 75:69–88.

Griffiths, I.R., McCulloch, M.C., and Abrahams, S., 1985, Progressive axonopathy: An inherited neuropathy of boxer dogs. 2: The nature and distribution of the pathological changes, *Neuropathol. Appl. Neurobiol.* 11:431–446.

Gunnarsson, L-G., Lindberg, G., Soderfelt, B., and Axelson, O., 1990, The mortality of motor neuron disease in Sweden, *Arch, Neurol.* 47:42–46.

Hamburger, V., and Levi-Montalcini, R., 1949, Proliferation, differentiation and degeneration in the spinal ganglia of the chick embryo under normal and experimental conditions, *J. Exp. Zool.* 111:457–501.

Hamburger, V., 1939, Motor and sensory hyperplasia following limb-bud transplantation in chick embryos, *Physiol. Zool.* 12:268–284.

Hamburger, V., 1975, Cell death in the development of the lateral motor column of the chick embryo, *J. Comp. Neurol.* 160:535–546.

Harris-Flanagan, A.E., 1969, Differentiation and degeneration in the motor horn of the foetal mouse, *J. Morphol.* 129:281–306.
Hartley, W.J., 1963, Lower motor neurone disease in dogs. *Acta Neuropathol.* 2:334–342.
Hartmann, H.A., Lalich, J.J., and Akert, K., 1958, Lesions in the anterior motor horn cells of rats after administratin of BIS-β-cyanoethylamine, *J. Neuropathol. Exp. Neurol.* 17:298–304.
Hausmanowa-Petrusewicz, I., 1991, Spinal muscular atrophies: How many types?, *in* "Advances in Neurology: Amyotrophic Lateral Sclerosis and other Motor Neuron Diseases," Vol. 56 (L.P. Rowland, ed.), pp. 157–167, Raven Press, New York.
Higgins, R.J., Vandevelde, M., Hoff, E.J., Jagar, J.E., Cork, L.C., and Silberman, M.S., 1977, Neurofibrillary accumulation in the zebra (*Equus burchelli*), *Acta Neuropathol.* (Berl.) 37:1–5.
Higgins, R.J., Rings, D.M., Fenner, W.R., and Stevenson, S., 1983, Spontaneous lower motor neuron disease with neurofibrillary accumulation in young pigs. *Acta Neuropathol* (Berl.) 59:288–294.
Hirano, A., 1973, Progress in the pathology of motor neuron diseases, *in* "Progress in Neuropathology, Vol. 2 (H.M. Zimmerman, ed.), pp. 181–215, Grune & Stratton, New York.
Hirano, A., 1977, Fine structural changes in the mutant hamster with hind leg paralysis, *Acta Neuropathol.* (Berl.) 39:225–230.
Hirano, A., 1978, A possible mechanism of demyelination in the Syrian hamster with hind leg paralysis, *Lab. Invest.* 38:115–121.
Hirano, A., 1991, Cytopathology of amyotrophic lateral sclerosis, *in* "Advances in Neurology, Amyotrophic Lateral Sclerosis and Other Motor Neuron Diseases, Vol. 56, (L.P. Rowland, ed.), pp. 91–101, Raven Press, New York.
Hirano, A., and Inoue, K., 1980, Early pathological changes in amyotrophic lateral sclerosis: Electron microscopic study of chromatolysis, spheroids and Bunina bodies, *Neurol. Med.* 13:148–160.
Hirano, A., Kurland, L.T., and Sayre, G.P., 1967, Familial amyotrophic lateral sclerosis, *Arch. Neurol.* 16:232–242.
Hirano, A., Llena, J.F., Streifler, M., and Cohn, F., 1976, Anterior horn cell changes in a case of neurolathyrism, *Acta. Neuropathol.* (Berl.) 35:277–283.
Hoffman, P.M., Pitts, O.M., Bilello, J.A., and Cimino, E.F., 1988, Retrovirus induced motor neuron degeneration, *Rev. Neurol.* (Paris), 144:676–679.
Homburger, F., and Bajusz, E., 1970, New models of human disease in Syrian hamsters, *JAMA* 212:604–610.
Hudson, A., Davenport, A., and Hader, W.J., 1986, The incidence of amyotrophic lateral sclerosis in southwestern Ontario, Canada, *Neurology* 36:1524–1528.
Hughes, A.F.W., 1968, Aspects of neural ontogeny, Academic Press, New York.
Hugon, J., and Vallat, J.M., 1990, Abnormal distribution of phosphorylated neurofilaments in neuronal degeneration induced by kainic acid, *Neurosci. Lett.* 119:45–48.
Hugon, J., Vallat, J.M., Spencer, P.S., Leboutet, M.J., and Barthe, D., 1989, Kainic acid induces early and delayed degenerative neuronal changes in rat spinal cord, *Neurosci. Lett.* 104:258–262.
Igata, A., Osame, M., and Inada, S., 1979, Hereditary amyotrophic dogs, *in* "Amyotrophic Lateral Sclerosis" (T. Tsubaki and Y. Toyokura, eds.), pp. 351–362, University of Tokyo Press, Tokyo.
Inada, S., Sakamoto, H., Haruta, K., Miyazono, Y., Sakaki, M., Yamauchi, C., Igata, A., Osame, M., and Fukunaga, H., 1978, A clinical study on hereditary progressive neurogenic muscular atrophy in pointer dogs, *Jpn. J. Vet. Sci.* 40:539–547.
Iwata, S., Sasajima, K., Yase, Y., and Chen, K.M., 1978, Report of investigation of the environmental factors related to occurrence of amyotrophic lateral sclerosis in Guam Island. Japanese Ministry of Education, Tokyo.
Izumo, S., Ikuta, F., Igata, A., Osame, M., Yamauchi, C., and Inada, S., 1983, Morphological

study on the hereditary neurogenic amyotrophic dogs: Accumulation of lipid compound-like structures in the lower motor neuron, *Acta Neuropathol.* (Berl.) 61:270–276.

Jakobsen, J., and Brimijoin, S., 1981, Axonal transport of enzymes and labeled proteins in experimental axonopathy induced by ρ-bromophenylace-tylurea, *Brain Res.* 229:103–122.

Jolicoeur, P., 1990, Retrovirus-induced lower motor neuron disease in mice: A model for amyotrophic lateral sclerosis and human spongiform neurological diseases, *in* "Amyotrophic Lateral Sclerosis: Concepts in Pathogenesis and Etiology" (Arthur J. Hudson, ed.), pp. 53–82, University of Toronto Press, Toronto.

Jolicoeur, P., Rassart, E., DesGroseillers, L., Robitaille, Y., Paquette, Y., and Kay, D.G., 1991, Retrovirus-induced motor neuron disease of mice: Molecular basis of neurotropism and paralysis, *in* "Advances in Neurology, Amyotrophic Lateral Sclerosis and other Motor Neuron Diseases" (L.P. Rowland, ed.), pp. 481–494, Raven Press, New York.

Jones, H.B., 1989, Dithiobiuret neurotoxicity: An ultrastructural investigation of the lesion in preterminal axons and motor endplates in the rat lumbrical muscle, *Acta Neuropathol.* 78:72–85.

Jubelt, B., 1991, Viruses and motor neuron diseases, *in* "Advances in Neurology: Amyotrophic Lateral Sclerosis and other Motor Neuron Diseases," Vol. 56 (L.P. Rowland, ed.), pp. 463–472, Raven Press, New York.

Jubelt, B., and Meagher, J.B., 1984, Poliovirus infectin of cyclophosphamide-treated mice results in persistence and late paralysis. I: Clinical, pathologic, and immunologic studies, *Neurology* 34:486–493.

Juergens, S.M., Kurland, L.T., Okazaki, H., and Mulder, D.W., 1980, ALS in Rochester, Minnesota, 1925–1977, *Neurology* 30:463–470.

Kaupmann, K., Simon-Chazottes, D., Guénet, J.-L., and Jockusch, H., 1992, Wobbler, a mutation affecting motoneuron survival and gonadal functions in the mouse, maps to proximal chromosome 11, *Genomics* 13:39–43.

Kawamura, R., O'Brien, P., Odazaki, H., and Dyck, P.J., 1977, Lumbar motoneurons of man. II: The number of diameter distribution of large- and intermediate-diamter cytons in "motoneuron columns" of spinal cord of man, *J. Neuropathol. Exp. Neurol.* 36:861–870.

Kohn, R., 1968, Postmortem findings in a case of Wohlfart-Kugelberg-Welander disease, *Confinia Neurol.* 30:253–260.

Krekoski, C.A., Mathew, A., and Parhad, I.M., 1988, Neuronal gene transcription is decreased with aluminum treatment, *J. Cell Biol.* 9:123–138.

Kugelberg, E., and Welander, L., 1956, Heredofamilial juvenile muscular atrophy stimulating muscular dystrophy, *Arch, Neurol. Psychiatry* 75:500–509.

Lamb, A.H., 1974, The timing of the earliest motor innervation in the hind-limb in the xenopus tadpole, *Brain Res.* 67:527–530.

Lamb, A.H., 1977, Neuronal death in the development of the somatotopic projections of the ventral horn in *Xenopus, Brain Res.* 134:145–150.

Lawyer, T. Jr., and Netsky, M.G., 1963, Amyotrophic lateral sclerosis, *Arch. Neurol.* 8:117–127.

Lee, C.-C., Surtees, R., and Duchen, L.W., 1992, Distal motor axonopathy and central nervous system myelin vacuolation caused by cycloleucine, an inhibitor of methionine adenosyltransferase, *Brain* 115:935–955.

Lee, S., Park, Y.D., Yen, S-H.C., Ksiezak-Reding, H., Goldman, J.E., and Dickson, D.W., 1989, A study of infantile motor neuron disease with neurofilament and ubiquitin immunocytochemistry, *Neuropediatrics* 20:107–111.

Leigh, P., and Swash, M., 1991, Cytoskeletal pathology in motor neuron diseases, *in* "Advances in Neurology, Amyotrophic Lateral Sclerosis and Other Motor Neuron Diseases, Vol. 56, (L.P. Rowland, ed.), pp. 115–124, Raven Press, New York.

Leigh, P.N., Dodson, A., Swash, M., Brion, J-P., and Anderton, B.H., 1989a, Cytoskeletal abnormalities in motor neuron disease, *Brain* 112:521–535.

Leigh, P.N., Swash, M., Whitwell, H., Garofalo, O., Dodson, A., and Anderton, B.H., 1989b, Ubiquitin inclusions in motor neurons in amyotrophic lateral sclerosis, *Neurology* 39:401.

Lilienfeld, D.E., Ehland, J., Landrigan, P.J., Chan, E., Godbold, J., Marsh, G., and Perl, D.P., 1989, Rising mortality from motoneuron disease in the USA, 1962–84, *Lancet* (April) 1:710–712.

Linton, R., Bryan, S.R., Cox, X.B., Griffis, D.P., Shelburne, J.D., Fiori, C.E., and Garruto, R.M., 1987, Digital imaging studies of aluminum and calcium in neurofibrillary tangle-bearing neurons using SIMS (secondary ion mass spectrometry), *Trace Elements Med.* 4:99–104.

Lippa, C.F., and Smith, T.W., 1988, Chromatolytic neurons in Werdnig-Hoffmann disease contain phsophorylated neurofilaments, *Acta Neuropathol.* 77:91–94.

Longo, F.M., Manthorpe, M., and Varon, S., 1982, Spinal cord neuronotrophic factors (SCNTFs). I: Bioassay of schwannoma and other conditioned media, *Dev. Brain Res.* 3:277–294.

Lowe, J., Aldridge, F., Lennox, G., Doherty, F., Jefferson, D., Landon, M., and Mayer, R.J., 1989, Inclusion bodies in motor cortex and brainstem of patients with motor neurone disease are detected by immunocytochemical localisation of ubiquitin, *Neurosci. Lett.* 105:7–13.

Ludolph, A.C., Hugon, J., Dwived, M.P., Schaumburg, H.H., and Spencer, P.S., 1987, Studies on the aetiology and pathogenesis of motor neuron diseases, *Brain* 110:149–165.

Lumsden, A., 1990, The cellular basis of segmentation in the developing hindbrain, *Trends Neurosci.* 3:329–335.

Lutsep, H.L., and Rodriguez, M., 1989, Ultrastructural, morphometric, and immunocytochemical study of anterior horn cells in mice with "wasted" mutation, *J. Neuropathol. Exp. Neurol.* 48:519–531.

Ma, W., and Vacca-Galloway, L.L., 1991, Reduced branching and length of dendrites detected in cervical spinal cord motoneurons of wobbler mouse, a model of inherited motoneuron disease, *J. Comp. Neurol.* 311:210–222.

Manetto, V., Sternberger, N.J., Perry, G., Sternberger, L.A., and Gambetti, P., 1988, Phosphorylation of neurofilaments is altered in amyotrophic lateral sclerosis, *J. Neuropathol. Exp. Neurol.* 47:642–653.

Masuko, S., Kuromi, H., and Shimada, Y., 1979, Isolation and culture of motoneurons from embryonic chicken spinal cords, *Proc. Natl. Acad. Sci. USA* 76:3537–3541.

Matsumoto, S., Hirano, A., and Goto, S., 1990, Ubiquitin-immunoreactive filamentous inclusions in anterior horn cells of Guamanian and non-Guamanian amyotrophic lateral sclerosis, *Acta Neuropathol.* 80:233–238.

Mazurkiewicz, J.E., 1990, Ubiquitin deposits in spinal motoneurons of the MND (motor neuron degeneration) mouse, *J. Neurol. Sci.* 985:349–350.

Mazurkiewcz, J.E., 1991, Ubiquitin deposits are present in spinal motor neurons in all stages of the disease in the motor neuron degeneration (*Mnd*) mutant of the mouse, *Neurosci. Lett.* 128:182–186.

Mazurkiewcz, J.E., Strominger, N.L., and Messer, A., 1986, Morphological analysis of the motor neuron degeneration (Mnd) mouse, *Muscle and Nerve* 9:125.

McManaman, J., Crawford, F., Clark, R., Richker, J., and Fuller, F., 1989, Multiple neurotrophic factors from skeletal muscle: Demonstration of effects of basic fibroblast growth factor and comparisons with the 22-kilodalton choline acetyltransferase development factor, *J. Neurochem.* 53:1763–1771.

McManaman, J.L., Crawford, F. G., Stewart, S.S., and Appel, S.H., 1988, Purification of a skeletal muscle polypeptide which stimulates choline acetyltransferase activity in cultured spinal cord neurons, *J. Biol. Chem.* 263:5890–5897.

McManaman, J.L., Oppenheim, R.W., Prevette, D., and Marchetti, D., 1990, Rescue of motoneurons from cell death by a purified skeletal muscle polypeptide: Effects of the ChAT development factor, CDF, *Neuron* 4:891–898.

Meadows, J.C., Marsden, C.D., and Harriman, D.G., 1969a, Chronic spinal muscular atrophy in adults. Part 1: The Kugelberg-Welander syndrome, *J. Neurol. Sci.* 9:527–550.

Meadows, J.C., Marsden, C.D., and Harriman, D.G., 1969b, Chronic spinal muscular atrophy in adults. Part 2: Other forms, *J. Neurol. Sci.* 9:551–566.

Meldrum, B., and Garthwaite, J., 1990, Excitatory amino acid neurotoxicity and neurodegenerative disease, *Trends Pharm. Sci.* 11:379–387.

Messer, A., 1992, Animal models of amyotrophic lateral sclerosis, *in* "Handbook of Amyotrophic Lateral Sclerosis (R.A. Smith, ed.), pp. 433–451, Marcel Dekker Inc., New York.

Messer, A., and Flaherty, L., 1986, Autosomal dominant in a late-onset motor neuron disease in the mouse, *J. Neurogenetics* 3:345–355.

Messer, A., Strominger, N.L., and Mazurkiewicz, J.E., 1987, Histopathology of the late-onset motor neuron degeneration (*Mnd*) mutant in the mouse, *J. Neurogenetics* 4:201–213.

Messer, A., Plummer, J., Maskin, P., Coffin, J.M., and Frankel, W.N., 1992, Mapping of the motor neuron degeneration (Mnd) gene, a mouse model of amyotrophic lateral sclerosis (ALS), *Genomics* 13:797–802.

Mitsumoto, H., and Bradley, W.G., 1982, Murine motor neuron disease (the Wobbler mouse) degeneration and regeneration of the lower motor neuron, *Brain* 105:811–834.

Mitsumoto, H., and Gambetti, P., 1986, Impaired slow axonal transport in wobbler mouse motor neuron disease, *Ann. Neurol.* 19:36–43.

Mitsumoto, H., Hanson, M.R., and Chad, D.A., 1988, Amyotrophic lateral sclerosis: Recent advances in pathogenesis and therapeutic trials, *Arch. Neurol.* 45:189–202.

Mitsumoto, H., McQuarrie, I.G., Kurahashi, K., and Sunohara, N., 1990, Histometric characteristics and regenerative capacity in wobbler mouse motor neuron disease, *Brain* 113:497–507.

Mohammed, H.O., Cummings, J.F., Divers, T.J., Valentine, B., deLahunta, A., Summers, B., Farrow, B.R.H., Trembicki-Graves, K., and Mauskopf, A., (1993), Risk factors associated with equine motor neuron disease: a possible model for human MND, *Neurology* 43:966–971.

Muma, N.A., Troncoso, J.C., Hoffman, P.N., Koo, E.H., and Price, D.L., 1988, Aluminum neurotoxicity: Altered expression of cytoskeletal genes, *Mol. Brain Res.* 3:115–122.

Munoz, D.G., Greene, C., Perl, D.P., and Selkoe, D.J., 1988, Accumulation of phosphorylated neurofilaments in anterior horn motoneurons of amyotrophic lateral sclerosis patients, *J. Neuropathol. Exp. Neurol.* 47:9–18.

Munsat, T.L., Skerry, L., Korf, B., Pober, B., Schapira, Y., and Gascon, G.G., Al-Rajeh, S.M., Dubowitz, V., Davies, K., Brzustowicz, L.M., Penchaszadeh, G.K., and Gilliam, T.C., 1990, Phenotypic heterogeneity of spinal muscular atrophy mapping to chromosome 5q11.2–13.3 (SMA 5q), *Neurology* 40:1831–1836.

Munsat, T.L., Woods, R., Fowler, W., and Pearson, C.M., 1969, Neurogenic muscular atrophy of infancy with prolonged survival, *Brain* 92:9–24.

Murakami, T., Mastaglia, F.L., and Bradley, W.G., 1980, Reduced protein synthesis in spinal anterior horn neurons in Wobbler mouse mutant, *Exp. Neurol.* 67:423–432.

Murakami, T., Mastaglia, F.L., Mann, D.M.A., and Bradley, W.G., 1981, Abnormal RNA metabolism in spinal motor neurons in the Wobbler mouse, *Muscle Nerve* 4:407–412.

Murayama, S., Bouldin, T.W., and Suzuki, K., 1990a, Immunocytochemical study of primary motor area in motor neuron disease, *J. Neuropathol. Exp. Neurol.* 49:280.

Murayama, S., Bouldin, T.W., and Suzuki, K., 1991, Immunocytochemical and ultrastructural studies of Werdnig-Hoffmann disease, *Acta Neuropathol.* 81:408–417.

Murayama, S., Mori, H., Ihara, Y., Bouldin, W., Suzuki, K., and Tomonaga, M., 1990b, Immunocytochemical and ultrastructural studies of lower motor neurons in amyotrophic lateral sclerosis, *Ann. Neurol.* 27:137–148.

Murayama, S., Ookawa, Y., Mori, H., Nakano, I., Ihara, Y., Kuzuhara, S., and Tomonaga, M.,

1989, Immunocytochemical and ultrastructural study of Lewy body-like hyaline inclusions in familial amyotrophic lateral sclerosis, *Acta Neuropathol.* 78:143–152.

Nag, S., and Riopelle, R.J., 1990, Spinal neuronal pathology associated with continuous intrathecal infusion on N-methyl-D-aspartate in the rat, *Acta Neuropathol.* 81:7–13.

Nagata, H., and Brimijoin, S., 1986a, Axonal transport in the motor neurons of rats with neuropathy induced by ρ-bromophenylacetylurea, *Ann. Neurol.* 19:458–464.

Nagata, H., and Brimijoin, S., 1986b, Neurotoxicity in halogenated phenylacetylureas is linked to abnormal onset of rapid axonal transport, *Brain Res.* 385:136–142.

Nagata, H., Brimijoin, S., Low, P., and Schmelzer, J.D., 1987, Slow axonal transport in experimental hypoxia and in neuropathy induced by ρ-bromophenylacetylurea, *Brain Res.* 422:319–326.

Namba, T., Aberfeld, D.C., and Grob, D., 1970, Chronic proximal spinal muscular atrophy, *J. Neurol. Sci.* 11:401–423.

Nixon, C.W., and Connelly, M.E., 1968, Hind-leg paralysis: A new sex-linked mutation in the Syrian hamster, *J. Heredity* 59:276–278.

Nunn, P.B., Seelig, M., Zogoren, J.C., and Spencer, P.S., 1987, Stereospecific acute neuronotoxicity of "uncommon" plant amino acids linked to human motor-system diseases, *Brain Res.* 410:375–379.

O'Brien, R.J., and Fischbach, G.D., 1986, Isolation of embryonic chick motoneurons and their survival *in vitro, J. Neurosci.* 6:3265–3274.

Officer, J.E., Teeson, N., Estes, J.D., Fontanilla, E., Rongey, R.W., and Gardner, M.B., 1973, Isolation of a neurotropic type C virus, *Science* 181:945–947.

Ohnishi, A., and Ikeda, M., 1980, Morphometric evaluation of primary sensory neurons in experimental ρ-bromophenylacetylurea intoxication, *Acta Neuropathol.* (Berl.) 52:111–118.

Okado, N., and Oppenheim, R.W., 1984, Cell death of motoneurons in the chick embryo spinal cord: The loss of motoneurons following removal of afferent inputs, *J. Neurosci.* 4:1639–1652.

Oldstone, M.B.A., Jensen, F., Dixon, F.J., and Lampert, P.W., 1980, Pathogenesis of the slow disease of the central nervous system associated with wild mouse virus, *Virology* 107:180–193.

Oppenheim, R.W., Cope, T., and Prevette, D., 1989, Early regional variations in motoneuron numbers arise by differential proliferation in the chick embryo spinal cord, *Dev. Biol.* 133:468–474.

Oppenheim, R.W., Haverkamp, L.J., Prevette, D., McManaman, J.L., and Appel, S.H., 1988, Reduction of naturally occurring motoneuron death *in vivo* by a target-derived neurotrophic factor, *Science* 240:919–922.

Papapetropoulos, T.A., and Bradley, W.G., 1972, Spinal motor neurones in murine muscular dystrophy and spinal muscular atrophy: A quantiative histological study, *J. Neurol. Neurosurg. Psych.* 35:60–65.

Parhad, I.M., Krekoski, C.A., Mathew, A., and Tran, P.M., 1989, Neuronal gene expression in aluminum myelopathy, *Cell Mol. Neurobiol.* 9:123–138.

Pearn, J., 1978, Incidence, prevalence, and gene frequency studies of chronic childhood spinal muscular atrophy, *J. Med. Genet.* 15:409–413.

Perl, D.P., Gajdusek, D.C., Garruto, R.M., Yanagihara, R.T., and Gibbs, C.J., Jr., 1982, Intraneuronal aluminum accumulation in amyotrophic lateral sclerosis and parkinsonism-dementia of Guam, *Science* 217:1053–1055.

Pisharodi, M., and Nauta, H.J., 1985, An animal model for neuron-specific cord lesions by the microinjection of N-methylaspartate, kainic acid, and quisqualic acid, *Appl. Neurophysiol.* 48:226–233.

Plato, C.C., Garruto, R.M., Yanagihara, R., Chen, K.M., Wood, J.L., Gajdusek, D.C., and Norris, A.N., 1982, Cortical bone loss and measurements of the second metacarpal bone. I: Comparisons between adult Guamanian Chamorros and American Caucasians, *Am. J. Phys. Anthropol.* 59:461–465.

Prestige, M.C., 1967, The control of cell number in the lumbar ventral horns during the development of *Xenopus laevis* tadpoles, *J. Embryol. Exper. Morphol.* 18:359–387.
Pringle, C.E., Hudson, A.J., Munoz, D.G., Kiernan, J.A., Brown, W.F., and Ebers, G.C., 1992, Primary lateral sclerosis, *Brain* 115:495–520.
Pullen, A.H., Martin, J.E., and Swash, M., 1992, Ultrastructure of pre-synaptic input to motor neurons in Onuf's nucleus: Controls and motor neuron disease, *Neuropathol. Appl. Neurobiol.* 18:213–231.
Rao, S.L.N., Adiga, P.R., and Sarma, P.S., 1964, The isolation and characterization of β-N-Oxalyl-L-α,β-Diaminopropionic acid: A neurotoxin from the seeds of *Lathyris sativus, Biochemistry* 3:432–436.
Reed, D.M., Torres, J.M., and Brody, J.A., 1975, Amyotrophic lateral sclerosis and parkinsonism-dementia on Guam, 1945–1972. II: Familial and genetic studies, *Am. J. Epidemiol.* 101:302–310.
Rodgers-Johnson, P., Garruto, R.M., Yanagihara, R., Chen, K.M., Gajdusek, D.C., and Gibbs, C.J., Jr., 1986, Amyotrophic lateral sclerosis and parkinsonism-dementia on Guam: A 30-year evaluation of clinical and neuropathological trends, *Neurology* 36:7–13.
Rosen, D.R., Siddique, T., Patterson, D., Figlewicz, D.A., Sapp, P., Hentati, A., Donaldson, D., Goto, J., O'Regan, J.P., Deng, H.-X., Rahmani, Z., Krizus, A., McKenna-Yasek, D., Cayabyab, A., Gaston, S.M., Berger, R., Tanzi. R.E., Halperin, J.J., Herzfeldt, B., Van den Bergh, R., Hung, W.-Y., Bird, T., Deng, G., Mulder, D.W., Syth, C., Laing, N.G., Soriano, E., Pericak-Vance, M.A., Haines, J., Rouleau, G.A., Gusella, J.S., Horvitz, H.R., and Brown, R.H., Jr., 1993, Mutations in Cu/Zn superoxide dismutase gene are associated with familial amyotrophic lateral sclerosis, *Nature* 362:59–62.
Ross, S.M., and Spencer, P.S., 1987, Specific antagonism of behavioral action of "uncommon" amino acids linked to motor-system diseases, *Synapse* 1:248–253.
Rothman, S.M., and Olney, J.W., 1987, Excitotoxicity and the NMDA receptor, *TINS* 10:299–302.
Rothman, S.M., Thurston, J.H., and Hauhart, R.E., 1987, Delayed neurotoxicity of excitatory amino acids *in vitro, Neuroscience* 22:471–480.
Russo, L.S., 1982, Clinical and electrophysiological studies in primary lateral sclerosis, *Arch. Neruol.* 39:662–664.
Sahenk, Z., 1990, Distal terminal axonopathy produced by 2,4-dithiobiuret: Effects of long-term intoxication in rats, *Acta Neuropathol.* 81:141–147.
Sack, G.H., Cork, L.C., Morris, J.M., Griffin, J.W., and Price, D.L., 1984, Autosomal dominant inheritance of hereditary canine spinal muscular atrophy, *Ann. Neurol.* 15:369–373.
Sanger, V.L., Yang, M., and Mickelsen, O., 1972, Cycasin-induced central nervous system lesions in postnatal mice, *Fed. Proc.* 31:1524–1528.
Sasaki, S., and Maruyama, S., 1992, Increase in diameter of the axonal initial segment is an early change in amyotrophic lateral sclerosis, *J. Neurol. Sci.* 110:114–120.
Schaffner, A.E., St. John, P.A., and Barker, J.L., 1987, Fluorescence-activated cell sorting of embryonic mouse and rat motoneurons and their long-term survival *in vitro, J. Neurosci.* 7:3088–3104.
Schmalbruch, H., Skovgaard Jensen, H.-J., Bjerg, B., Kamieniecka, Z., and Kurland, L., 1991, A new mouse mutant with progresive motor neuropathy, *J. Neuropathol. Exp. Neurol.* 50:192–204.
Schnarr, R.L., and Schaffner, A.E., 1981, Separation of cell types from embryonic chicken and rat spinal cord: Characterization of motoneuron-enriched fractions, *J. Neurosci.* 1:204–217.
Selye, H., 1957, Lathyrism, *Rev. Can. de Biol.* 16:2–82.
Sendtner, M., Schmalbruch, H., Stock, K.A., Carroll, P., Kreutzberg, G.W., and Thoenen, H., 1992, Ciliary neurotrophic factor prevents degeneration of motor neurons in mouse mutant progressive motor neuronopathy, *Nature* 358:502–504.
Shields, R.P., and Vandevelde, M., 1978, Spontaneous lower motor neuron disease in rabbits (Oryctolagus cuniculus), *Acta Neuropathol.* (Berl.) 44:85–90.

Shiraki, H.K., and Yase, Y., 1975, Amyotrophic lateral sclerosis in Japan, *in* "Handbook of Clinical Neurology," Vol. 22. (P.J. Vinken and G.W. Bruyn, eds.), pp. 353–419, Elsevier Publishing, New York.

Shishikura, K., Hara, M., Sasaki, Y., and Misugi, K., 1983, A neuropathologic study of Werdnig-Hoffmann disease with special reference to the thalamus and posterior roots, *Acta Neuropathol.* (Berl.) 60:99–106.

Shultz, L.D., Sweet, H.O., Davisson, M.T., and Coman, D.R., 1982, 'Wasted' a new mutant of the mouse with abnormalities characteristic of ataxia telangiectasia, *Nature* 297:402–404.

Stockard, C.R., 1936, A hereditary lethal for localized motor and preganglionic neurones with a resulting paralysis in the dog, *Am. J. Anat.* 1:1–53.

Siddique, T., Figlewicz, D.A., Pericak-Vance, M.A., Haines, J.L., Rouleau, G., Jeffers, A.J., Sapp, P., Hung, W.-Y., Bebout, J., McKenna-Yasek, D., Deng, G., Horvitz, H.R., Gusella, J.F., Brown, R.H., and Roses, A.D., 1991, Linkage of a gene causing familial amyotrophic lateral sclerosis to chromosome 21 and evidence of genetic-locus heterogeneity, *N. Engl. J. Med.* 324:1381–1384.

Sobue, G., Hashizume, Y., Yasuda, T., Mukai, E., Kumagai, T., Mitsuma, T., and Trojanowski, J.Q., 1990, Phosphorylated high molecular weight neurofilament protein in lower motor neurons in amyotrophic lateral sclerosis and other neurodegenerative diseases involving ventral horns, *Acta Neuropathol.* 79:402–408.

Sotaniemi, K.A., and Myllyia, V.V., 1985, Primary lateral sclerosis: A debated entity, *Acta Neurol. Scand.* 71:334–336.

Spencer, P.S., 1987, Guam ALS/Parkinsonism-dementia: A long-latency neurotoxic disorder caused by "slow toxin(s)" in food? *Can. J. Neurol. Sci.* 14:347–357.

Spencer, P.S., Ludolph, A., Dwived, M.P., Roy, D.N., Hogon, J., and Schaumburg, H.H., 1986, Lathyrism: Evidence for role of the neuroexcitatory aminoacid BOAA, *Lancet* (November) 8:1066–1067.

Spencer, P.S., Nunn, P.B., Hugon, J., Ludolph, A.C., Ross, S.M., Roy, D.N., and Robertson, R.C., 1987, Guam amyotrophic lateral sclerosis-parkinsonism-dementia linked to a plant excitant neurotoxin, *Science* 237:517–522.

Sternberger, N.A., Sternberger, L.A., and Ulrich, J., 1985, Aberrant neurofilament phosphorylation in Alzheimer disease, *Proc. Natl. Acad. Sci. USA* 82:4274–4276.

Striefler, M., Cohn, D.F., Hirano, A., and Schujman, E., 1977, The central nervous system in a case of neurolathyrism, *Neurology* 37:1176–1178.

Strong, M.J., and Garruto, R.M., 1889, Isolation of fetal mouse motor neurons on discontinuous percoll density gradients, *In Vitro Cell. Devel. Biol.* 25:939–945.

Strong, M.J., and Garruto, R.M., 1991a, Potentiation in the neurotoxic induction of experimental chronic neurodegenerative disorders: N-butyl benzenesulfonamide and aluminum chloride, *Neurotoxicology* 12:426–426.

Strong, M.J., and Garruto, R.M., 1991b, Chronic aluminum-induced motor neuron degeneration: Clinical neuropathological and molecular biological aspects, *Can. J. Neurol. Sci.* 18:428–431.

Strong, M.J., and Garruto, R.M., 1991c, Aluminum neurotoxicity *in vitro:* A comparative analysis of dissociated fetal rabbit hippocampal and motor neuron-enriched cultures, *Lab. Invest.* 65:243–249.

Strong, M.J., Garruto, R.M., Wolff, A.V., Chou, S.M., Fox, S.D., and Yanagihara, R., 1991a, N-butyl benzenesulfonamide: A neurotoxic plasticizer inducing a spastic myelopathy in rabbits, *Acta Neuropathol.* 81:235–241.

Strong, M.J., Hudson, A.J., and Alvord, W.G., 1991b, Familial amyotrophic lateral sclerosis, 1850–1989: A statistical analysis of the world literature, *Can. J. Neurol. Sci.* 18:45–58.

Strong, M.J., Wolff, A.V., Wakayama, I., and Garruto, R.M., 1991c, Aluminum-induced chronic myelopathy in rabbits, *Neurotoxicology* 12:9–22.

Strong, M.J., Yanagihara, R., Wolff, A.V., Shankar, S.H., and Garruto, R.M., 1990, Experimental neurofilamentous aggregates: Acute and chronic models of aluminum-induced encephalomyelopathy in rabbits, *in* "The Etiology of Amyotrophic Lateral Sclerosis: Epidemiological and Neurotoxicological Aspects" (F.B. Norris, and F.C. Rose, eds.), pp. 157–173, Smith-Gordon, London.

Swartz, J.R., Brooks, B.R., and Johnson, R.T., 1981, Spongiform polioencephalomyelopathy caused by a murine retrovirus. II: Ultrastructural localization of virus replication and spongiform changes in the central nervous system, *Neuropathol. Appl. Neurobiol* 7:365–380.

Tada, K., Ohshita, S., Yonenobu, K., Ono, K., Satoh, K., and Shimizu, N., 1979, Development of spinal motoneuron innvervation of the upper limb muscle in the rat, *Exp. Brain Res.* 35:287–293.

Takeda, M., Tatebayashi, Y., Tanimukai, S., Nakamura, Y., Tanaka, T., and Nishimuca, T., 1991, Immunohistochemical study of microtubule-associated protein 2 and ubiquitin in chronically aluminum-intoxicated rabbit brain, *Acta Neuropathol.* 82:346–352.

Tanaka, H., and Obata, K., 1982, Survival and neurite growth of chick embroy spinal cord cells in serum-free culture, *Dev. Brain Res.* 4:313–321.

Tomonaga, M., Saito, M., Yoshimura, M., Shimada, H., and Tohgi, H., 1978, Ultrastructure of the Bunina bodies in anterior horn cells of amyotrophic lateral sclerosis, *Acta Neuropathol.* 42:81–86.

Tomlinson, B.E., and Irving, D., 1977, The number of limb motor neurons in the human lumbosacral cord throughout life, *J. Neurol. Sci.* 34:213–219.

Towfighi, J., Young, R.S.K., and Ward, R.M., 1985, Is Werdnig-Hoffmann disease a pure lower motor neuron disorder?, *Acta Neuropathol.* 65:270–280.

Tracey, J.I., Jr., Stensland, C.H., Doan, D.B., May, H.G., Schlanger, S.O., and Stark, Jr., 1959, Military geology of Gaum, Mariana Islands. Part I: Description of terrain and environment. Part II: Engineering aspects of geology and soils, pp. 1–282. Intelligence Division, Office of the Engineer Headquarters, United States Army of the Pacific, Washington DC.

Tsukagoshi, H., Yanagisawa, N., and Ogushi, K., 1980, Morphometric quantification of the cervical limb motor cells in various neuromuscular diseases, *J. Neurol. Sci.* 47:463–472.

Tsukagoshi, H., Yanagisawa, N., Ogushi, K., Nagashima, K., and Murakami, T., 1979, Morphometric quantification of the cervical limb motor cells in controls and in amyotrophic lateral sclerosis, *J. Neurol. Sci.* 41:287–297.

Uemura, E., 1984, Intranuclear aluminum accumulation in chronic animals with experimental neurofibrillary changes, *Exp. Neurol.* 85:10–18.

Uemura, E., and Ireland, W.P., 1984, Synaptic density in chronic animals with experimental neurofibrillary change, *Exp. Neurol.* 85:1–9.

Ungar-Sargon, J.Y., Lovelace, R.E., and Brust, J.C., 1980, Spastic paraplegia-paraparesis, *J. Neurol. Sci.* 46:1–12.

Wakayama, I., 1992, Morphometry of spinal motor neurons in amyotrophic lateral sclerosis with special reference to chromatolysis and intracytoplasmic inclusion bodies, *Brain Res.* 586:12–18.

Watkins, J.C., and Olverman, H.J., 1987, Agonists and antagonists for excitatory amino acid receptors, *TINS* 10:265–272.

Weiss, J.H., Koh, J.-Y., and Choi, D.W., 1989, Neurotoxicity of β-N-Oxalylamino-L-alanine (BOAA) on cultured cortical neurons, *Brain Res.* 497:64–71.

Whiting, M.G., 1964, Food practices in ALS foci in Japan, the Marianas, and New Guinea, *Fed. Proc.* 23:1343–1345.

Wisniewski, H.M., Sturman, J.A., Shek, J.W., 1980, Aluminum chloride induced neurofibrillary changes in the developing rabbit: A chronic animal model. *Ann. Neurol.* 8:479–490.

Wisniewski, H.M., Sturman, J.A., and Shek, J.W., 1982, Chronic model of neurofibrillary changes induced in mature rabbits by metallic aluminum, *Neurobiol. Aging* 3:11–22.

Yamamoto, T., Iwasaki, Y., and Konno, H., 1984, Retrograde axoplasmic transport of adriamycin, *Neurology* 34:1299–1304.

Yamamoto, T., Iwasaki, Y., Konno, H., and Kudo, H., 1985, Primary degeneration of motor neurons by toxic lectins conveyed from the peripheral nerve, *J. Neurol. Sci.* 70:327–337.

Yamasaki, H., Itakura, C., and Mizutani, M., 1991, Hereditary hypotrophic axonopathy with neurofilament deficiency in a mutant strain of the Japanese quail, *Acta Neuropathol.* (Berl.) 82:427–434.

Yanagihara, R., Garruto, R.M., and Gajdusek, D.C., 1984, Calcium and vitamin D metabolism in Guamanian Chamorros with amyotrophic lateral sclerosis and parkinsonism-dementia, *Ann. Neurol.* 15:42–48.

Yano, I., Yoshida, S., Uebayashi, Y., Yoshimasu, F., and Yase, Y., 1989, Degenerative changes in the central nervous system of Japanese monkeys induced by oral administration of aluminum salt, *Biomed. Res.* 10:33–41.

Yase, Y., 1988, Metal studies in ALS—further development, *in* "Amyotrophic Lateral Sclerosis: Recent Advances in Research and Treatment" (T. Tsubaki and Y. Yase, eds.), Elsevier Science Publishers, New York.

Yase, Y., 1978, The basic process of amyotrophic lateral sclerosis as reflected in the Kii Peninsula and Guam, *in* "Proceedings of the Eleventh World Congress of Neurology: (G.W. Bryun and A.P.J. Heijstee, eds.), pp. 413–427, Excerpta Medica, Amsterdam.

Yoshida, S., 1977, X-ray microanalytical studies on amyotrophic lateral sclerosis. I: Metal distribution compared with neuropathological findings in cervical spinal cord, *Clin. Neurol. (Japan)* 17:299–309.

Yoshimasu, F., Yasui, M., Yase, Y., Iwata, S., Gajdusek, D.C., Gibbs, C.J., Jr., and Chen, K.M., 1980, Studies on amyotrophic lateral sclerosis by neutron activation analysis. 2: Comparative study of analytical results on Guam PD, Japanese ALS, and Alzheimer disease cases, *Folia Psychiatry Neurol. Jpn.* 34:75–85.

Younger, D.S., Chou, S., Hays, A.P., Lange, D.J., Emerson, R., Brin, M., Thompson, H., and Rowland, L.P., 1988, Primary lateral sclerosis, *Arch. Neurol.* 45:1304–1307.

Zachary, J.F., Knupp, C.J., and Wong, P.K.Y., 1986, Noninflammatory spongiform polioencephalomyelopathy caused by a neurotropic temperature-sensitive mutant of Moloney murine leukemia virus TB, *Am. J. Pathol.* 124:457–468.

Zagoren, J.C., Seelig, M., Roy, D.N., Bornstein, M.B., and Spencer, P.S., 1985, Pathological changes in spinal cord induced by lathyrus neurotoxin (Beta-(*N*)-oxalylamino L-alanine), *J. Neuropathol. Exp. Neurol.* 44:124.

Zwellweger, H., Schneider, H.J., Schuldt, D.R., and Mergner, W., 1969, Heritable spinal muscular atrophies, *Helv. Paed. Acta* 24:92–102.

4

6-Hydroxydopamine Lesions of Nigrostriatal Neurons as an Animal Model of Parkinson's Disease

PAUL J. READING and STEPHEN B. DUNNETT

1. Why Have an Animal Model of Parkinson's Disease?

Parkinson's disease (PD) is a uniquely human affliction that profoundly affects all aspects of daily living as it progresses. In addition to the obvious and characteristic physical disability the sufferer displays to an outside observer, it is becoming increasingly evident that the inner worlds of emotion and thought can also be devastated during the advance of the disease.

Animal models have at least two direct roles to play in developing a better understanding and treatment of the complex symptomatology of PD. First, the models provide an experimental basis for trials of novel therapeutic strategies. These include the development of new drugs, novel delivery mechanisms, and surgical interventions, notably using the revolutionary technique of embryonic neural transplantation. Second, by virtue of their anatomical and biochemical selectivity, animal models allow various aspects of the disease to be fractionated, raising the ultimate hope of more specific treatments for particular symptoms.

A further aim of animal models is to provide insight into the normal functions of the basal ganglia by examining the nature of parkinsonian-type deficits. Particularly useful comparisons can be made when some of the more enigmatic features of the disease are considered. Examples of these include the unpredictable daytime variability of symptoms seen in many patients (Marsden and Parkes, 1976) and the progressive hypokinesia, observed as a susceptibility to fatigue when patients are asked to repeat a fixed motion for prolonged periods

PAUL J. READING and STEPHEN B. DUNNETT. • Department of Experimental Psychology, University of Cambridge, Downing Street, Cambridge CB2 3EB, England.
Toxin-Induced Models of Neurological Disorders, edited by Michael L. Woodruff and Arthur J. Nonneman. Plenum Press, New York, 1994.

(Selby, 1990). This latter class of symptoms is well demonstrated by progressive micrographia seen on a written page, which cannot be explained in terms of simple muscle fatigue. Conversely and more dramatically, it is well documented that advanced parkinsonian patients may temporarily overcome their seemingly profound hypokinesia in response to an intense external stimulus or internal effort (Schwab, 1972). As will be discussed later, such stress-induced "kinesia paradoxica" is analogous to features of the animal model.

An additional role of animal models centers on the notion of functional recovery of damaged dopamine systems. Idiopathic parkinsonism is asymptomatic until the striatal concentration of dopamine is reduced beyond 60%–80% depletion in various studies (Bernheimer *et al.*, 1973; Lloyd, 1976). This suggests that compensatory processes can accommodate for dopamine loss so that the patients manifest no detectable deficits in early stages of the disease. Nevertheless, the slow and subsequent progression of symptoms following diagnosis suggests that the effects of continuing neural degeneration overtake compensatory mechanisms as the disease advances. Even though most animal models have relied on observing the effects of dopamine depletions induced by acute experimental lesions (rather than by progressive disease), similar compensatory processes can be observed in the animals' subsequent recovery. The study of these aspects can be expected to provide information on why parkinsonian patients reach a point of functional decompensation typically in their later years and on how that progression may be slower or reversed.

In addition to demonstrating the utility of experimental parkinsonism in animals, this chapter will also attempt to offer a critique of the most studied animal model of PD—that induced by the toxin 6-hydroxydopamine (6-OHDA). In particular, neuropathological and behavioral aspects of the syndrome induced by 6-OHDA lesions of experimental animals will be compared to the idiopathic form of the disease seen in humans, highlighting both the similarities between the two conditions and the important differences.

2. Dopamine and Parkinson's Disease

An early landmark in the study of PD was the prescient demonstration that destruction of a region of midbrain by a tuberculoma produced a hemiparkinsonian syndrome (Blocq and Marinesco, 1893). Brissaud, however, is generally credited with suggesting that the crucial pathological damage involved the substantia nigra (cited by Greenfiled, 1955), an area of brain first described over a century earlier by Soemmering (cited by Stern, 1966). Tretiakoff (1919) is remembered for correlating nigral degeneration seen both in postencephalitic parkinsonism and the idiopathic form of the disease. However, the view that discrete and limited bilateral nigral cell loss might represent pathognomic neuropathology for such a devastating disease was not readily accepted. Indeed, it was not until 1953 that Greenfield and Bosanquet emphasized the importance of the substantia nigra in PD and established the modern view that this nucleus was central to the disease process.

Identification of the role of dopamine and the subsequent biochemical characterization of PD developed as a direct result of an animal model. While studying the functional consequences of central catecholamine depletion, Carlsson (1957, 1959) described a characteristic akinetic syndrome induced by reserpine treatment in rabbits, which was reversible by the administration of dihydroxyphenylalanine (DOPA) (Carlsson, 1959). Although reserpine treatment depleted the brain of all catecholamines, Carlsson speculated that the akinetic effects were attributable to a depletion of forebrain dopamine rather than of noradrenaline. First, since DOPA treatment increased brain levels of dopamine with little effect on noradrenaline, Carlsson proposed that the recovery from akinesia was dependent upon dopamine repletion. Second, since the intact corpus striatum was rich in dopamine but virtually devoid of noradrenaline, and this system was known from classical neurology to be implicated in motor control, it was plausible that the reserpine-induced akinesia was a consequence of striatal depletion of dopamine. Moreover, Carlsson noted the similarity between akinesia in his reserpine-treated rabbits and the motor deficits of Parkinson's disease. These basic studies quickly led to the prediction that PD itself may be due to dopamine depletion, and that its symptoms, as in the reserpine-treated rabbit, may be alleviated by DOPA treatment.

The first of these predictions was tested in 1960 by Ehringer and Hornykiewicz, who reported a marked decline of dopamine in both the striatum and the substantia nigra of postmortem parkinsonian brains. They suggested that these two structures were interconnected and proposed that PD involves degeneration in a hypothesized striato-nigral dopamine pathway. At the same time the anatomical technique of catecholamine histofluorescence was under development, and Andén *et al.* (1964) were able to confirm the long-suspected presence of a forebrain dopamine pathway but projecting from the substantia nigra up to the caudate nucleus and putamen of the striatum, rather than in the descending direction as first proposed.

The second prediction of Carlsson's animal studies was that if PD is also attributable to dopamine depletions, then DOPA may provide an effective replacement therapy in the disease. Birkmayer and Hornykiewicz (1961) first tested peripheral injection of d,l-DOPA and reported a remarkable alleviation of symptoms, a result rapidly replicated in other centers (Cotzias *et al.*, 1967). With the development of soluble tablet forms of delivery, the use of the active levo-isomer of DOPA and its combination with a peripheral decarboxylase inhibitor to block the major side effects of raw l-DOPA, this drug has not only revolutionized but also remains the most widespread treatment for Parkinson's disease.

3. 6-Hydroxydopamine Lesions in Rats

3.1. 6-OHDA as a Catecholamine Neurotoxin

With the emerging detailed characterization of PD, the scene was set for the potential development of animal models that would mimic the human condition faithfully in many, if not all, of its aspects. Although relatively large electrolytic lesions of the ventral tegmentum had already been found to produce tremor, rigidity, and hypokinesia in primates (Poirier, 1960), these were nonspecific, invariably involving other regions of midbrain, notably the parvocellular portion of the red nucleus. However, it was the discovery of the catecholamine neurotoxin, 6-hydroxydopamine (6-OHDA), that enabled more selective lesions to be made, and thereby provided the basis for major advances in studying the functional organization and contribution of forebrain dopamine systems of the brain and in developing more selective animal models of PD. Developed as a means of destroying peripheral catecholamine-containing neurons (Tranzer and Thoenen, 1968), 6-OHDA was soon applied in the central nervous system to produce large effective depletions of forebrain dopamine and noradrenaline (Ungerstedt, 1968; Uretsky and Iversen, 1969). The mechanism of toxicity appears to be due to active uptake of 6-OHDA by catecholaminergic neurons, which are then destroyed by cytotoxic auto-oxidation products following intracellular metabolism (Sachs and Jonsson, 1975).

In general, all catecholamine neurons are subject to 6-OHDA toxicity, noradrenaline as well as dopamine, in the forebrain. Various alternative strategies have therefore been adopted to block the toxicity in other nondopaminergic pathways. First, if delivered as an intraventricular or intracisternal infusion, 6-OHDA will diffuse throughout the brain, but accumulation in noradrenaline neurons can be blocked by systemic administration of the selective uptake inhibitor des-methylimipramine (DMI; Jacks *et al.*, 1972; Kostrzewa and Jacobowitz, 1974; Breese and Traylor, 1970). Second, precise stereotaxic injections of 6-OHDA can be given to produce local and discrete lesions of individual forebrain pathways. This strategy is particularly convenient for lesioning noradrenergic neurons in the brainstem, where the dorsal and ventral ascending bundles diverge caudal to (and remote from) the dopaminergic neurons of the nigra (Robbins and Everitt, 1982). However, this does not work particularly well for selective lesions of the nigrostriatal pathway, because of the proximity of forebrain noradrenaline axons coursing together with the dopaminergic axons in the medial forebrain bundle. Consequently, nigrostriatal 6-OHDA lesions should be combined with DMI if the dopaminergic specificity of the lesion is an important constraint of the experiment. A third strategy has been to administer 6-OHDA to neonatal animals. Because the blood-brain barrier does not become fully formed until after the first weeks of life, 6-OHDA administered peripherally in neonatal animals will penetrate the brain. It has been found that effective targeting of the toxin depends on the protocol for its administration: noradrenaline neurons being particularly sensitive to repeated small injections, whereas a single large injection in combination with DMI preferentially destroys forebrain dopamine neurons (Cooper *et al.*, 1973, Luthman *et al.*, 1989).

3.2. Neurochemistry of 6-OHDA Nigrostriatal Lesions

The introduction of techniques enabling accurate biochemical assay of the affected neural tissue, notably using high-performance liquid chromatography (HPLC), has confirmed the 6-OHDA model as efficient and selective for studying the neurobiology of central dopamine systems. The majority of work has used the rat as the experimental subject, but relevant observations have also been made with other species (see Zigmond and Stricker, 1989), including cats, dogs, mice, and primates (Redmond *et al.*, 1973; Annett *et al.*, 1992).

After an acute lesion of the nigrostriatal dopamine system, measurable changes of dopamine (DA) neurochemistry in the striatum follow a complex time-course. Initially, an increase in DA levels occurs within 15–30 min, reaching a maximum at 150%–180% of control values (Bell *et al.*, 1970; Walters *et al.*, 1973) by 60 min after injection. This peak is maintained for about 24 hr before falling precipitously to reach a nadir 2 to 3 days after the lesion (Faull and Laverty, 1969), a level that is maintained at a stable low for at least 180 days (Uretsky and Iversen, 1970; Mishra *et al.*, 1980). This asymptotic level of DA depletion correlates well with the degree of cell loss within the substantia nigra (Range and Ungerstedt, 1977).

The initial elevation of DA levels probably reflects a combination of temporarily increased synthesis and decreased DA reuptake and metabolism combined with nonspecific leakage of DA from degerenating terminals. The first is implied by measurements of DOPA following a lesion in the presence of a dopa-decarboxylase (DDC) inhibitor revealing 2- to 3-fold increases (Carlsson *et al.*, 1972). The mechanism underlying this probably reflects two effects. First, reduced autorecptor stimulation by extracellular DA occurs since impulse flow is abolished. The resulting disinhibited DA synthesis increases intracellular DA, which normally would lead to end-product inhibition of tyrosine hydroxylase (TH) and normalized levels of synthesis. However, this fails to occur owing to a reduced affinity of TH for DA (Roth *et al.*, 1975). This phenomenon is probably also due to inhibited impulse flow and is explainable by acute reductions in intraneuronal calcium concentrations (Morgrenroth *et al.*, 1976). The net result is a build-up of intraneuronal DA in agonal neurons until the synthesizing enzymes are deactivated. It has been established that TH activities diminish *pari passu* with declining DA levels following intraventricular 6-OHDA (Bell *et al.*, 1970; Uretsky and Iversen, 1970).

In addition to increased synthesis, decreased metabolism of DA is suggested by the appearance of greatly reduced homovanillic acid (HVA) and dihydroxyphenylacetic acid (DOPAC) levels only 20–30 min after lesioning (Westerink *et al.*, 1978). In the absence of *in vivo* studies investigating the extracellular proportion of the early lesion-induced DA increase, however, it is impossible to deduce its behavioral consequences in the early postoperative animal. It is generally assumed that nonspecific leakage of the increased stores of DA lead to the behavioral hyperactivity seen in animals on the day of their lesion. Indeed, the phenomenon can be abolished by co-administration of α-methyl-para-tyrosine, which blocks synthesis, but not by reserpine, which interferes with the storage mechanisms (Fibiger *et al.*, 1972).

3.3. Biochemical Plasticity after 6-OHDA Nigrostriatal Lesions

After moderate depletions of DA using 6-OHDA, animals show no obvious behavioral deficits (Bloom *et al.*, 1969). Even after stable extensive depletions, made possible by the development of more refined lesioning techniques, varying degrees of functional recovery are possible (Marshall *et al.*, 1974; Zigmond and Stricker, 1972). Stricker and Zigmond (1976) hypothesized that the functional recovery may be attributable to compensatory changes in the biochemical effectiveness of remaining DA neurons at a number of different levels: in the firing rate of the dopaminergic neurons; in the synthesis, storage, release, and reuptake of the dopamine transmitter; and in the responsiveness of the postsynaptic receptor to what dopamine is released into the synapse. Although based largely on speculation at the time it was proposed, the biochemical compensatory hypothesis has gained much support in subsequent studies (e.g., Zigmond *et al.*, 1984). In addition to identifying a degree of redundancy within the nigrostriatal DA system, these studies have illuminated the longer time-course of compensatory changes evident in human parkinsonism.

Surviving DA neurons are able to increase their metabolic activity after an acute lesion. This has been measured between 6 days and 12 weeks following the initial damage (Westerink *et al.*, 1978; Hefti *et al.*, 1980). The hyperactivity may be reflected electrophysiologically by increased neuronal firing rates and biochemically by a rise in DA turnover. The former mechanism only occurs if depletions are in excess of 95% of control values (Hollerman and Grace, 1988), perhaps, in part, reflecting the removal of neighboring DA neurons, which can provide reciprocal inhibition via dendro-dendritic connections (Aghajanian and Bunney, 1977; Groves *et al.*, 1975). The latter mechanism is apparent, however, when as much as 75% of DA remains (Agid *et al.*, 1973) and has been variously documented as increased rates of formation or disappearance of radiolabeled DA in the striatum (Agid *et al.*, 1973), a higher ratio of DA metabolites to DA (Hefti *et al.*, 1980), or increased extracellular concentration of DA and its metabolites (Robinson and Whishaw, 1988; Zhang *et al.*, 1988).

The above adaptions are also generally accompanied by an increase in TH activity (Zigmond *et al.*, 1984; Onn *et al.*, 1986), measured as a short-term activation of existing molecules or as an increase in their number by a process of enzymatic induction. The latter process can be expected to enhance the functional capacity of residual DA neurons. Indeed, the time-course of increases in DA turnover correlates with that of TH activity (Chiueh, 1988). The underlying mechanism remains speculative but may involve decreased stimulation of presynaptic DA receptors, located on DA terminals, as a consequence of lower concentrations of released DA. These receptors influence the regulation of DA synthesis (Nowycky and Roth, 1978), and a reduction in their activation leads to increased metabolic activity in the remaining DA fibers. This is supported by the observation that DA overflow from unlesioned striatal slice preparations is enhanced by the presence of the dopamine antagonist sulpiride, but this effect is abolished if the slices are taken from 6-OHDA-lesioned rats (Snyder *et al.*, 1986).

Studies using electrical field stimulation have also indicated that lesioned striatal slices have an enhanced ability to release DA. This has been recorded as an increased fractional DA efflux, a measure of released DA as a fraction of remaining DA content (Stachowiak *et al.*, 1987). Because this study showed that the index was also raised by the DA uptake inhibitor nomifensine, decreased uptake by fewer available DA terminals may have accounted for the relatively increased concentrations of extracellular DA released in the lesioned preparation. It has been known for a long time that destruction of nigrostriatal fibers reduces the uptake of catecholamines in the striatum (Goldstein *et al.*, 1969; Bell *et al.*, 1970; Uretsky and Iversen, 1970; Schoenfield and Uretsky, 1973). This mechanism can be termed *presynaptic supersensitivity,* and support for its existence comes from its rapid onset (Trendelenberg, 1966). Reduced uptake may also allow DA to diffuse over longer distances (Kelly and Wightman, 1987), increasing the field of influence of DA released from a given terminal. The ramifying nature of the nigrostriatal projection, each neuron capable of releasing DA from about 500,000 varicosities over a length of about 50 cm (Andén *et al.*, 1966), increases the chances of this nonspecific mechanism having functional significance.

Changes such as enzyme induction and transport to nerve terminals clearly do not occur immediately. Another longer-term compensatory change of importance relates to the development of supersensitive postsynaptic receptors. Evidence for the existence of this mechanism comes from a variety of sources. First, sensitivity to DOPA increases over a period of several weeks (Thornburg and Moore, 1975), whereas degeneration, and the extent of reduced uptake, is more or less complete after 48 hr (Hökfelt and Ungerstedt, 1973). Further, supersensitivity to apomorphine, a DA agonist not susceptible to the reuptake process, is usually taken as direct evidence for receptor upregulation (Ungerstedt, 1971b; Iversen *et al.*, 1976; Zigmond and Stricker, 1980). The DA receptor assays

(Creese *et al.*, 1977) and supersensitivity to DA antagonists (Muller and Seeman, 1977) also corroborate the existence of increased numbers of postsynaptic receptors. The increased receptors also produce an enhanced electrophysiological response by the striatal cells containing them (Feltz and DeChamplain, 1972; Fibiger and Grewaal, 1974; Ohye *et al.*, 1970).

Postsynaptic supersensitivity is not apparent unless 90% of DA neurons have been destroyed. It is the number of receptors rather than their affinity for DA that becomes increased (Creese and Snyder, 1979). Moreover, it is the nonadenylate cyclase D_2 subtype that appears to be particularly responsive to development of the phenomenon (Nagy *et al.*, 1978). This is implied by the existence of a correlation between apomorphine-induced behaviors and spiroperidol (D_2) binding but not DA-stimulated adenylate activity (Waddington *et al.*, 1979). However, some studies have implied that the D_1 receptor may also be upregulated (Mishra *et al.*, 1974; Zigmond and Stricker, 1980), although, as adenylate cyclase activity was used as an assay, this may be an epiphenomenon secondary to reduced DA uptake mechanisms (Krueger *et al.*, 1976).

4. The Specificity of 6-OHDA as a Model of Parkinson's Disease

4.1. Neurochemical Specificity in Parkinson's Disease

The relative specificity of the 6-OHDA model in destroying DA neurons may, ironically, limit its ability to mimic the human disease. Recent neurochemical studies have emphasized that other parts of the so-called isodendritic core also degenerate in PD to varying degrees. Notably, the locus coeruleus, areas of the reticular formation (raphe nuclei), and the nucleus basalis of Meynert are all significantly affected in the idiopathic form of the disease (Jellinger, 1986).

Of these nuclei, the locus coeruleus has been most studied. Associated changes in the level of the neurotransmitter noradrenaline, its metabolite, 3-methoxy-4-hydroxyphenol-glycol and its main synthetic enzyme, dopamine-β-hydroxylase, have been documented in various cortical and subcortical parts of the parkinsonian brain (Farley and Hornykiewicz, 1976; Scatton *et al.*, 1983). Most loss is evident from the pigmented locus coeruleus, which is the source of the dorsal noradrenergic bundle, innervating wide areas of neocortex and limbic areas such as nucleus accumbens, amygdala, and hippocampus. The extent of the depletion is 40%–70% (see Agid *et al.*, 1987). Cortical receptor changes are also seen in the form of lowered, presynaptic, α_2 receptors and increased numbers of the postsynaptic β_1 subtype (Cash *et al.*, 1987). In contrast to the extensive changes in the dorsal system, the ventral noradrenergic system that originates from the lower brainstem is relatively spared (Kott *et al.*, 1982), explaining the less extensive transmitter depletions from the hypothalamus, which receives the majority of its innervation from this ventral source. Noradrenaline neurons descending to the spinal cord are also affected, with a 50% reduction in the dorsal and ventral horns of the lumbar spinal cord (Scatton *et al.*, 1986). The extent to which 6-OHDA models also deplete the subject of noradrenaline is largely dependent on where the neurotoxin is infused and whether it is administered with DMI or another noradrenergic uptake inhibitor. Thus, if it is injected (without any other pharmacological agents) just rostral to the dopaminergic cell bodies in the substantia nigra, where the axons converge to form part of the medial forebrain bundle, the ascending axons of the dorsal noradrenergic bundle are also reliably involved in the lesion (Tassin *et al.*, 1991). Such a treatment may provide a closer model of the human condition neurochemically, but the involvement of more than one population of catecholamine neurons makes it correspondingly harder to disentangle the different functional components in the behaving animal.

The ascending and descending serotonergic systems are only moderately and nonhomogeneously affected in PD, probably reflecting selective damage to cell groups in the raphe nuclei (Agid *et al.*, 1987; Forno, 1982). Limbic areas such as cingulate and entorhinal cortex, nucleus accumbens, amygdala, and the ventral tegmental area are unaffected with regards to biochemical indices of depletion. Frontal cortex, caudate-putamen, and hippocampus, however, have in the order of 50% reductions in serotonin concentrations (see Agid *et al.*, 1987). This presynaptic depletion is not accompanied by changes in serotonin receptors, and the density of both striatal and cortical 5-HT_1 (Rinne *et al.*, 1980) and 5-HT_2 (Perry *et al.*, 1984; Maloteaux *et al.*, 1985) receptors appear normal in the parkinsonian brain. However, the moderate serotonin deficiency may account for the reduced levels of 5-HIAA reported in the cerebrospinal fluid of patients (Chase, 1974).

The functional significance of impaired serotonin transmission remains largely speculative. Early theorizing that said reduced serotonin lay behind parkinsonian symptoms of tremor (Goldstein *et al.*, 1969) has not been confirmed by subsequent drug trials, which have failed to find any motoric effects of serotonergic drugs (Chase, 1972). More recently, the serotonin theory of depression (van Praag, 1982) has led to the suggestion that its depletion may explain the depressive symptoms in PD patients, especially since 5-HIAA levels are particularly low in the cerebrospinal fluid of depressed parkinsonian patients (Mayeux *et al.*, 1984). Some commentators have even suggested that serotonin depletion may offset some of the functional consequences of severe dopamine depletion (e.g., Zigmond and Stricker, 1989, p. 13). This is based on the notion deduced from early formulations that serotonin has functions generally antagonistic to dopamine (Kostowski *et al.*, 1974; Mabry and Campbell, 1973). The 6-OHDA models do not routinely affect serotonin neurons, although this transmitter system is sensitive to the related neurotoxins, 5,6- and 5,7-dihydroxytryptamine (Baumgarten *et al.*, 1983). Nevertheless, the proposed inhibitory interactions of serotonin on dopamine neurons in the substantia nigra and/or striatum remain as obscure in the animal models as they do in PD itself.

The situation regarding cholinergic depletion in PD is equally confused. Considerable attention has recently been directed to the function of the long cholinergic fiber systems projecting from basal forebrain and brainstem to the neo- and allo-cortices and hippocampus, following the proposal that dysfunction in this system is critically involved in the learning and memory deficits seen in aging and dementia (Drachman and Sahakian, 1980; Bartus *et al.*, 1982; Coyle *et al.*, 1983). It is a remarkable fact that the Lewy body, often considered to be the definitive neuropathological hallmark in the substantia nigra of Parkinson's disease, was first described by Lewy in the cholinergic basal forebrain nucleus of Meynert. Moreover, reductions in the activity of the cholinergic marker choline acetyltransferase (CAT) in the neocortex can be as great in PD as is typically observed in Alzheimer's disease (AD) (Ruberg *et al.*, 1982, Perry, E.K., *et al.*, 1985). The associated cell loss seen particularly in the nucleus basalis of Meynert (Arendt *et al.*, 1985; Whitehouse *et al.*, 1982; Rogers *et al.*, 1985) has substantiated the presence of degeneration in this pathway. Because neocortical losses of cholinergic markers have been purported to correlate with the extent of intellectual deterioration in patients dying of PD (Dubois *et al.*, 1983; T.L. Perry *et al.*, 1983; E. K. Perry, 1985), it has been suggested that the relationship may be causative. However, evidence for cholinergic depletion (especially in frontal and occipital lobes) (E.K. Perry *et al.*, 1985) and cell loss (Nakano and Hirano, 1984) is present even in patients with no intellectual impairments (E.K. Perry *et al.*, 1985; Aubert *et al.*, 1992) so degeneration of this pathway may reflect general advance of the disease rather than representing a particular demented subgroup. The 6-OHDA lesions in animals spare the cholinergic neurons of the basal forebrain, although various other experimental toxins are available for making experimental lesions in this system (Dunnett *et al.*, 1991; Wiley *et al.*, 1991).

Although 6-OHDA can be used effectively to deplete the experimental subject of dopamine, the resulting gross transmitter loss fails to match the increasingly detailed and complex topography of depletion evident in the human disease. The importance of these differences depends on the experimental issues under investigation. Thus, experimental 6-OHDA-treated rats typically have a relatively "clean" lesion, reflecting relatively complete dopamine denervation, with other transmitter systems spared. This may be ideal if the goal of the experimental study is to identify dopaminergic processes in the disease or improve on a dopaminergic-replacement therapy. Similarly, other experimental toxins are available to manipulate noradrenergic, serotonergic, or cholinergic components selectively, as fits a controlled experimental design. However, in other situations (e.g., investigating the natural time-course and functional progression of the disease process), it may be preferable to mimic the progressive nature of a partial but developing dopamine denervation, accompanied by parallel (albeit less severe) decline in other neurotransmitter systems. In such circumstances, the acute injection of 6-OHDA is less suitable, and other neurotoxins, such as MPTP in monkeys (Burns *et al.*, 1983; Jenner and Marsden, 1986), or the genetic mutation that leads to progressively developing neurological impairment in *wv/wv* Weaver mice (Triarhou *et al.*, 1986; Lalonde, 1987), although more complex, may be more appropriate.

A further aspect of the human situation in this respect is that the dopamine depletion is uneven in the various subregions of the striatal complex. The anterolateral part of the striatum, particularly the putamen, is affected to the greatest degree, followed by the caudate nucleus, and relatively less or later-developing depletions in the ventral striatum, including the nucleus accumbens. The neurochemical heterogeneity is matched by evidence of a gradient of cell loss in the substantia nigra, with the greatest loss in the nigral subregions that project to the most affected areas of the striatum. Thus, nigral cell loss is most prominent in the caudal and internal portions of the nigra that project preferentially to the putamen (Agid *et al.*, 1987).

Recent evidence has added a further layer of complexity to the picture. The substantia nigra can be divided into subregions (including dorsal and ventral tiers) depending on the orientation of the dendritic trees and the presence or absence, respectively, of calcium-binding protein. In PD, it is the ventral tier, particularly laterally, that is most degenerate (Gibb and Lees, 1991). Parenthetically, the ventral tier projects to the patch or striosome compartment in the caudate-putamen (Gerfen, 1992). By contrast, the cellular degeneration that occurs during normal aging preferentially affects the dorsal tier (Fearnley and Lees, 1991), arguing against the view that PD is simply an accelerated aging process.

Given the heterogeneous functions associated with the subregions of the striatal complex (Robbins and Everitt, 1982; Dunnett and Robbins, 1992), animal models should ideally mimic the gradient of dopamine depletion seen in the disease. Such subtle aspects of the PD profile have not been addressed when using 6-OHDA-induced dopamine depletions so that finer aspects of the clinical syndrome have not been explicitly modeled. Nevertheless, the techniques for making circumscribed striatal injections of 6-OHDA are well established, and the functional effects of differential denervation of distinct striatal areas have been investigated in other contexts (Dunnett and Iversen, 1982).

4.2. Compensatory Plasticity in Parkinson's Disease

Because the level of DA depletion needed to produce behavioral deficits is similar in both 6-OHDA experimental models and the disease itself, at about 70%, it is quite plausible that similar compensatory mechanisms apply in PD as have been demonstrated in the experimental animal studies (see Section 3.3). The limited data that are available from postmortem material uphold this speculation. Although DA may degrade after death, the HVA/DA ratio can be used as an index of amine turnover in the remaining neurons, especially if careful controls are used (Hornykiewicz, 1979). The HVA/DA ratio is significantly increased in the putamen, caudate nucleus, nucleus accumbens, and substantia nigra of the parkinsonian brain (Scatton *et al.*, 1984). Further studies have confirmed the existence of increased metabolic activity in the partially degenerated human nigrostriatal DA system (Davidson *et al.*, 1971; Lloyd, 1976).

By contrast, the question of receptor hypersensitivity in PD patients is more contenious and a consensus has yet to materialize. The D_2 receptors in the striatum of parkinsonian patients have been reported to increase (Lee *et al.*, 1978) or decrease (Reisine *et al.*, 1977). A later study found no overall significant modulation of receptor density or affinity constant (Rinne *et al.*, 1981), although some patients had hypersensitivity and others hyposensitivity of striatal D_2 receptors. Such discrepancies may be partly explained by the uneven severity of DA depletion throughout the striatal complex (see below). Of note is that D_1 and D_2 receptors specifically located in the putamen are increased when the results of several studies are amalgamated (Agid *et al.*, 1987). Also, taking into account the suspected presence of D_2 autoreceptor sites located presynaptically on degenerated terminals, near-normal gross measurements of D_2 receptors may underestimate the degree of postsynaptic hypersensitivity. An increased density of D_2 receptors has been seen particularly in those who responded well to DA replacement therapy and those who developed dyskinesias or fluctuations in treatment response in life (Rinne *et al.*, 1981). Further, the increased density is observed preferentially in untreated patients (Lee *et al.*, 1978) or in those with a lengthy evolution of the disease (Bokobza *et al.*, 1984). These conclusions have been challenged, however, by a study that found no correlation between l-DOPA dose or duration of either treatment or disease and the density of D_2 receptors (Guttman *et al.*, 1986) and another study that found no relation between receptor density and l-DOPA side effects (Pierot *et al.*, 1988).

The D_1 receptor changes are even more controversial. Adenylate cyclase activity measurements have implied either increased (Nagatsu *et al.*, 1978) or decreased (Riederer *et al.*, 1978; Shibuya, 1979) D_1 receptors. When more direct specific binding methods are used, uncertainties still prevail. Labeling with the selective D_1 ligand [^{3}H]-SCH 23390 has implied increases (Raisman *et al.*, 1985) or no change (Pimoule *et al.*, 1985) in putamen receptor numbers. Another study using [^{3}H]-flupenthixol as a D_1 ligand also suggested receptor upregulation but, confusingly, this was further potentiated by l-DOPA treatment (Rinne *et al.*, 1985).

In summary, the pre- and postsynaptic compensations that occur in acutely lesioned experimental animals have pointed the way forward for identification of several similar compensatory processes in the human condition. In both situations, the degree of depletion is predictive of the predominant compensatory mechanism. When less than 70% of DA neurons are lost, increased activity in the remaining population of spared neurons probably keeps parkinsonian symptoms latent unless the system becomes transiently decompensated. This might occur after stress (Snyder *et al.*, 1985) or short-term administration of neuroleptic drugs (Rajput *et al.*, 1982). With 70%–90% of DA neurons lost, symptoms emerge, controlled by l-DOPA therapy, and with few drug-related abnormal involuntary movements, probably because receptor sensitivity is still within the normal range (Agid *et al.*, 1985). Postsynaptic (D_2) supersensitivity is only apparent at late stages of functional decompensation when 90% or more of the neurons undergo degeneration. At this stage, l-DOPA has declining duration of effectiveness, wears off more rapidly, and at the peak of its action is prone to induce pronounced dyskinetic side effects. Striatal DA transmission may be partly re-established by agonist administration, but these newer compounds may also be associated with severe fluctuations in performance and dyskinesias.

5. Functional Insights from 6-OHDA Lesions

As outlined above, and perhaps not surprisingly, the 6-OHDA model fails to match the fine details of the neurochemical abnormalities seen in the human condition. Nevertheless, it is indisputable that the most significant neurochemical aspect of the disease, namely destruction of the nigrostriatal pathway, is reflected in the animal model. Detailed reviews of the functional consequences of nigrostriatal destruction in the experimental rat are provided elsewhere (Robbins and Everitt, 1982; LeMoal and Simon, 1990; Dunnett and Robbins, 1992). The present review then focuses on the next question that arises: To what extent do the functional effects of this experimental destruction correspond to and/or elucidate clinical features of the disease?

5.1. Parkinsonian Symptoms after Bilateral 6-OHDA Lesions in Rats

The early studies of Ungerstedt showed that bilateral infusions of 6-OHDA into the substantia nigra pars compacta of the rat had profound effects on the experimental subject (Ungerstedt, 1971c). The dramatic behavioral deficits included catalepsy (the maintenance of an unusual posture), behavioral inactivity, and life-threatening aphagia and adipsia. Subsequent work has shown that some of these features, such as catalepsy, represent an unmasking of primitive motor reflexes (Schallert *et al.*, 1978). Furthermore, the profound aphagia and adipsia, not characteristic of the clinical condition, may reflect motivational deficits consequent to the near total depletion of dopamine from the hypothalamus. However, the akinesia and problems in initiating actions—for example, in escape situations (Price and Fibiger, 1975)—are reminiscent of the core movement disorder seen in PD. Indeed, Marsden argues with some force that the hypokinesia seen in PD reflects the direct consequences of basal ganglia damage (Marsden, 1984). In contrast, the occasionally more obvious symptoms of rigidity and tremor represent secondary effects or "release phenomena" in a Jacksonian sense.

The possibility of accelerating recovery in bilaterally lesioned animals also correlates to some aspects of the human situation. More specifically, if the animals are exposed to "activating" environmental challenges such as social interaction, cold, palatable food, and tail pinch, their spontaneous recovery from the debilitating effects of the lesion is subsequently enhanced (Antelman *et al.*, 1975; Marshall *et al.*, 1976; Wolgin and Teitelbaum, 1978). The concept that the lesioned animals retain the capacity to move but need exceptionally motivating circumstances or stimuli to elicit the actions finds a possible parallel in the phenomenon of "kinesia paradoxica" (Schwab, 1972; Schwab and Zieper, 1965). This describes the remarkable ability of severe parkinsonian patients to overcome their obvious hypokinesia under conditions of extreme or intense arousal. In this respect, anecdotal accounts abound of galvanized patients leaping to avoid incipient danger such as oncoming trains or buses.

Because of the gross debilitating effects of experimental bilateral 6-OHDA lesions and the concomitant need for tube-feeding, at least in the early postoperative period, few studies have investigated the possibility of the existence of more subtle lesion-induced behavioral or learning deficits. An early study by Ranje and Ungerstedt (1977) demonstrated that lesioned animals could swim with vigor to escape from an underwater T-maze. Spatial or brightness cues were used to discriminate the correct arm of the maze, and although no deficits were seen if the discrimination was learned preoperatively, problems arose if the animals had to learn it postoperatively. Another learning task involving aversive training is the Morris water maze where rats have to locate the presence of a submerged platform in order to escape from deep water. Excluding the most severely depleted animals, one study found no lesion-induced deficit, aside from an increased latency to escape, that was explainable as a motoric problem (Hagan *et al.*, 1983). In contrast, a comparable study by Whishaw and Dunnett (1985) did find lesion-induced deficits. However, given the nature of the task, any interpretation regarding the precise nature of the underlying psychomotor deficit is necessarily speculative. For example, is the problem one of learning a spatial discrimination or is it due to neglect of the distal spatial stimuli that are essential for expedient solution of the task? In any case, attempts to tie in such behavioral effects with the symptomatology of PD can appear even more contrived.

A recent study by Robbins *et al.* (1990) compared terminal 6-OHDA-induced dopamine depletions of the rat dorsal striatum (caudate-putamen) and ventral striatum (centered on the nucleus accumbens) on the acquisition and performance of a difficult conditional discrimination paradigm. The task necessitated rats learning a procedural rule (push the right or left lever in a Skinner box) according to whether a flashing visual stimulus was "fast" or "slow." Those rats with specifically dorsal striatal depletions were dramatically impaired in the speed of acquisition of the task and were also deficient if they had learned the task postoperatively. Neither motor nor motivational factors were sufficient to account for the magnitude of the effects seen. The relationship of this apparently "cognitive" deficit to PD remains speculative, but the putative concept of bradyphrenia or "slow-thinking" may be relevant. This refers to the slowed mental processing that occurs in subcortical dementing conditions, in which accuracy of responding in a variety of situations is maintained at the cost of speed (Rogers, 1986). Of particular note is that the experimentally parkinsonian rats of the study by Robbins and colleagues were found to be significantly biased to respond on the side predetermined to be correct for the "slow" stimulus. In other words, the errors were systematic; the rats were treating an abnormally high proportion of the stimuli as being "slow." Whether this reflected slowed internal processing of the visual stimuli needs further study. However, the notion that this type of deficit may reflect an analogous inability of parkinsonian patients to gauge temporal stimuli finds support in recent investigations. More specifically, parkinsonian patients have an increased threshold for detecting two temporally spaced stimuli compared to controls (Artieda *et al.*, 1992). Importantly, this was reliably seen for a variety of sensory modalities and, intriguingly, was partly ameliorated by systemic dopaminergic therapy. A companion study also showed that parkinsonian patients consistently underestimated the length of passage of time (Pastor *et al.*, 1992). Both these investigations provide superficial support for a slowed "internal clock" in the patient group.

5.2. Parkinsonian Symptoms after Unilateral 6-OHDA Lesions in Rats

Unilaterally depleting nigrostriatal dopamine from experimental animals has been much more widely used as a model of hemiparkinsonism. The advantages of the unilateral model include a relative lack of postoperative morbidity and the possibility of comparing the influences of the intact and lesioned dopamine pathways by observing aspects of behavioral asymmetry in individual animals. Thus, in the quiescent state, the animal initially manifests a pronounced postural bias with head and trunk turned toward the side of the lesion, but this typically recovers within about 1 week after lesion. However, more marked asymmetries can be reestablished if the animal is activated, either behaviorally by a strong stimulus such as novelty or tail-pinch, or pharmacologically. In the activated animal, the postural asymmetry combined with locomotor activation becomes manifest as "rotation," head to tail turning of the animal, and this phenomenon is long lasting and not subject to the same degree of spontaneous recovery. The ease of measuring this gross rotational behavior and the potential for "amplifying" asymmetries with dopaminergic drugs once spontaneous lateralized turning behavior has recovered have stimulated many studies on the neurobiology of rotation, and the phenomenon has been the subject of several reviews (Glick *et al.*, 1976; Dankova *et al.*, 1978; Pycock, 1980).

From the early studies (Ungerstedt 1971a,b), a virtual dogma has arisen that an animal will turn away from that side of the brain with greater overall dopaminergic receptor occupancy. However, the complexity and fractionation of the rotational response in terms of underlying psychomotor processes, of the relation and significance of rotation to idiopathic PD, and of recent unexplained anomalies in the direction of turning after various manipulations, possibly reflecting the heterogeneity of striatal subregions in the initiation of circling, have all severely compromised the utility of this simple model (Miller and Beninger, 1991). Largely due to our lack of knowledge of the processes distal to the dopamine receptor that lead to overt rotation, useful applications of this model remain confined primarily to the field of neuropharmacology.

First, it has become clear that rotation is not a unitary behavioral measure and that its extent and even direction can depend on how it is measured. For example, a recent study (Schwarting *et al.*, 1991) used video analysis of the spontaneous turning behavior of unilaterally lesioned rats in a relatively large "open field" (60 × 60 cm) as opposed to the more widely used rotometer bowls, first introduced by Ungerstedt and Arbuthnott (1970). Moderate depletions of dopamine (74%) produced no ipsiversive circling when "tight" ($<$30 cm in diameter) turns were considered but did produce a *contraversive* asymmetry of "wide" ($>$ 55 cm) turns. This apparent contradiction to Ungerstedt's principle was termed "thigmotactic" circling and was explained as a sensory imbalance in the animals. More specifically, the animals were tending to explore preferentially the side of space contralateral to the intact side of the brain, which, given the geometry of the open field, was recorded as contralateral turning. More severe depletions of dopamine ($>$ 90%) produced only ipsiversive asymmetries, seen as "tight" circling. This type of study serves to illustrate the heterogeneous nature of a turning response and hints at the possible differing psychomotor explanations behind various types of circling.

Along similar lines, a further recent study has shown the unreliability of correlating lesion extent with degree of rotational asymmetry (Carman *et al.*, 1991). Using partial lesions of the rat substantia nigra as a more realistic model of hemiparkinsonism, the authors attempted to correlate the number of remaining TH-positive neurons with the extent of rotational behavior as measured in an automated rotometer. Predictably, amphetamine produced ipsiversive turning as a consequence of inducing release and inhibiting reuptake of dopamine in the intact striatum whereas low-dose apomorphine led to contralateral turning, presumably as a result of the stimulation of supersensitive dopamine receptors in the lesioned striatum. However, when the extent of rotation was considered, somewhat surprising data were found. Amphetamine-induced circling showed no correlation whatsoever with the number of TH-positive neurons remaining in the substantia nigra. In fact, when the percentage of the dorsal striatum containing dopaminergic neurons was considered, a seemingly *positive* correlation was seen between the extent of lesions and the rate of rotation under amphetamine. This contrasted with the results for apomorphine rotation, which showed good inverse correlations between the extent of (contralateral) turning and the two measures of dopaminergic neuronal loss. This and other studies also showing discrepancies between the types of turning induced by different dopaminergic agents (Cools *et al.*, 1989; Casas *et al.*, 1988) serve to highlight how little is still known about this deceptively simple behavioral response following lateralized stimulation of dopamine receptors.

Given the above examples, which highlight the complexities of the rotating rodent model, its relation to clinical aspects of parkinsonism appear to be even more obscure. A pertinent conundrum arises from a superficial comparison of the animal and human situations. In human patients who have a degree of asymmetry in their motor symptoms, the hypokinesia and variable degrees of tremor and rigidity are referred to the limbs contralateral to the more affected striatum. Extrapolating to the rat model, therefore, one might expect to see turning *toward* the side of increased dopaminergic stimulation if, indeed, the limbs contralateral to the intact striatum are more activated than those ipsilaterally. The fact that the opposite of this simple pivotal behavior occurs and that the rat uses all four limbs in an apparently purposeful way implies the existence of a different mechanism behind the movement deficit in the animal model when compared to the human situation. Nevertheless, some support is found for the verisimilitude of the animal model in a report by Bracha (1987), who demonstrated that patients with significantly lateralized symptoms did, in fact, show a turning tendency contralateral to their more disabled side (i.e., ipsilateral to the side of their presumably more severely depleted striata). This was measured over the course of a day without the patient's knowledge of the aims of the study. Perhaps importantly, no interaction occurred between direction of circling and measures of handedness in the patient group. Nevertheless, the comparison of this type of rotational bias with the situation of drug-induced circling in rodents remains contrived at the very least.

One final aspect of rotational studies that has implications for PD and, indeed, the neurobiology of striatal dopamine concerns the heterogeneity of the subregions of the striatum in producing circling. As mentioned previously, the uneven but predictable depletion of dopamine throughout the striatal complex in PD is important for understanding the nature of the disease given the differing functions of these subregions. In the context of circling behavior in rats, it is convenient to divide the striatum into dorsal and ventral regions, roughly corresponding to caudate-putamen and nucleus accumbens, respectively. A widely held view, based mostly on lesion studies, is that the ventral striatum has no direct role in determining the direction of motoric bias following unilateral lesions, but that dopamine activation within it can increase the vigor of circling produced by dorsal striatal dopamine asymmetries (Kelly and Moore, 1976; Pycock and Marsden, 1978). This, in turn, has led to speculation that ventral striatal depletions in PD, although less marked than those in dorsal areas, may be behind some of the activational deficits seen in the disease (Rose *et al.*, 1989). However, recent evidence implies that ventral striatal dopamine manipulations do have a directional role in behavioral asymmetries (Colle and Wise, 1988; Messier *et al.*, 1990; Koshikawa *et al.*, 1990). Moreover, this may reflect different underlying psychomotor mechanisms for rotation, alluded to above. Extrapolating to the potential treatment of PD by the implantation of embryonic neural grafts, animal models have consistently shown that reinnervation of dorsal areas of striatum often leads to an "overcompensation" of rotational behavior, that is, the animals turn contralaterally to the lesion-and-grafted side (Dunnett *et al.*, 1981; Herman *et al.*, 1985). This "overcompensation" is not readily explained by increased sensitivity to drugs, remaining receptor supersensitivity, nor to increased release of dopamine on the grafted side (Herman *et al.*, 1989). We have recently obtained evidence that subsequent dopamine grafts to areas of ventral striatum in such overcompensated animals can partly reverse the contralateral turning (Reading and Dunnett, unpublished results). In a potentially human context, this may imply that reinnervation by grafts of a single area of striatum may produce abnormal behavioral results whereas a more even distribution of implantation sites into areas with differing and even opposing functions might be more effective at normalizing final behavioral output.

The rotation model in rats involves spontaneous and drug-induced behaviors that are unlearned or unconditioned. Two other paradigms using the unilateral 6-OHDA nigrostriatal lesion that have potential relevance for PD have looked at circling behavior that is dependent on the prior experience of the animal. The first type of conditioned rotation used thirsty rats that were trained in the absence of any lesion to turn in circles for water reward (Dunnett and Björklund, 1983). Subsequent lesions were then made to the nigrostriatal pathway either contralateral or ipsilateral to the direction of learned circling. Ipsilateral lesions had no effect. In contrast, those made contralaterally dramatically disrupted the animal's ability to turn in the reinforced direction. This observation is important because it represents a purposeful behavior rather than random spontaneous circling and therefore gives more credence to its application to a clinical situation. Furthermore, the authors note that on the rare occasions a lesioned animal did make turns contralateral to the lesioned striatum, the speed and morphology of movement were normal, implying that the animals were impaired in the initiation of a goal-directed movement rather than in its execution per se. This notion finds correlation in human studies of PD that have emphasized the relatively profound deficits of motor program initiation especially when the action necessitates sequential movements (e.g., Benecke *et al.*, 1987).

The second type of conditioned rotation relates to an animal's ability to circle spontaneously in a direction contralateral to a lesion providing it has had prior experience of doing so under the influence of a drug, such as apomorphine (Carey, 1986). The contralateral turning is found to depend crucially on the animal being placed in the specific environment in which it had previously received the drug. Although speculative, this type of study may lead the way to novel strategies for PD patients, allowing them to function without the need for dopaminergic therapy providing that have had the necessary relevant "experience."

From observations of behavioral asymmetries in lesioned rats, the concept arose that the contralateral deficit was neither purely motor nor sensory in nature but "sensorimotor." Marshall and colleagues developed a neurological test battery in which they assessed the animal's orientation to spatially localized visual, tactile, and olfactory stimuli, and a variety of limb placing, grasping, and withdrawal reflexes (Marshall *et al.*, 1974). Because the animals would readily make a nonlateralized response to contralateral stimuli, such as a vocalization to pinprick, the lack of orientation or biting of the eliciting stimulus was not a sensory loss. Similarly, the contralateral limbs were used normally in automatic acts of grooming or locomotion. Therefore, the term "sensorimotor" was used to imply that the animals' symptoms reflected an impairment in their ability to organize an integrated lateralized response to a lateralized stimulus.

However, it is not clear to what extent the polymodal neglect documented by Marshall extends to the human condition of PD. Superficially, it would appear that the neglect exhibited by the animals was more analogous to that seen after parietal rather than striatal lesions in man (Critchley, 1953). In concordance with the animal model, however, one study of a subject with unilateral striatal damage observed contralateral neglect in a sensory extinction task when an arm movement was required but not when it was a verbal response (Valenstein and Heilman, 1981). Another study has demonstrated increased omissions in a cancellation task by a parkinsonian patient group, but these were not lateralized in those subjects with clear left-sided symptomatology (Villardita *et al.*, 1983). To resolve the apparent discrepancies between clinical studies and experimental models of parkinsonism, recent studies have analyzed the hemi-neglect syndrome in much greater detail, extending the paradigms to involve measurements of reaction times for which there exists a considerable comparable human literature.

A landmark was an early experiment by Turner (1973), which showed that unilateral hypothalamic lesions led specifically to deficits of contralateral head turning, as a learned escape response, when a shock was administered contralaterally and not ipsilaterally. This fitted with the concept of sensorimotor integration as a function of the ascending dopamine pathway that was involved in the lesion. When an appetitive paradigm was used, however, similar lesions produced contralateral response deficits independent of which side the stimulus was presented, implying the existence of a lateralized response-related deficit (Hoyman *et. al.*, 1979). This has been extended in a series of experiments using unilateral dopamine depletions of the striatum (Carli *et al.*, 1985, 1989; Brown and Robbins, 1989). These studies were based on a novel lateralized vigilance task involving training rats to hold their heads in a central location monitored by a photocell beam. Unpredictable lateralized visual stimuli are then presented to the left or right of the animal, to which it must respond for food reward either by turning the head to the same side as the stimulus in one group or to the opposite side in a second group. Unilateral striatal or nigrostriatal lesions induce consistent biases in trained animals and slow their reaction times, and such deficits in this lateralized vigilance test are stable over many months following the lesion (Robbins and Brown, 1990). In brief, the deficit exhibited by unilaterally lesioned rats appears to be one of response initiation in space contralateral to the lesion, regardless of whether the trained response is ipsilateral or contralateral to the eliciting stimulus. This pattern of deficit is analogous to the human condition of hemi-akinesia that has been documented in a variety of neuropathological conditions, including PD (Heilman *et al.*, 1976). Where correct responses were made contralateral to the lesion, reaction times were much slower. However, of importance is the finding that it was the nonlateralized initial withdrawal of the head from the central location rather than the contralateral movement itself that was retarded. This conclusion is remarkably similar to that reached by Dunnett and Björklund (1983), namely, that nigrostriatal dopamine lesions selectively disrupt initiation but not execution of contralateral responding in their conditioned rotation studies.

Recently, the lateralized vigilance paradigm has been extended further to compare reaction rimes of unilaterally depleted rats depending on whether they had advance knowledge of the correct direction of movement (Brown and Robbins, 1991). This is highly relevant to the clinical study of PD because of the reported evidence that patients are relatively unable to improve their performance on reaction time tasks when advance information is available. In other words, they are more impaired on simple versus choice reaction time tasks (SRT *vs.* CRT), perhaps as a result of being unable to prepare or select a response in advance (Bloxham *et al.*, 1984). The animal study trained rats to perform a visual spatial discrimination where stimulus luminance (bright *vs.* dim) provided the cue to turn right or left. The animals were trained to wait for an imperative auditory cue before making a response. The visual stimuli were presented either in advance of this, with an unpredictable variable delay (simple reaction time condition), or simultaneously (choice reaction time condition). Following the lesion, the expected spatial bias to the ipsilateral side was seen. However, the rats did show a benefit in performance of the SRT compared to the CRT condition, indicating that they were able to use advance information to prepare a response, regardless of the side of response.

In contrast, on responses contralateral to the lesion, the rats failed to show a delay-dependent quickening of reaction times as the temporal interval between the visual cue and auditory stimulus increased. This delay-dependent speeding of reaction times, however, was seen on the side where responses were ipsilateral to the lesion. It was termed "motor readiness," and the phenomenon is well documented in human subjects and, moreover, appears deficient in PD patients (Bloxham *et al.*, 1987). The overall conclusion reached by the study was that the SRT/CRT dichotomy seen in parkinsonian patients may be epiphenomenal to a deficit of "internal activation" in which cues generated internally fail to influence motor output appropriately. This accords with other descriptions of the fundament deficit in PD involving an impairment in the control of action (Brown and Marsden, 1988).

Clearly, the approach of using complex paradigms in which the behavior of 6-OHDA-lesioned rats can be analyzed and compared to equivalent situations in parkinsonian patients has potential for confirming and/or generating theories of basal ganglia function.

In a further attempt to make animal experiments more relevant to the human motor deficits in PD, many studies have investigated paw-reaching deficits. Nigrostriatal lesions have been found to disrupt rats' abilities to use the contralateral limb to lever-press or to reach, grasp, manipulate, and retrieve pellets from tubes, food wells, or moving conveyor belts (Evenden and Robbins, 1984; Sabol *et al.*, 1985; Siegfried and Bures, 1980; Whishaw *et al.*, 1986). The most reliable effect of the lesion was to switch paw preference solely to that side ipsilateral to the lesion. Whenever the animal was forced to use the contralateral paw (e.g., using an ipsilateral bracelet) precision and accuracy of movement were comprised. However, although this approach may be used to model the human motoric deficit in PD semi-quantitatively and may generate behavioral assays to assess therapies, it is not clear what additional information it can provide into the nature of the deficit other than describing "motor difficulty."

6. Summary

In summary, the question arises as to what extent the vast amount of neurobiological data generated by the 6-OHDA model of parkinsonism has had useful applications to the human disease. From the discussion above, it is clear that the differences between the experimental and pathological condition far outnumber the similarities when anything other than superficial observations are considered. It is inevitable that the profound deficits of posture, gait, and speech will never be adequately modeled in subhuman species. However, perhaps counterintuitively, the less obvious features of the disease, termed "cognitive" for want of a better terminology, may have analogous representation in dopamine-depleted animals. The crucial emphasis for future studies will be to investigate the behavior of animals in paradigms that can be credibly compared with those that have generated data from human performance. The reaction-time studies mentioned above come close to this ideal.

Needless to say there remain virtually limitless possibilities for further investigations using 6-OHDA-treated animals to shed light on the human disease. Many aspects of PD that have profound implications for the individual patient are overlooked in animal studies. Nevertheless, useful answers may yet be found if the right questions are asked. For example, it is unknown if the autonomic dysfunction seen in PD, such as the universally common problems of constipation, has its origins centrally or peripherally. Animal models would seem to be appropriate for this type of analysis, especially as therapeutic implications are direct.

ACKNOWLEDGMENTS. We thank the Medical Research Council and Parkinson's Disease Society for support of our own studies in this research area.

7. References

Aghajanian, G.K., and Bunney, B.S., 1977, Dopamine "autoreceptors": Pharmacological characterization by microiontophoretic single cell recording studies, *Naunym Schmiedebergs Arch. Exp. Pathol. Pharmak.* 297:1–7.

Agid, Y., Javoy, F., and Glowinski, J., 1973, Hyperactivity of remaining dopaminergic neurones after partial destruction of the nigro-striatal dopaminergic system in the rat, *Nature* 245:150–151.

Agid, Y., Bonnett, A.M., Ruberg, M., and Javoy-Agid, F., 1985, Pathophysiology of levodopa induced abnormal involuntary movements, *in* "Dyskinesia, Research and Treatment: Psychopharmacology," Supplementum 2 (D. Casey, T.N. Chase, J. Christensen, and J. Gerlach, eds.), pp. 145–159, Springer-Verlag, Berlin.

Agid, Y., Javoy-Agid, F., and Ruberg. M., 1987, Biochemistry of transmitters in Parkinson's disease, *in* "Movement Disorders" (Vol. 2, C.D. Marsden and S. Fahn, eds.), pp. 166–230, Butterworth, London.

Andén, N.E., Carlsson, A., Dahlström, A., Fuxe, K., Hillarp, N.A., and Larsson, K., 1964, Demonstration and mapping of nigro-neostriatal dopamine neurons, *Life Sci.* 3:523–530.

Andén, N.E., Dahlström, A., Fuxe, K., and Larsson, K., 1966, Functional role of the nigro-neostriatal dopamine neurons, *Acta Pharmacol. Toxicol.* 24:263–274.

Annett, L.E., Rogers, D.C., Hernandez, T.D., and Dunnett, S.B., 1992, Behavioural analysis of unilateral dopamine depletion in the marmoset, *Brain,* 115:825–856.

Antelman, S.M., Szechtman, H., Chin, P., and Fisher, A.E., 1975, Tail pinch-induced eating, gnawing and licking behavior in rats: Dependence on the nigrostriatal dopamine system, *Brain Res.* 99:319–337.

Arendt, T., Bigl, V., Tennstedt, A., and Arendt, A., 1985, Neuronal loss in different parts of the nucleus basalis is related to neuritic plaque formation in cortical target areas in Alzheimer's disease, *Neuroscience* 14:1–14.

Artieda, J., Pastor, M.A., Lacruz, F., and Obeso, J.A., 1992, Temporal discrimination is abnormal in Parkinson's disease, *Brain* 115:199–210.

Aubert, I., Araujo, D.M., Cecyre, D., Robitaille, Y., Gauthier, S., and Quirion, R., 1992, Comparative alterations of nicotinic and muscarinic binding sites in Alzheimer's and Parkinson's diseases, *J. Neurochem.* 58:529–541.

Bartus, R.T., Dean, R.L., Beer, B., and Lippa, A.S., 1982, The cholinergic hypothesis of geriatric memory dysfunction, *Science* 217:408–417.

Baumgarten, H.G., Björklund, A., and Lachenmayer, L., 1983, Evaluation of the effects of 5,7-dihydroxytryptamine on serotonin and catecholamine neurons in the rat CNS, *Acta Physiol. Scand. Suppl.* 391:1–19.

Bell, L.J., Iversen, L.L., and Uretsky, N.J., 1970, Time course of the effects of 6-hydroxydopamine on catecholamine-containing neurons in rat hypothalamus and striatum, *Br. J. Pharmacol.* 40:790–799.

Benecke, R., Rothwell, J.C., Dick, J.P.R., Day, B.L., and Marsden, C.D., 1987, Disturbance of sequential movements in patients with Parkinson's disease, *Brain* 110:361–379.

Bernheimer, H., Birkmayer, W., Hornykiewicz, O., Jellinger, K., and Seitelberger, F., 1973, Brain dopamine and the syndromes of Parkinson and Huntington: Clinical, morphological and neurochemical correlations, *J. Neurol. Sci.* 20:415–455.

Birkmayer, W., and Hornykiewicz, O., 1961, Der L-dioxyphenylalanin (=L-dopa) effekt bei der Parkinson-akinese, *Wien. Klin. Wochenschr.* 73:787.

Blocq, P., and Marinesco, G., 1893, Sur un cas de tremblement parkisonien hemiplegique: Symptomatique d'une tumeur du pedoncule cerebral, *C.R. Soc. Biol. (Paris)* 5:105–111.

Bloom, F.E., Algeria, S., Gropetti, S., Revuelta, A., and Costa E., 1969, Lesions of central norepinephrine terminals with 6-OH-dopamine: Biochemistry and fine structure, *Science* 166:1284–1286.

Bloxham, C.A., Mindel, T.A., and Frith, C.D., 1984, Initiation of predictable and unpredictable movements in Parkinson's disease, *Brain* 107:371–384.

Bloxham, C.A., Dick, D.J., and Moore, M., 1987, Reaction time and attention in Parkinson's disease, *Brain* 107:371–384.

Bokobza, B., Ruberg, M., Scatton, B., Javoy-Agid, F., and Agid, Y., 1984, [^{3}H] spiperone binding, dopamine and HVA concentrations in Parkinson's disease and supranuclear palsy, *Eur. J. Pharmacol.* 99:167–175.

Bracha, H.S., Shults, C., Glick, S.D., and Kleinman, J.E., 1987, Spontaneous asymmetric circling behavior in hemi-parkinsonism: A human equivalent of the lesioned-circling rodent behavior, *Life Sci.* 40:1127–1130.

Breese, G.R., and Traylor, T.D., 1970, Effect of 6-hydroxydopamine on brain norepinephrine and dopamine: Evidence for selective degeneration of catecholamine neurons, *J. Pharmacol. Exp. Ther.* 174:413–420.

Brown, R.G., and Marsden, C.D., 1988, Internal versus external cues and the control of attention in Parkinson's disease, *Brain* 111:23–45.

Brown, V.J., and Robbins, T.W., 1989, Elementary processes of response selection mediated by distinct regions of the striatum, *J. Neurosci.* 9:3760–3765.

Brown, V.J., and Robbins, T.W., 1991, Simple and choice reaction time performance following unilateral striatal dopamine depletion in the rat: Impaired motor readiness but preserved response preparation, *Brain* 114:513–525.

Burns, R.S., Chiueh, C.C., Markey, S.P., Ebert, M.H., Jacobowitz, D.M., and Kopin, I.J., 1983, A primate model of parkinsonism: Selective destruction of dopaminergic neurons in the pars compacta of the substantia nigra by *N*-methyl-4-phenyl-1,2,3,6-tetrahydropyridine. *Proc. Natl. Acad. Sci.* USA 80:4546–4550.

Carey, R.J., 1986, Conditioned rotational behavior in rats with unilateral 6-hydroxydopamine lesions of the substantia nigra, *Brain Res.* 365:379–382.

Carli, M., Evenden, J.L., and Robbins, T.W., 1985, Depletion of unilateral striatal dopamine impairs initiation of contralateral actions and not sensory attention, *Nature* 313:679–682.

Carli, M., Jones, G.H., and Robbins, T.W., 1989, Effects of unilateral dopamine depletion from the dorsal and ventral striatum on visual neglect in the rat: A neural and behavioural analysis, *Neuroscience* 29:309–327.

Carlsson, A., 1959, The occurrence, distribution and physiological role of catecholamines in the nervous system, *Pharmacol. Rev.* 11:490–493.

Carlsson, A., Lindqvist, M., and Magnusson, T., 1957, 3,4-dihydroxytryptamine and 5-hydroxytryptophan as reserpine antagonists, *Nature* 180:1200.

Carlsson, A., Kehr, W., Lindquist, M., Magnusson, T., and Atack, C.V., 1972, Regulation of monoamine metabolism in the central nervous system, *Pharmacol. Rev.* 24:371–384.

Carman, L.S., Gage, F.H., and Shults, C.W., 1991, Partial lesion of the substantia nigra: Relation between extent of lesion and rotational behaviour, *Brain Res.* 553:275–283.

Casas, M., Ferre, S., Cobos, A., Cadafalch, J., Grau, J.M., and Jane, F., 1988, Comparison between apomorphine and amphetamine-induced rotational behaviour in rats with a unilateral nigrostriatal pathway lesion, *Neuropharmacology* 27:657–659.

Cash, R., Dennis, T., L'Heureux, Raisman, R., Javoy-Agid F., and Scatton, B., 1987, Parkinson's disease and dementia—norepinephrine and dopamine in locus ceruleus, *Neurology* 37:42–46.

Chase, T.N., 1972, Parkinson's disease—modification by 5-hydroxytryptophan, *Neurology* 22:479–484.

Chase, T.N., 1974, Serotoninergic mechanisms and extrapyramidal function in man, *Adv. Neurol.* 5:31–39.

Chiueh, C.C., 1988, Dopamine in the extrapyramidal motor function: A study based upon the MPTP-induced primate model of parkinsonism, *Ann. N.Y. Acad. Sci.* 515:226–238.

Colle, L.M., and Wise, R.A., 1988, The nucleus accumbens contributes to the direction of dopamine-dependent circling, *Soc. Neurosci. Abstr.* 14:663.

Cools, A.R., Scheenen, W., Eilam, D., and Golani, I., 1989, Evidence that apomorphine and (+)-amphetamine produce different types of circling in rats, *Brain Res.* 34:111–116.

Cooper, B.R., Breese, G.R., Grant, L.D., and Howard, J.L., 1973, Effects of 6-hydroxydopamine treatments on active avoidance responding: Evidence for involvement of brain dopamine, *J. Pharmacol. Exp. Ther.* 185:358–370.

Cotzias, G.C., van Woert, M.H., and Schiffer, L.M., 1967, Aromatic amino acids and modification of Parkinsonism, *N. Engl. J. Med.* 276:374–379.

Coyle, J.T., Price, D.L., and DeLong, M.R., 1983, Alzheimer's disease: A disorder of cortical cholinergic innervation, *Science* 219:1184–1190.

Creese, I., Burt, D., and Snyder, S.H., 1977, Dopamine receptor binding enhancement accompanies lesion-induced behavioral supersensitivity, *Science* 197:596–598.

Creese, I., and Snyder, S.H., 1979, Nigrostriatal lesions enhance striatal (^{3}H) apomorphine and (^{3}H) spiroperidol binding, *Eur. J. Pharmacol* 56:277–281.

Critchley, M., 1953, "The Parietal Lobes," Haffner, New York.

Dankova, J., Bedard, P., Langelier, P., and Poirier, L.J., 1978, Dopaminergic agents and circling behaviour, *Gen. Pharmacol.* 9:295–302.

Davidson, J., Lloyd, K., Dankova, K., and Hornykiewicz, O., 1971, L-Dopa treatment in Parkinson's disease: Effect on dopamine and related substances in discrete brain regions, *Experientia* 27:1048–1049.

Drachman, D.A., and Sahakian, B.J., 1980, Memory, aging and pharmacosystems, *in* "The Psychobiology of Aging: Problems and Perspectives" (D. Stein, ed.), pp. 347–368, Elsevier/North Holland, Amsterdam.

Dubois, B., Ruberg, M., Javoy-Agid, F., Ploska, A., and Agid, Y., 1983, A subcortico-cortical cholinergic system is affected in Parkinson's disease, *Brain Res.* 288:213–218.

Dunnett, S.B., and Björklund, A., 1983, Conditioned turning in rats: Dopaminergic involvement in the initiation of movement rather than the movement itself, *Neurosci. Lett.* 41:173–178.

Dunnett, S.B., and Iversen, S.D., 1982, Sensorimotor impairments following localized kainic acid and 6-hydroxydopamine lesions of the neostriatum, *Brain Res.* 248:121–127.

Dunnett, S.B., Björklund, A., Stenevi, U., and Iversen, S.D., 1981, Behavioural recovery following transplantation of substantia nigra in rats subjected to 6-OHDA lesions of the nigrostriatal pathway. I: Unilateral lesions, *Brain Res.* 215:147–161.

Dunnett, S.B., Everitt, B.J., and Robbins, T.W., 1991, The basal forebrain-cortical cholinergic system: Interpreting the functional consequences of excitotoxic lesions, *Trends Neurosci.* 14:494–501.

Dunnett, S.B., and Robbins, T.W., 1992, The functional role of mesotelencephalic dopamine systems, *Biol. Rev.* 67:491–518.

Ehringer, H., and Hornykiewicz, O., 1960, Verteilung von noradrenalin und dopamin (3-hydroxytyramin) im Gehirn des Menschen und ihr Verhalten bei Erkrankungen, *Wien. Klin. Wochenschr.* 38:1236–1239.

Evenden, J.L., and Robbins, T.W., 1984, Effects of unilateral 6-hydroxydopamine lesions of the caudate-putamen on skilled forepaw use in the rat, *Behav. Brain Res.* 14:61–68.

Farley, I.J., and Hornykiewicz, O., 1976, Noradrenaline in subcortical brain regions of patients with Parkinson's disease and control subjects, *in* "Advances in Parkinsonism (W. Birkmayer and O. Hornykiewicz, eds.), pp. 178–185, Roche, Basel.

Faull, R.L.M., and Laverty, R., 1969, Changes in dopamine levels in the corpus striatum following lesions in the substantia nigra, *Exp. Neurol.* 23:332–340.

Fearnley, J.M., and Lees, A.J., 1991, Ageing and Parkinson's disease: Substantia nigra regional specificity, *Brain* 114:2283–2301.

Feltz, P., and DeChamplain, J., 1972, Enhanced sensitivity of caudate neurons to microiontophoretic injections of dopamine in 6-hydroxydopamine treated rats, *Brain Res.* 43:601–605.

Fibiger, H.C., Lonsbury, B., Cooper, H.P., and Lytle, L.D., 1972, Early behavioural effects of intraventricular administration of 6-hydroxydopamine in the rat, *Nature* 236:209–211.

Fibiger, H.C., and Grewaal, D.S., 1974, Neurochemical evidence for denervation supersensitivity: The effect of unilateral substantia lesions on apomorphine-induced increases in neostriatal acetylcholine levels, *Life Sci.* 15:57–63.

Forno, L.S., 1982, Pathology of Parkinson's disease, *in* "Movement Disorders" (C.D. Marsden and S. Fahn, eds.), pp. 25–40, Butterworths, London.

Gerfen, C.R., 1992, The neostriatal mosaic: Multiple levels of compartmental organization in the basal ganglia, *Ann. Rev. Neurosci.* 15:285–320.

Gibb, W.R.G., and Lees, A.J., 1991, Anatomy, pigmentation, ventral and dorsal subpopulations of the substantia nigra, and differential cell death in Parkinson's disease, *J. Neurol. Neurosurg. Psychiatry* 54:388–396.

Glick, S.D., Jerussi, T.P., and Fleisher, L.N., 1976, Turning in circles: The neuropharmacology of rotation, *Life Sci.* 18:889–896.

Goldstein, M., Anagnoste, G., Battista, A.F., Owen, W.S., and Nakatani, S., 1969, Studies of amines in the striatum monkeys with nigral lesions, *J. Neurochem.* 16:645–653.

Greenfield, J.G., 1955, The pathology of Parkinson's disease, *in* "James Parkinson 1755–1824" (ed.), pp. 219–243, M. Critchley, Macmillan, London.

Groves, P.M., Wilson, C.J., Young, S.J., and Rebec, G.V., 1975, Self-inhibition in dopamine neurons, *Science* 190:522–529.

Guttman, M., Seeman, P., Reynolds, G.P., Riederer, P., Jellinger, K., and Tourtelotte, W.W., 1986. Dopamine D2 receptor density remains constant in treated Parkinson's disease, *Ann. Neurol.* 19:487–492.

Hagan, J.J., Alpert, J.E., Morris, R.G.M., and Iversen, S.D., 1983, The effects of central catecholamine depletions on spatial learning in rats, *Behav. Brain Res.* 9:83–104.

Hefti, F., Melamed, E., and Wurtman, R.J., 1980, Partial lesions of the dopaminergic nigrostriatal system in rat brain: Biochemical characterization, *Brain Res.* 195:123–137.

Heilman, K.M., Bowers, D., Watson, R.T., and Greer, M., 1976, Reaction time in Parkinson's disease, *Arch. Neurol.* 33:139–140.

Herman, J.P., Choulli, K., and Le Moal, M, 1985, Hyper-reactivity to amphetamine in rats with dopaminergic grafts, *Exp. Brain Res.* 60:521–526.

Herman, J.P., Choulli, K., Abrous, N., and Le Moal, M., 1989, Intracerebral grafts of dopaminergic neurons: A discussion of their functional effects and mechanisms of action, *in* "Neuronal Grafting and Alzheimer's Disease" (F. Gage, A. Privat, and Y. Christen, eds.), pp. 21–33, Springer-Verlag, Berlin.

Hökfelt, T., and Ungerstedt, U., 1973, Specificity of 6-hydroxydopamine induced degeneration of central monoamine neurones: An electron and fluorescence microscopic study with special reference to intracerebral injection on the nigro-striatal dopamine system, *Brain Res.* 60:269–297.

Hollerman, J.R., and Grace, A.A., 1988, Nigral DA cell recruitment as a compensatory mechanism, *Soc. Neurosci. Abstr.* 14:116.

Hornykiewicz, O., 1979, Compensatory biochemical changes at the striatal dopamine synapse in Parkinson's disease—limitations of L-DOPA therapy, *Adv. Neurol.* 24:275–281.

Hoyman, L., Weese, G.D., and Frommer, G.P., 1979, Tactile discrimination performance following neglect-producing unilateral lateral hypothalamic lesions in the rat, *Physiol. Behav.* 22:139–147.

Iversen, L.L., Rogawski, M.A., and Miller, R.J., 1976, Comparison of the effects of neuroleptic drugs on pre- and postsynaptic dopaminergic mechanisms in the rat striatum, *Mol. Pharmacol.* 12:251–262.

Jacks, B.R., De Champlain, J., and Cordeau, J.P., 1972, Effects of 6-hydroxydopamine on putative transmitter systems in the central nervous system, *Eur. J. Pharmacol.* 18:353–360.

Jellinger, K., 1986, Pathology of Parkinsonism, *in* "Recent Developments in Parkinson's Disease" (S. Fahn, C.D. Marsden, P. Jenner, and P. Teychenn, eds.), pp. 33–66, Raven Press, New York.

Jenner, P.D., and Marsden, C.D., 1986, The actions of 1-methyl-4phenyl-1,2,3,6-tetrahydropyridinein animals as a model of Parkinson's disease. *J. Neural Transm. Suppl.* 20:11–39.

Kelly, P.H., and Moore, R.Y., 1976, Mesolimbic dopaminergic neurones in the rotational model of nigrostriatal function, *Nature* 263:695–696.

Kelly, R.S., and Wightman, R.M., 1987, Detection of dopamine overflow and diffusion with voltammetry in slices of rat brain, *Brain Res.* 423:79–87.

Kopp, N., Denoroy, L., Thomasi, M., Gay, N., Chazot, G., and Renaud, B., 1982, Increase in noradrenaline synthesizing enzyme activity in medulla oblongata of Parkinson's disease, *Acta Neuropathol.* 56:17–21.

Koshikawa, N., Mori, E., Maruyama, Y., Yatsushige, N., and Kobayashi, M., 1990, Role of dopamine D1 and D2 receptors in the ventral striatum in the turning behaviour of rats, *Eur. J. Pharmacol.* 178:233–237.

Kostowski, W., Samanin, R., Bareggie, S.R., Marc, V., Garratini, S., and Valzelli, L., 1974, Biochemical aspects of the interaction between midbrain raphe and locus ceruleus in the rat, *Brain Res.* 82:178–182.

Kostrzewa, R.M., and Jacobowitz, D.M., 1974, Pharmacological actions of 6-hydroxydopamine, *Pharamocol. Rev.* 26:199–288.

Krueger, B.K., Forn, J., Walters, J.R., Roth, R.H., and Greengard, P, 1976, Stimulation by dopamine of cyclic 3,5-monophospate formation in rat caudate nucleus: Effect of lesions of the nigoneostriatal pathway, *Mol. Pharmacol.* 12:639–648.

Lalonde, R., 1987, Motor abnormalities in Weaver mutant mice, *Exp. Brain Res.* 65:479–481.
Le Moal, M., and Simon, H., 1990, Mesocorticolimbic dopaminergic network: Functional and regulatory roles, *Physiol. Rev.* 71:155–234.
Lee, T., Seeman, P., Rajput, A., Farley, I.J., and Hornykiewicz, O., 1978, Receptor basis for dopaminergic supersensitivity in Parkinson's disease, *Nature* 273:59–60.
Lloyd, K.G., 1976, Neurochemical compensation in Parkinson's disease, *in* "Parkinson's Disease: Concepts and Prospects (J.P. Laake, W.F. Korf, and J. Wesseling, eds.), pp. 61–72, Excerpta Medica, Amsterdam.
Luthman, J., Fredriksson, A., Sundström, E. Jonsson, G., and Archer, T., 1989, Selective lesion of central dopamine or noradrenaline neuron systems in the neonatal rat: Motor behavior and monoamine alterations at adult stage, *Behav. Brain Res.* 33:267–277.
Mabry, P.D., and Campbell, B.A., 1973, Serotonergic inhibition of catecholamine-induced behavioral arousal, *Brain Res.* 49:381–391.
Maloteaux, J.M., Lyabeya, M.K., Laterre, E.C., Javoy-Agid, F., Agid, Y., and Laduron, P.M., 1985, S2-serotonin receptors in frontal cortex of parkinsonian patients, *J. Neurol. Suppl.* 32:108.
Marsden, C.D., 1984, Which motor disorder in Parkinson's disease indicates the true motor function of the basal ganglia? *in* "Ciba Foundation Symposium 107: Functions of the Basal Ganglia," pp. 225–237, Pitman, London.
Marsden, C.D., and Parkes, J.D., 1976, "On-off" effects in patients with Parkinson's disease on chronic levodopa therapy, *Lancet* i:292–296.
Marshall, J.F., Richardson, J.S., and Teitelbaum, P., 1974, Nigrostriatal bundle damage and the lateral hypothalamic syndrome, *J. Comp. Physiol. Psychol.* 87:808–830.
Marshall, J.F., Levitan, D., and Stricker, E.M., 1976, Activation-induced restoration of sensorimotor functions in rats with dopamine depleting brain lesions, *J. Comp. Physiol. Psychol.* 90:536–546.
Mayeux, R., Stern, Y., Cote, L., and Williams, J.B.W., 1984, Altered serotonin metabolism in depressed patients with Parkinson's disease, *Neurology* 34:642–646.
Messier, C., Mrabet, O., and Destrade, C., 1990, Dopamine D1 and D2 agonists and antagonists produce turning when injected into the nucleus accumbens, *Soc. Neurosci. Abstr.* 15: 1053.
Miller, R., and Beninger, R.J., 1991, On the interpretation of asymmetries of posture and locomotion produced with dopamine agonists in animals with unilateral depletion of striatal dopamine, *Prog. Neurobiol.* 36:229–256.
Mishra, R.K., Gardner, E.L., Katzman, R., and Makman, M.H., 1974, Enhancement of dopamine-stimulated adenylate cyclase activity in rat caudate after lesions in substantia nigra: Evidence for degeneration supersensitivity, *Proc. Natl. Acad. Sci. USA* 71:3883–3887.
Mishra, R.K., Marshall, A.M., and Varmuza, S.L., 1980 Supersensitivity in rat caudate nucleus: Effects of 6-hydroxydopamine on the time course of dopamine receptor and cyclic AMP changes, *Brain Res.* 200:47–57.
Morgenroth, V.H. III, Walters, J.R., and Roth, R.H., 1976, Dopaminergic neurons: Alteration in the kinetic properties of tyrosine hydroxylase after cessation of impulse flow, *Biochem. Pharmacol.* 25:655–661.
Muller, P., and Seeman, P., 1977, Brain neurotransmitter systems after long-term haloperidol: Dopamine, acetylcholine, serotonin, noradrenergic and naloxone receptors, *Life Sci.* 21:1751–1758.
Nagatsu, T., Kanamori, T., Kato, T., Iizuka, R., and Narabayashi, H., 1978, Dopamine stimulated adenylate cyclase activity in the human brain changes in parkinsonism, *Biochem. Med.* 19:360–365.
Nagy, J.I., Lee, T., Seeman, P., and Fibiger, H.C., 1978, Direct evidence for presynaptic and postsynaptic dopamine receptors in brain, *Nature* 274:278–281.

Nakano, I., and Hirano, A., 1984, Parkinson's disease: Neuron loss in the nucleus basalis without concomitant Alzheimer's disease, *Ann. Neurol.* 15:415–418.

Nowycky, M.C., and Roth, R.H., 1978, Dopaminergic neurons: Role of presynaptic receptors in the regulation of transmitter biosynthesis, *Prog. Neuropsychopharmacol.* 2:139–158.

Ohye, C., Bouchard, R., Boucher, R., and Poirier, L.J., 1970, Spontaneous activity of the putamen after chronic interruption of the dopaminergic pathway: Effect of L-Dopa, *J. Pharmacol. Exp. Ther.* 175:700–708.

Onn, S.P., Berger, T.W., Stricker, E.M., and Zigmond, M.J., 1986, Effects of intraventricular 6-hydroxydopamine on the dopaminergic innervation of striatum: Histochemical and neurochemical analysis, *Brain Res.* 376:8–19.

Pastor, M.A., Artieda, J., Jahanshahi, M., and Obeso, J.A., 1992, Time estimation is abnormal in Parkinson's disease, *Brain* 115:211–225.

Perry, E.K., Curtis, M., Dick, D.J., Candy, J.M., Atack, J.R., Bloxham, C.A., Fairbairn, A., Tomlinson, B.E., and Perry, R.H., 1985, Cholinergic correlates of cognitive impairment in Parkinson's disease: Comparisons with Alzheimer's disease, *J. Neurol. Neurosurg. Pyschiatry.* 48:413–442.

Perry, E.K., Perry, R.H., Candy, R.M., Fairbairn, A.F., Blessed, G., Dick, D.J., and Tomlinson, B.E., 1984, Cortical serotonin-S2 receptor binding abnormalities in patients with Alzheimer's disease: Comparisons with Parkinson's disease, *Neurosci. Lett.* 51:353–357.

Perry, T.L., Javoy-Agid, F., Agid, T., and Fibiger, H.C., 1983, Is striatal GABAegic neuronal activity reduced in Parkinson's disease? *J. Neurochem.* 40:1120–1123.

Pierot, L., Desnos, C., Blin, J., Raisman, R., Scherman, D., Javoy-Agid, F., Ruberg, M., and Agid, Y., 1988, D1 and D2-type dopamine receptors in patients with Parkinson's disease and progressive supranuclear palsy, *J. Neurol. Sci.* 86:291–301.

Pimoule, C., Schoemaker, H., Reynolds, G.P., and Langer, S.Z., 1985, [^{3}H]SCH 23390 labeled D1 dopamine receptors are unchanged in schizophrenia and Parkinson's disease, *Eur. J. Pharmacol.* 114:235–237.

Poirier, L.J., 1960, Experimental and histological study of midbrain dyskinesias, *J. Neurophysiol.* 23:534–551.

Price, M.T.C., and Fibiger, H.C., 1975, Discrimination escape learning and response to electric shock after 6-hydroxydopamine lesions of the nigro-neostriatal dopaminergic projection, *Pharmacol Biochem. Behav.* 3:285–290.

Pycock, C.J., 1980, Turning behaviour in animals, *Neuroscience* 5:461–514.

Pycock, C.J., and Marsden, C.D., 1978, The rotating rat: A two-component system? *Eur. J. Pharmacol.* 47:167–175.

Raisman, R., Cash, R., Ruberg, M., Javoy-Agid, F., and Agid, Y., 1985, Binding of [^{3}H]SCH 23390 to D-1 receptors in the putamen of control and parkinsonian subjects, *Eur. J. Pharmacol.* 113:467–468.

Rajput, A.H., Rozdilsky, B., Hornykiewicz, O., Shannak, K., Lee, T., and Seeman, P., 1982, Reversible drug-induced parkinsonism: Clinicopathologic study of two cases, *Arch. Neurol.* 39:644–646.

Ranje, C., and Ungerstedt, U., 1977, Lack of acquisition of dopamine denervated animals tested in an underwater maze, *Brain Res.* 134:95–111.

Redmond, D.E., Jr., Hinrichs, R.L., Maas, J.W., and Kling, A., 1973, Behavior of free-ranging macaques after intraventricular 6-hydroxydopamine, *Science* 181:1256–1258.

Reisine, T.D., Fields, J.Z., Yamamura, H.I., Bird, E.D., Spokes, E., Schreiner, P.S., and Enna, S.J., 1977, Neurotransmitter receptor alterations in Parkinson's disease, *Life Sci.* 21:335–344.

Riederer, P., Rausch, W.D., Birkmayer, W., Jellinger, K., and Danielczyk, W., 1978, Dopamine-sensitive adenylate cyclase in the caudate nucleus and metabolic encephalopathies, *J. Neural Transm. Suppl* 14:153.

Rinne, U.K., Koskinen, V., and Lönnberg, P., 1980, Neurotransmitter receptors in the parkinsonian brain, *in* "parkinson's Disease: Current Progress, Problems and Management" (U.K. Rinne, M. Klinger, and G. Stamm, eds.), pp. 93–107, Elsevier, Amsterdam.

Rinne, U.K., Lönnberg, P., and Koskinen, V., 1981, Dopamine receptors in the parkinsonian brain, *J. Neural Transm.* 51:97–106.

Rinne, J.O., Rinne, J.K., Laakso, K., Lönnberg, P., and Rinne, U.K., 1985, Dopamine D-1 receptors in the parkinsonian brain, *Brain Res.* 359:306–310.

Robbins, T.W., and Everitt, B.J., 1982, Functional studies of central catecholamines, *Int. J. Neurobiol.* 23:303–365.

Robbins, T.W., Giardini, V., Jones, G.H., Reading, P., and Sahakian, B.J., 1990, Effects of dopamine depletion from the caudate-putamen and nucleus accumbens septi on the acquisition and performance of a conditional discrimination task, *Behav. Brain Res.* 38:243–261.

Robbins, T.W., and Brown, V.J., 1990, The role of the striatum in the mental chronometry of action: A theoretical review, *Rev. Neurosci.* 2:181–213.

Robinson, T.W., and Whishaw, I.Q., 1988, Normalization of extracellular dopamine in striatum following recovery from a partial unilateral 6-OHDA lesion of the substantia nigra: A microdialysis study in freely moving rat, *Brain Res.* 450:209–224.

Rogers, D., 1986, Bradyphrenia in parkinsonism: A historical review, *Psychol. Med.* 16:257–265.

Rogers, J.D., Brogan, D., and Mirra, S.S., 1985, The nucleus basalis of Meynert in neurological disease: A quantitative morphological study, *Ann. Neurol.* 17:163–170.

Rose, S., Nomoto, M., Jenner, P., and Marsden, C.D., 1989, Transient depletion of nucleus accumbens dopamine content may contribute to initial akinesia induced by MPTP in common marmosets, *Biochem, Pharmacol.* 38:3677–3681.

Roth, R.H., Walter, J.R., Murrin, L.C., and Morgenroth, V.H. III, 1975, Dopamine neurons: Role of impulse flow and presynaptic receptors in the regulation of tyrosine hydroxylase, *in* "Pre- and Postsynaptic Receptors" (E. Usdin and W.E. Bunney, eds.), pp. 5–46, Marcel Dekker, New York.

Ruberg, M., Ploska, A., Javoy-Agid, F., and Agid, Y., 1982, Muscarinic binding and choline acetyltranferase activity in parkinsonian subjects with reference to dementia, *Brain Res.* 232:129–139.

Sabol, K.E., Neill, D.B., Wages, S.A., Church, W.H., and Justice, J.B., 1985, Dopamine depletion in a striatal subregion disrupts performance of a skilled motor task in the rat, *Brain Res.* 335:33–43.

Sachs, C.H., and Jonsson, G., 1975, Mechanisms of action of 6-hydroxydopamine, *Biochem. Pharmacol.* 24:1–8.

Scatton, B., Javoy-Agid, F., Rouquier, L., Dubois, B., and Agid, Y., 1983, Reduction of cortical dopamine, noradrenaline, serotonin and their metabolites in Parkinson's disease, *Brain Res.* 275:321–328.

Scatton, B., Monfort, J.C., Javoy-Agid, F., and Agid, Y., 1984, Neurochemistry of monoaminergic neurones in Parkinson's disease, *in* "Catecholamines Part C: Neuropharmacology and Central Nervous System—Therapeutic Aspects," (E. Usdin, A. Carlsson, A. Dahlström, and J. Engel, eds.), pp. 43–52, Alan R. Liss, New York.

Scatton, B., Dennis, T., Lheureux, R., Monfort, J.C., Duyckaerts, C., and Javoy-Agid, F., 1986, Degeneration of noradrenergic and serotoninergic but not dopaminergic neurons in the lumbar spinal cord of parkinsonian patients, *Brain Res.* 380:181–185.

Schallert, T., Whishaw, I.Q., DeRyck, M., and Teitelbaum, P., 1978, The postures of catecholamine-depletion catalepsy: Their possible adaptive value in thermoregulation, *Physiol. Behav.* 21:817–820.

Schoenfield, R.I., and Uretsky, N.J., 1973, Enhancement by 6-hydroxydopamine of the effects of dopa upon motor activity of rats, *J. Pharmacol. Exp. Ther.* 186:616–624.

Schwab, R.S., 1972, Akinesia paradoxica, *Electroencephalogr. Clin. Neurophysiol.* 31:87–92.
Schwab, R.S., and Zieper, I., 1965, Effect of mood, motivation, stress, and alertness on performance in Parkinson's disease, *Psychiatry Neurol.* 150:345–357.
Schwarting, R.K.W., Bonatz, A.E., Carey, R.J., and Huston, J.P., 1991, Relationships between indices of behavioral asymmetries and neurochemical changes following mesencephalic 6-hydroxydopamine injections, *Brain Res.* 554:46–55.
Selby, G., 1990, Clinical features, *in "Parkinson's Disease"* (G. Stern, ed.) Chapman and Hall, London.
Shibuya, M., 1979, Dopamine sensitive adenylate cyclase activity in the striatum of Parkinson's disease, *J. Neural Transm.* 44:287.
Siegfried, B., and Bures, J., 1980, Handedness in rats: Blockade of reaching behavior by unilateral 6-OHDA injections into substantia nigra and caudate nucleus, *Physiol. Psychol.* 8:360–368.
Snyder, A.M., Stricker, E.M., and Zigmond, M.J., 1985, Stress-induced neurological impairments in an animal model of parkinsonism, *Ann Neurol.* 18:544–551.
Snyder, A.M., Zigmond, M.J., and Lund, R.D., 1986, Sprouting of serotinergic afferents into striatum after dopamine-depleting lesions in infant rats—a retrograde transport and immunocytochemical study, *J. Comp. Neurol.* 245:274–281.
Stachowiak, M.K., Keller, R.J., Stricker, E.M., and Zigmond, M.J., 1987, Increased dopamine efflux from striatal slices during development and after nigrostriatal bundle damage, *J. Neurosci.* 7:1648–1654.
Stern, G., 1966, The effects of lesions in the substantia nigra, *Brain* 89:449–478.
Stricker, E.M., and Zigmond, M.J., 1976, Recovery of function after damage to central catecholamine-containing neurons: A neurochemical model for the lateral hypothalamic syndrome, *in* "Progress in Physiological Psychology and Psychobiology" (J.M. Sprague and A.E. Epstein, eds.), pp. 121–189, Academic Press, New York.
Tassin, J.P., Hervé, D., Vezina, P., Trovero, F., Blanc, G., and Glowinski, J., 1991, Relationships between mesocortical and mesolimbic dopamine neurons: Functional correlates of D1 receptor heteroregulation, *in* "The Mesolimbic Dopamine System: From Motivation to Action (P. Willner and J. Scheel-Kruger, eds.), pp. 175–196, Wiley, Chichester.
Thornburg, J.E., and Moore, K. E., 1975, Supersensitivity to dopamine agonists following unilateral, 6-hydroxydopamine-induced striatal lesions in mice, *J. Pharmacol. Exp. Ther.* 192:42–49.
Tranzer, J.P., and Thoenen, H., 1968, An electron microscopic study of selective, acute degeneration of sympathetic nerve terminals after the administration of 6-hydroxydopamine, *Experientia* 24:155–156.
Trendelenberg, U., 1966, Mechanisms of supersensitivity and subsensitivity to sympathomimetic amines. *Pharmacol. Rev.* 18:629.
Tretiakoff, C., 1919, Contribution a l'etude de l'anatomie pathologique du locus niger de Soemmering avec quelques deductions relatives a la pathogenie des troubles du tonus musculaire et de la maladie de Parkinson, *These de Paris.*
Triarhou, L.C., Low, W.C., and Ghetti, B., 1986, Transplantation of ventral mesencephalic anlagen to hosts with genetic nigrostriatal dopamine deficiency, *Proc. Natl. Acad. Sci. USA* 83:8789–8793.
Turner, B.H., 1973, Sensorimotor syndrome produced by lesions of the amygdala and lateral hypothalamus, *J. Comp. Physiol. Psychol.* 82:37–47.
Ungerstedt, U., 1968, 6-Hydroxydopamine induced degeneration of central monoamine neurons, *Eur. J. Pharmacol.* 5:107–110.
Ungerstedt, U., 1971a, Striatal dopamine release after amphetamine or nerve degeneration revealed by rotational behaviour, *Acta Physiol Scand. Suppl.* 367:49–68.
Ungerstedt, U., 1971b, Postsynaptic supersensitivity after 6-hydroxydopamine-induced degeneration of the nigro-striatal dopamine system, *Acta Physiol. Scand. Suppl.* 367:69–93.

Ungerstedt, U., 1971c, Adipsia and aphagia after 6-hydroxydopamine-induced degeneration of the nigro-striatal dopamine system, *Acta Physio. Scand. Suppl.* 367:96–122.

Ungerstedt, U., and Arbuthnott, G.W., 1970, Quantitative recording of rotational behaviour in rats after 6-hydroxydopamine lesions of the nigrostriatal dopamine system, *Brain Res.* 24:485–493.

Uretsky, N.J., and Iversen, L.L., 1969, Effects of 6-hydroxydopamine on catecholamine containing neurones in the brain, *Nature* 221:557.

Uretsky, N.J., and Iversen, L.L., 1970, Effects of 6-hydroxydopamine on catecholamine containing neurones in the rat brain, *J. Neurochem.* 17:269–278.

Valenstein, E., and Heilman, K.M., 1981, Unilateral hypokinesia and motor extinction. *Neurology* 31:445–448.

Van Praag, 1982, Depression, *Lancet* ii:1259–1264.

Villardita, C., Smirni, P., and Zappala, G., 1983, Visual neglect in Parkinson's disease, *Arch. Neurol.* 40:737–739.

Waddington, J.L., Cross, A.J., Longden, A., Owen, F., and Poulter, M., 1979, Functional distinction between DA-stimulated adenylate cyclase and ^{3}H-spiperone binding sites in rat striatum, *Eur. J. Pharmacol.* 58:341–342.

Walters, J.R., Roth, R.H., and Aghajanian, G.K., 1973, Dopaminergic neurons: Similar biochemical and histochemical effects of gamma-hydroxybutyrate and acute lesions of the nigrostriatal pathway, *J. Pharmacol. Exp. Ther.* 186:630–639.

Westerink, B.H.C., Van der Hayden, J.A.M., and Korf, J., 1978, Enhanced dopamine metabolism after small lesions in the midbrain of the rat, *Life Sci.* 22:749–756.

Whishaw, I.Q., and Dunnett, S.B., 1985, Dopamine depletion, stimulation or blockade in the rat disrupts spatial navigation and locomotion dependent upon beacon or distal cues, *Behav. Brain Res.* 18:11–29.

Whishaw, I.Q., O'Connor, W.T., and Dunnett, S.B., 1986, The contributions of motor cortex, nigrostriatal dopamine and caudate-putamen to skilled forelimb use in the rat, *Brain* 109:805–843.

Whitehouse, P.J., Price, D.L., Struble, R.G., Clark, A.W., Coyle, J.T., and DeLong, M.R., 1982, Alzheimer's disease and senile dementia: Loss of neurons in the basal forebrain, *Science* 215:1237–1239.

Wiley, R.G., Oeltmann, T.N., and Lappi, D.A., 1991, Immunolesioning: Selective destruction of neurons using immunotoxin to rat NGF receptor, *Brain Res.* 562:149–153.

Wolgin, D.L., and Teitelbaum, P., 1978, Role of activation and sensory stimuli in recovery from lateral hypothalamic damage in the cat, *J. Comp. Physiol. Psychol.* 92:474–500.

Zhang, W.Q., Tilson, H.A., Nanry, K.P., Hudson, P.M., Hong, K.S., and Stachowiak, M.K., 1988, Increased dopamine release from striata of rats after unilateral nigrostriatal bundle damage, *Brain Res.* 461:335–342.

Zigmond, M.J., and Stricker, E.M., 1972, Deficits in feeding behavior after intraventricular injection of 6-hydroxydopamine in rats, *Science* 177:1211–1213.

Zigmond, M.J., and Stricker, E.M., 1980, Supersensitivity after intraventricular 6-hydroxydopamine: Relation to dopamine depletion, *Experientia* 36:436–438.

Zigmond, M.J., and Stricker, E.M., 1989, Animal models of parkinsonism using selective neurotoxins: Clinical and basic implications, *Int. Rev. Neurobiol.* 31:1–79.

Zigmond, M.J., Acheson, A.L., Stachiowiak, M.K., and Stricker, E.M., 1984, Neurochemical compensation after nigrostriatal bundle injury in an animal model of preclinical Pakinsonism, *Arch. Neurol.* 41:856–861.

5

MPTP-Treated Mice: A Useful Model for Parkinson's Disease?

ERIK SUNDSTRÖM, BENGT G. HENRIKSSON, ABDUL H. MOHAMMED, and FRÉDÉRIC SOUVERBIE

1. Introduction

In this chapter we consider whether mice treated with MPTP (1-methyl-4-phenyl-1,2,3,6-tetrahydropyridine) represent a suitable model for Parkinson's disease (PD). We classify the disease model system as one that is capable of reproducing certain aspects of the human disease ranging from pathology to clinical symptoms. Sometimes the development of animal models also aims toward reproducing or at least understanding the etiology and/or the pathophysiology of the disease. In PD, the most apparent neuronal lesion is the degeneration of the dopaminergic neurons in substantia nigra, which terminate in nucleus caudatus and putamen, while the classical triad of symptoms consists of rigidity, hypokinesia, and tremor. As is well known, other neuronal systems also degenerate, and the parkinsonian patient displays a number of clinical symptoms in addition to the three mentioned above.

The link between PD and dopamine was established by studies of the behavioral and biochemical effects of the drug reserpine, which provide the first established model for PD. Reserpine treatment, often used in combination with the catecholamine synthesis inhibitor α-methyl-p-tyrosine, results in an acute/subchronic depletion of monoamine neurotransmitters. This means that the system under study is probably more or less constantly changing. Experiments in this model are restricted to studies of short-term effects of dopamine depletion on biochemical or behavioral parameters and the influence of various drugs on these

Erik Sundström, Bengt G. Henriksson, Abdul H. Mohammed, and Frédéric Souverbie
• Department of Geriatric Medicine, Karolinska Institutet, Huddinge University Hospital, Huddinge, Sweden

Toxin-Induced Models of Neurological Disorders, edited by Michael L. Woodruff and Arthur J. Nonneman. Plenum Press, New York, 1994.

parameters. Another important drawback is that reserpine affects all catecholamine neurons. Nevertheless, an impressive amount of important data has been gathered using this model.

Another neurotoxin that is widely used to study various aspects of PD is 6-hydroxydopamine (6-OHDA). Selective lesions of different monoaminergic neurons can be achieved using intracerebral injections of 6-OHDA. If injections are made in substantia nigra or among ascending axons in the medial forebrain bundle, the nigrostriatal dopamine neurons are irreversibly lost, which is a necessary prerequisite for various kinds of studies such as neural transplantation. However, stereotaxic injections are by nature time-consuming and somewhat variable. If bilateral lesions are performed, the animals generally develop aphagia and adipsia, and tube-feeding has to be maintained for some time. To avoid this, unilateral lesions can be made. This leads to rotation of the animal upon stimulation of the dopamine receptors (Ungerstedt, 1971) and has been used widely as a model for PD.

Both of these models have been used extensively in studies examining the biochemical and behavioral consequences of dopamine depletion and the adaptive mechanisms involved. One particularly important finding in 6-OHDA–lesioned rats, which reflects the situation in the patient with PD, is the existence of a threshold of degeneration below which symptoms appear. In rats, more than 95% depletion is necessary for expression of behavioral effects as well as certain biochemical changes. However, given the disadvantages of these models, a substance that could cause selective, reproducible, and severe lesions of the nigrostriatal dopamine neurons would be ideal. As we will see, MPTP-treated mice may fulfill these demands.

2. Mechanisms of Toxicity of MPTP

Research on PD grew considerably with the discovery of 1-methyl-4-phenyl-4-propionoxypiperidine (MPTP) as a dopamine neurotoxin. This neurotoxin was inadvertently discovered when it was found that self-administration of the so-called designer drug, rapidly caused a severe state of parkinsonism (Davis *et al.*, 1979; Langston *et al.*, 1983). Subsequently, MPTP was identified as a contaminant, and administration of this substance to monkeys induced a behavioral syndrome resembling PD (Burns *et al.*, 1983). In contrast, rats were found to be insensitive to MPTP (Chiueh *et al.*, 1984), whereas catecholamine neurons in mice could be lesioned with MPTP although using considerably higher doses than is necessary in primates (Hallman *et al.*, 1984; Heikkila *et al.*, 1984a).

The first step in unraveling the mechanism of MPTP was taken upon discovery that MPTP is metabolized to MPP^+ (1-methyl-4-phenylpyridinium) via the intermediate $MPDP^+$ (1-methyl-4-phenyl-2,3-dihydropyridine) in the brain, and that this oxidation is catalyzed by monoamine oxidase type B (MAO B) (Chiba *et al.*, 1984; Markey *et al.*, 1984). That this is a crucial sequence in the mechanism of toxicity was shown by the protective effect of deprenyl and other MAO B inhibitors administered prior to MPTP (Heikkila *et al.*, 1984b, Langston *et al.*, 1984). MAO B is localized in astroglia among other sites, which can accumulate MPTP *in vitro* and oxidize it to $MPDP^+$ and MPP^+ that readily cross the cell membrane (Di Monte *et al.*, 1992). Because serotonin lesions do not occur, the MAO B that is present in serotonergic neurons does not seem to be involved in MPTP toxicity (Melamed *et al.*, 1986).

The occurence of MAO B in many regions of the brain and the preferential location of MAO B in noncatecholaminergic cells indicated that the local formation of MPP^+ does not determine the selectivity of MPTP. The demonstration that catecholaminergic neurons accumulate MPP^+ (Javitch *et al.*, 1985) suggested that this mechanism was responsible for the selectivity of MPTP. This was proven when a number of different catecholamine uptake inhibitors were shown to prevent MPTP toxicity to dopamine and noradrenaline neurons in mice (Pileblad and Carlsson, 1985; Ricaurte *et al.*, 1985; Sundström and Jonsson, 1985) as well as toxicity of intracerebroventricularly administered MPP^+ (Sundström *et al.*, 1986). It was also demonstrated that oxidation of MPTP to MPP^+ precedes the uptake process in both dopaminergic and noradrenergic neurons (Sundström and Jonsson, 1986). In contrast to NMRI mice, the MPTP-induced degeneration of nigrostriatal dopamine neurons of C57 BL/6 mice seems to be anterograde with an initial loss of nigral TH-neurons during a 48-hr period following MPTP administration and degeneration of the striatal terminals during the first week (Sundström *et al.*, 1988). Still the uptake of MPP^+ could occur in striatum with a subsequent retograde transport to the cell soma. Such a transport has been demonstrated in the rat (Campbell *et al.*, 1990).

The mechanism whereby catecholamine neurons are killed (or damaged) by MPP^+ is less clear. Theoretically, some steps in the bioactivation of MPTP could produce free radicals. However, *in vitro* studies have given conflicting results. Although it was initially reported by Perry *et al.* (1985) that a number of free radical scavengers lessened the MPTP-induced lesion, several investigators failed to find any protective effects of free radical scavengers. Overall, studies using different approaches to this problem have generated an equal number of opposing findings (see Bloem *et al.*, 1990).

The predominant opinion is that MPTP causes its degenerative effects by interfering with mitochondrial function. Nicklas *et al.* (1985) originally showed that MPP^+ inhibits NADH ubiquinone (CoQ) reductase (complex I), thereby depleting the cells of ATP. This inhibition was preceded by accumulation of MPP^+ in mitochondria (Ramsay *et al.*, 1986), and this accumulation was shown to depend on the transmembrane potential (Ramsay and Singer, 1986). In a preparation of synaptosomes, MPP^+ was shown to induce a decrease of ATP content in a dose-dependent manner (Scotchner *et al.*, 1990). However, the decrease induced by MPP^+ was not inhibited by a dopamine uptake inhibitor, indicating that MPP^+ had nonspecific effects on all neurons. Subsequently, however, it was shown that the MPP^+-induced decrease in oxygen consumption observed *in vitro* appears selectively in striatum at low concentrations, whereas nonselective effects occur at high MPP^+ concentrations (Martin *et al.*, 1991). Chan *et al.* (1991) recently demonstrated that the MPTP-induced decrease of ATP *in vitro* is restricted to the ventral mesencephalon and striatum, and that it is prevented by pretreatment with the dopamine uptake inhibitor mazindol. Thus, nonspecific effects of MPP^+ appear only at high concentrations that are not found after systemic administration of MPTP, but are often reached during intracerebral injections of MPP^+.

A possible link between the involvement of free radicals in MPTP toxicity and inhibition of complex I of the mitochondrial respiratory chain has been suggested. Thus, inhibition of complex I by rotenone or MPP^+ (as well as inhibition of complex III) causes an increase in free radical production (Hasegawa *et al.*, 1990). This could lead to the irreversible decrease of complex I activity reported by Cleeter *et al.* (1992) following prolonged incubation with MPP^+. Still, the evidence for free radical involvement in MPTP is not convincing.

Recently, researchers suggested involvement of NMDA-receptors in the toxicity of MPTP. Turski *et al.* (1991) reported that both competitive and noncompetitive NMDA antagonists protect neurons in substantia nigra from degeneration induced by infusions of MPP^+. This study may be criticized for its use of rats, a species that is not sensitive to MPTP toxicity and for the very high concentrations of MPP^+ injected into the brain. As mentioned later in this chapter, this leads in general to non-specific cell death. This study was repeated by Sonsalla *et al.* (1992) who failed to see any protection by the non-competitive NMDA antagonist MK-801. They, as well as others (Kupsch *et al.*, 1992) did not observe any protective effect of repeated injections of NMDA antagonists against MPTP-toxicity. However, others have found such a protection (Tabatabaei *et al.*, 1992; Brouillet and Beal, 1993) and Zuddas *et al.* (1992) reported on the protective effect of MK-801 against MPTP-induced lesion of dopamine neurons in monkeys. MK-801 does not prevent degeneration of cultured dopamine neurons by MPP^+ (Michel and Agid, 1992; Finiels-Marlier *et al.*, 1993) although this could be due to the fact that MK-801 only gets access to its binding site if the neuron is depolarized. We found recently that MPP^+ is not an agonist at striatal NMDA receptors (manuscript in preparation), so any involvement of NMDA receptors in MPTP-toxicity is indirect. A possible mechanism is that the energy failure caused by MPTP depolarizes the neurons and relieves the Mg^{2+}-block of the NMDA receptor making it possible for endogenous excitatory amino acids to contribute to the toxicity. However, the role of NMDA receptors in MPTP-toxicity is still very unclear.

3. Acute Effects of MPTP

When discussing the effects of MPTP, its similarity to PD, and the relevance of MPTP-treated mice as a model for PD, it is very important to distinguish between acute and chronic, degenerative effects. MPTP induces a number of acute, biochemical effects. The local administration MPP^+ in rats has been reported to trigger a rapid release of dopamine possibly owing to a direct effect on the dopamine neurons (Rollema *et al.*, 1986). MPTP will indirectly affect most of the monoamine neurons by inhibiting MAO and thereby monoamine metabolism. This combination of the direct and indirect effects is seen in general as acute reductions in tissue levels of dopamine and noradrenaline as well as their deaminated metabolites. Serotonin levels are often increased with a simultaneous decrease in metabolite levels (Hallman *et al.*, 1985; Pileblad *et al.*, 1984; Tadano *et al.*, 1987). MPTP can also affect neurons acutely by interaction with neurotransmitter receptors. For example, MPTP is an antagonist at the α_2-adrenergic receptor (Enz *et al.*, 1984) and may also act as a cholinergic agonist (Hotchkiss *et al.*, 1984).

There is no evidence that dopamine release is involved in the mechanism of MPTP toxicity, and evidently acute effects occur in the MPTP-resistant rat. However, it is possible that some acute effects such as dopamine release reflects the initiated degenerative process. The indirect effects of MPTP will depend upon normal monoamine metabolism, the involvement of different types of MAO, and their respective anatomical distribution. Thus in our view, it is not possible to draw any conclusions about the susceptibility of different strains or species, or their suitability as a model for PD on the basis of acute effects of MPTP, especially not those caused by receptor interactions.

4. MPTP-Induced Degeneration in Mice

As mentioned previously, shortly after the discovery of MPTP it was found that MPTP induces degeneration of nigrostriatal and, to some extent, mesolimbic dopamine neurons in mice. However, the first two reports of this effect differed in that Hallman *et al.* (1984) found loss of striatal dopamine terminals without loss of nigral cell bodies, while Heikkila *et al.* (1984a) showed loss of the entire dopamine neurons. In the former study, noradrenaline terminals also degenerated, whereas these were not included in the latter study. These discrepant findings were subsequently clarified when a number of reports demonstrated that the extent of degeneration depends on the strain of mice used (Sundström *et al.*, 1987) with the nigrostriatal dopamine neurons of C57 BL/6 mice being more susceptible to MPTP than those of the NMRI, Swiss-Webster, and CAB/Ca strains. Moreover, the source of animals of a certain strain (Heikkila, 1985), the age of the animals (Ricaurte *et al.*, 1987), the route of administration, and the dosing interval (Sonsalla and Heikkila, 1986) are also important determinants for the extent of degeneration. To summarize these studies, the most efficient lesioning of dopamine neurons by MPTP is achieved by repeated, subcutaneous injections of MPTP to C57 BL/6 mice with a dose interval of a few hours. The increase of toxicity in aged mice may be an effect of decreased regeneration rather than more severe initial lesion. It was also shown that MPTP-induced degeneration of noradrenergic terminals only occurred in some strains, that the noradrenergic cell bodies in locus ceruleus are spared, and that in contrast to other strains, these neurons in C57 BL/6 mice are completely resistant to MPTP (Sundström *et al.*, 1987). Thus, one cannot rely on results from other laboratories concerning severity of the dopamine lesion, loss of dopaminergic cell bodies, degeneration of nondopaminergic neurons, and so forth, unless the animals are identical.

The most severe lesion by MPTP (2×50 mg/kg s.c., 16 hr interval) reported to date gave depletions of 90% to 95% of striatal dopamine levels with a loss of 70% of dopaminergic neurons in substantia nigra. However, this depletion is not severe enough to induce an up-regulation of dopamine receptors as seen in the 6-OHDA–lesioned rats.

Other methods that may be used to induce more severe dopaminergic degeneration include the use of more potent MPTP analogs or the administration of other substances in association with MPTP, which enhance toxicity. Many different analogs of MPTP have been identified of which 1-methyl-4-(2′-methylphenyl)-1,2,3,6-tetrahydropyridine (2′-CH_3-MPTP) is the most potent one discovered (Youngster *et al.*, 1986). Acetaldehyde, ethanol, and diethyldithiocarbamate apparently potentiate the toxicity of MPTP, the former partly acting by slowing the elimination of MPP^+ from the brain (Zuddas *et al.*, 1989). In addition, tetraphenylboron enhances MPTP-induced toxicity by increasing mitochondrial accumulation of MPP^+ (Heikkila *et al.*, 1990).

5. Local Injections of MPP^+

Intracerebral injections of MPP^+ have been used for two reasons—either to study the mechanism of action of MPTP/MPP^+ or to achieve localized dopamine lesions to study biochemical and behavioral changes. In both situations, nonspecific effects of MPP^+ compromise its use. It has been demonstrated that local MPP^+ injections cause degeneration of noncatecholaminergic cells (Jonsson *et al.*, 1986; Harik *et al.*, 1987) and several studies have failed to prevent neurotoxicity by pretreatment with dopamine uptake blockers (Jonsson *et al.*, 1986; Sun *et al.*, 1988). *In vitro* studies have also shown that MPP^+ is toxic to all cells at concentrations greater than 10 μM (Sanchez-Ramos *et al.*, 1988). Thus, if intracerebral administration is desirable, low concentrations of MPP^+ should be delivered by slow infusion and the specificity of the lesion should be ascertened biochemically or morphologically. Mere studies of dopamine-related behavior after intranigral injections does not exclude the MPP^+-induced lesions of other neurons. In this context, it should be mentioned that Marini and Paul (1992) have shown that the toxicity of reasonably low concentrations of MPP^+ towards cultured cerebellar granule cells could be prevented by preexposure of the cells to glutamate or NMDA as well as inhibitors of RNA and protein synthesis. This mechanism does not seem to be involved in the toxic effect of MPP^+ in cultured dopamine neurons (Finiels-Marlier *et al.*, 1993).

6. Species Differences in MPTP Toxicity

Because most of the events leading to MPTP-induced degeneration are now known, one would expect that the processes determining which species and neuronal systems are vulnerable to damage should be clear. However, this is not the case.

When comparing vulnerability to MPTP between species, the most pronounced difference occurs between highly susceptible species, like primates and dogs, and less susceptible ones like rodents. There seems to be a fundamental difference in the metabolism of MPTP between these species. The half-life of MPP^+ in mouse brain is 2–4 hr (Langston and Irwin, 1986; Sundström, 1988), whereas the half-life in the primate brain is more than 10 times longer (Langston and Irwin, 1986). This explains the inability of dopamine uptake inhibitor pretreatment to protect monkeys from MPTP toxicity, whereas prolonged treatment with an uptake inhibitor after administration of MPTP offers protection (Schultz *et al.*, 1986). Johannessen (1991) has proposed that the long retention of MPP^+ in striatum with deleterious inhibition of axonal transport is unique to highly susceptible species. However, the mechanism causing this retention is not clear. Other mechanisms for the high vulnerability of the primate brain such as binding of MPP^+ to neuromelanin have also been proposed and criticized (see Johannessen, 1991).

Giovanni *et al.* (1991) have shown a high correlation between sensitivity to MPTP and the concentration of MPP^+ in striatum of different mice strains. No difference in the affinity of MPP^+ for the dopamine uptake was found. In preliminary studies we have not found any dramatic differences between striatal concentrations of MPP^+ in rats and mice. At present, however, the data do not allow us to calculate the area under the curve (AUC).

In a series of papers, Harik and collaborators have suggested that the sensitivity to MPTP-induced degeneration in different species as well as strains of rodents correlates with the amount of MAO B in the capillary endothelium, thereby limiting the transport of MPTP into the brain parenchyma (Riachi *et al.*, 1991, and references therein). Even though these correlative studies are convincing, it is difficult to include results of other studies in this model. For example, the large variations in vulnerability to MPTP of different catecholamine neurons in several strains would imply regional differences in the blood–brain barrier. There is no reason per se to choose the sensitivity of nigrostriatal dopamine neurons for correlation with MAO B in whole-brain preparations. Thus, correlations between endothelial MAO B and vulnerability of, for example, noradrenaline neurons, would probably turn out completely different. Furthermore, if MPTP is prevented from entering the brain, it is difficult to explain the similar acute effects on dopamine metabolism in rats and mice (Pileblad *et al.*, 1984, 1985).

It is perhaps futile to search for one factor determining the susceptibility of a given species or strain. Differences between animals might very well be due to combinations of relatively small differences in various factors such as penetration of MPTP into the brain and astrocytes, formation and cellular distribution of MPP^+, uptake into catecholamine neurons and their vesicles, accumulation into mitochondria, and so forth. Still, identification of any process that significantly determines the susceptibility of an animal might lead to a better understanding of the etiology of Parkinson's disease.

7. Studies of Trophic Factors in MPTP-Treated Mice

In recent years, one of the areas of research in which MPTP-lesioned mice have been used most extensively as a model for PD is in studies of potentially neuroprotective substances, growth factors, and transplantation. This experimental situation is less complicated; thus, results can generally be accepted independent of the treatment regimen, strain used, etc. It is not necessary to achieve a maximal lesion of the nigrostriatal neurons to demonstrate a regenerative effect. It may even be advantageous to induce a moderate lesion, as that allows for easier detection of an eventual opposite effect—that is, if the treatment under study would potentiate toxicity. However, one must know whether MPTP causes degeneration of the complete neuron or if only the axon terminals are severed so as to interpret fully the results. Furthermore, if the aim is to demonstrate enhancement of regeneration, one should not initiate treatment earlier than 5–6 days after the last MPTP injection, which is the time the degenerative process takes in some species (Sundström *et al.*, 1988). In some of the studies mentioned below, treatment was initiated shortly after the MPTP-treatment, which limits the ability to differentiate between sparing of degenerating neurons and sprouting of nonlesioned fibers, truly an important difference.

Examples of these types of studies are the regenerative effects of several gangliosides, which have been tested in mice after MPTP lesions. It has been shown that both GM1 (Hadjiconstantinou *et al.*, 1991; Janson *et al.*, 1988; Date *et al.*, 1989; Schneider and Yuwiler, 1989) and GD1a (Date *et al.*, 1991) diminish the effect of an MPTP lesion. Basic fibroblast growth factor administered by implantation of a gel foam was shown to induce reinnervation after MPTP lesion to mice (Otto and Unsicker, 1990), and intracerebroventricular infusion of epidermal growth factor diminished MPTP-induced lesions (Hadjiconstantinou *et al.*, 1991).

Transplantation studies in MPTP- or $2'CH_3$-MPTP–treated mice, unfortunately of varying strain and age, have revealed that grafts of the adrenal medulla, even if nonviable, induce unilateral reinnervation of the lesioned striatum and a partial, bilateral recovery of striatal dopamine levels (Bohn *et al.*, 1988; Bohn and Kanuicki, 1990). The reinnervating fibers apparently were from the host animal. Similar results have been reported by Date *et al.* (1990) with demonstration of a decreased regenerative capacity in old mice.

8. Studies on Learning and Memory

In a clinical study, Hovestadt *et al.* (1988) claimed that patients with PD have an impaired spatial orientation. However, this defect was not ameliorated by L-dopa therapy, indicating involvement of nondopaminergic neuronal systems. Memory for temporal order and learning of conditional association have also been shown to be deficient in PD (Vriezen and Moscovitch, 1990). Thus, studies on involvement of dopaminegic neurons in learning and memory processes are of interest.

By using the Morris swim maze, it has been shown that rats with 6-OHDA lesions of their dopaminergic neurons are impaired both in acquisition of the spatial navigation task as well as retention—that is, the expression of learned behavior (Whishaw and Dunnett, 1985). However, concerning acquisition, the authors stressed that the sensorimotor deficits in these animals make it impossible to differentiate between a learning deficit and a deficit in the use of navigational strategies. At present we do not know which sensory deficits that MPTP-treated mice display. Although relatively mild motor deficits are found in these mice, the different changes in sensorimotor functions could, in the same manner, indirectly affect acquisition.

No studies of cognitive functions utilizing MPTP-treated mice have been published to our knowledge. Two studies have shown deficits in avoidance responding after intrastriatal injections of MPP^+ (Wagner *et al.*, 1986; Wagner and Walsh, 1987). L-dopa reversed the observed effects of MPP^+ in one of the studies, indicating that dopamine depletion caused the deficits. However, the nonspecific effect of local MPP^+ injections makes any interpretations uncertain.

We have performed preliminary studies on learning in MPTP-treated C57

BL/6 mice using a water T-maze. Mice (7 per group) were trained to escape by forcing them to swim to the correct arm during the first trial every day. They were then allowed free choice the following 8 trials. Intertrial interval was 5 min. The procedure was repeated daily until the animals reached a criterion of 7 out of 8 correct responses. They were then trained to choose the opposite arm (i.e., extinction of the previously learned behavior). Saline groups and MPTP-treated mice receiving L-dopa before the sessions learned both tasks on the second day. Three MPTP-treated mice took considerably longer time to learn. Extinction was also considerably delayed in three animals. One mouse performed poorly in both tests. No mouse had any difficulty to swim. However, because only half of the mice showed poor learning, these differences were not significant. Thus, it still remains to demonstrate eventual learning deficits in MPTP-treated mice.

9. Motor Behavior in MPTP-Treated Mice

One of the obvious goals in the use of MPTP-treated mice is the possibility of using these mice in tests of motor activity to obtain a behavioral model of PD. The lack of consistent effects of MPTP in mice is often referred to when dismissing them as a model for PD. Originally it was reported that MPTP-treated mice showed no permanent motor deficits. One report even found mild hyperactivity one week after MPTP treatment (Kordower *et al.*, 1986). This is not surprising since rats only display deficits in motor behavior after severe (> 95%) dopamine depletion with 6-OHDA. Sershen *et al.* (1987), however, reported significant decrease in spontaneous locomotor activity of MPTP-treated BALB/cBy mice placed three per cage and registered during the dark period. This hypoactivity persisted for at least 2 months. Willis and Donnan (1987a,b) observed no changes when studying individual MPTP-treated C57 mice in activity boxes or in a swim tests. Animals were probably studied immediately after being placed in the test box (not described in detail). MPTP-treated mice did, however, show deficits in initiation of step-down response after elevating their hind legs. Yurek *et al.* (1989) found significant MPTP-induced hyperactivity, especially during light hours when the mice were studied individually in their home cages. In all these studies striatal dopamine depletions were about 70% 1 to 2 weeks after i.p. injections of 30 mg MPTP/kg to Swiss-Webster and C57 black mice. In the study by Yurek *et al.* (1989) MPTP was administered with diethyldithiocarbamate.

We have also experienced discrepancies of this type. We reported that C57 BL/6 mice exhibited severe hypoactivity, tested individually and not habituated in activity boxes 4 weeks after MPTP treatment (Sundström *et al.*, 1990). This hypoactivity was dose-dependently reversed by L-dopa with a therapeutic interval between 20–40 mg/kg i.p. (Fredriksson *et al.*, 1990). Subsequently we performed several studies on the effects of dopamine agonists on MPTP-induced hypoactivity. Using subthreshold doses of the D_2 agonist bromocryptine and the D_1-agonist SKF 38393, we found significant synergistic effects on locomotor activity that were most prominent in MPTP-treated mice. In these studies, striatal dopamine levels were decreased by 85% to 90%.

However, when these studies were continued using similar activity boxes but in the laboratory where the animals were treated, MPTP-induced hypoactivity was not seen. The animals were delivered by the same breeder and the depletion of striatal dopamine was as severe as before. A similar pattern of locomotor activity was also found when mice were habituated 24 hr to their individual activity boxes. To minimize variation within test groups, in one experiment 50 mice were pretested for spontaneous motor activity to select two homogenous groups of 10 mice. Still, locomotion in the group treated with MPTP was almost identical to the control group. To exclude possible neophobic reactions or increased exploratory behavior in the test situation, MPTP-treated mice were compared to normal mice during exposure to a novel object. The animals were put in a test box with a red-and-white-striped block, approximately 1.5 × 2 × 2.5 inches, with a drop of cedar oil smeared on it. There were no differences in latency to first interaction (sniffing) with the object, or in the number of interactions (sniffing and touching) during 5 min.

One probable explanation for these discrepancies is that in the latter unsuccessful studies, animals were kept in the same laboratory throughout the studies, whereas in the initial, published experiments, animals were transported after MPTP treatment to the laboratory where the behavioral studies were performed. Although testing was started at least 5 days later, the stress of the transportation and the new environment might have provoked to MPTP-induced hypoactivity. Weihmuller *et al.* (1988) found that MPTP-treated mice without behavioral deficits expressed severe akinesia and catalepsy after acute stress or a low dose of the dopamine antagonist haloperidol. An interesting parallel is that these authors found that the haloperidol-induced deficits remained for 5 months although striatal dopamine levels recovered almost completely. We found that in the test situation where MPTP-induced hypoactivity was present, it remained the same for 12 weeks although striatal dopamine levels recovered from an 85% depletion at 4 weeks to less than 60% depletion at 12 weeks. Thus, it is probable that a nonsymptomatic dopamine depletion limits the ability of the remaining dopamine neurons of the animal to cope with a stressful situation, and that although recovery of dopamine levels occurs, the dopamine system is still severely restrained during stress.

Although measuring locomotor deficits in MPTP-treated mice is not as straightforward as one would wish, it is clear from observation of the animals that they display disturbances in motor behavior. By studying more complicated motor functions, it may be possible to detect and quantify MPTP-induced deficits. Accordingly, Colotla *et al.* (1990) described deficits in a rotorod test and a grip test 5 days after MPTP treatment. Ogawa *et al.* (1985) developed a test in which the animals are placed head up on a vertical pole, then turn around and climb down the pole. MPTP-treated C57 BL/6 mice also performed significantly worse than did control mice, and this deficit was reversed by L-dopa treatment.

10. Conclusions

Although MPTP has been used extensively to produce degeneration of the nigrostriatal dopamine neurons, the use of different mouse strains, MPTP dose regimens, and ways of administration have created some seemingly conflicting results, especially for the extent and distribution of degenerative changes, adaptive responses, and behavioral effects. In contrast, results from studies of the mechanism of MPTP toxicity in mice are on the whole unanimous, and the basic mechanisms of toxicity seem to be similar in mice and primates. Furthermore, a number of studies of MPTP-lesioned mice have given important results on regeneration of dopaminergic neurons following various treatments. However, an important disadvantage is that the majority of studies on neuroanatomy, pharmacology, and so forth have been performed in rats.

As described in this chapter, it should be possible to induce severe and selective lesions of nigrostriatal dopamine neurons by more effective treatment schedules, by administration of drugs enhancing MPTP toxicity, or by using more potent MPTP analogs. A combination of all three approaches might be a way to achieve dopamine depletions that are severe enough for robust behavioral changes to appear. However, given the variability between different mice, it must be emphasized that it is crucial in most studies to define the experimental model by determining the extent of the MPTP-induced lesion. With these considerations in mind, the MPTP-treated mouse is indeed a useful model for Parkinson's disease.

ACKNOWLEDGMENTS. This work was supported by grants from the Swedish MRC (06555), Axel och Margareta Ax:son-Johnsons Foundation, King Gustav V and Queen Victoria Foundation, Sigurd and Elsa Golje Foundation, Gun and Bertil Stohne Foundation, and the research funds of Karolinska Institutet. Drs. C. O'Neill and R. Cowburn are gratefully acknowledged for their helpful discussions.

11. References

Bloem, B.R., Irwin, I., Buruma, O.J.S., Haan, J., Roos, R.A.C., Tetrud, J.W., and Langston, J.W., 1990, The MPTP model: Versatile contributions to the treatment of idiopathic Parkinson's disease, *J. Neurol. Sci.* 97:273–293.

Bohn, M.C., and Kanuicki, M., 1990, Bilateral recovery of striatal dopamine after unilateral adrenal grafting into the striatum of the 1-methyl-4)(2′-methylphenyl)-1,2,3,6-tetrahydropyridine (2′CH_3-MPTP)-treated mouse, *J. Neurosci. Res.* 25:281–286.

Bohn, M.C., Cupit, L., Marciano, F., and Gash, D.M., 1988, Adrenal medullary grafts enhance recovery of striatal dopaminergic fibers, *Science* 237:913–916.

Brouillet, E. and Beal, M.F., 1993, NMDA antagonists partially protect against MPTP induced neurotoxicity in mice, *Neuroreport,* 4:387–390.

Burns, R.A., Chieuh, C.C., Markey, S.P., Elbert, M.H., Jacobowitz, D.M., and Kopin, I.J., 1983, A primate model of parkinsonism: Selective destruction of dopaminergic neurons in the pars compacta of the substantia nigra by N-methyl-4-phenyl-1,2,3,6-tetrahydropyridine, *Proc. Natl. Acad. Sci. U.S.A.* 80:4546–4550.

Campbell, K.J., Takada, M., and Hattori, T., 1990, Evidence for retrograde axonal transport of MPP^+ in the rat, *Neurosci. Lett.* 118:51–154.

Chan, P., DeLanney, L.E., Irwin, I., Langston, J.W., and Di Monte, D., 1991, Rapid ATP loss caused by 1-methyl-4-phenyl-1,2,3,6-tetrahydropyridine in mouse brain, *J. Neurochem.* 57:348–351.

Chiba, K., Trevor, A., and Castagnoli Jr., N., 1984, Metabolism of the neurotoxic tertiary amine, MPTP, by brain monoamine oxidase, *Biochem. Biophys. Res. Commun.* 120:1228–1232.

Chieuh, C.C., Markey, S.P., Burns, R.S., Johannessen, J.N., Pert, A., and Kopin, I.J., 1984, Neurochemical and behavioral effects of systemic and intranigral administration of N-methyl-4-phenyl-1,2,3,6-tetrahydropyridine in the rat, *Eur. J. Pharmacol.* 100:189–194.

Cleeter, M.W.J., Cooper, J.M., and Schapira, A.H.V., 1992, Irreversible inhibition of mitochondrial complex I by 1-methyl-4-phenylpyridinium: Evidence for free radical involvement, *J. Neurochem.* 58:786–789.

Colotla, V.A., Flores, E., Oscos, A., Meneses, A., and Tapia, R., 1990, Effects of MPTP on locomotor activity in mice, *Neurotoxicol. Teratol,* 12:405–407.

Date, I., Felten, S.Y., and Felten, D.L., 1989, Exogenous GM1 gangliosides induce partial recovery of the nigrostriatal dopaminergic system in MPTP-treated young mice but not in aging mice, *Neurosci. Lett.* 106:282–286.

Date, I., Felten, S.Y., Olschowka, J.A., and Felten, D.L., 1990, Limited recovery of striatal dopaminergic fibers by adrenal medullary grafts in MPTP-treated aging mice, *Exp. Neurol.* 107:197–207.

Date, I., Notter, M.F.D., Felten, S.Y., and Felten, D.L., 1991, Stereotaxic injection of GD1a ganglioside induces limited recovery of striatal dopaminergic system in MPTP-treated aging mice, *J. Neurosci. Res.* 28:525–530.

Davis, C.G., Williams, A.C., Markey, S.P., Elbert, M.H., Calne, E.D., Reichert, C.M., and Kopin, I.J., 1979, Chronic parkinsonism secondary to intravenous injection of meperidine analogues, *Psychiatry Res.* 1:249–254.

Di Monte, D.A., Wu, E.Y., Irwin, I., DeLanney, L.E., and Langston, J.W., 1992, Production and disposition of 1-methyl-4-phenylpyridinium in primary cultures of mouse astrocytes, *GLIA* 5:48–55.

Enz, H., Hefti, F., and Frick, W., 1984, Acute administration of 1-methyl-4-phenyl-1,2,3,6-tetrahydropyridine (MPTP) reduces dopamine and serotonin but accelerates norepinephrine metabolism in the rat brain: Effect of chronic pretreatment with MPTP, *Eur. J. Pharmacol.* 101:37–44.

Finiels-Marlier, F., Marini, A.M., Williams, P. and Paul, S.M., 1993, The N-methyl-D-aspartate antagonist MK-801 fails to protect dopaminergic neurons from 1-methyl-4-pyridinium toxicity in vitro, *J. Neurochem.* 60:1968–1971.

Fredriksson, A., Plaznik, A., Sundström, E., Jonsson, G., and Archer, T., 1990, MPTP-induced hypoactivity in mice: Reversal by L-dopa, *Pharmacol. Toxicol.* 67:295–301.

Giovanni, A., Sieber, B.-A., Heikkila, R.E., and Sonsalla, P.K., 1991, Correlation between the neostriatal content of the 1-methyl-4-phenylpyridinium species and dopaminergic neurotoxicity following 1-methyl-4-phenyl-1,2,3,6-tetrahydropyridine administration to several strains of mice, *J. Pharmacol. Exp. Ther.* 257:691–697.

Hadjiconstantinou, M., Fitkin, J.G., Dalia, A., and Neff, N.H., 1991, Epidermal growth factor enhances striatal dopaminergic parameters in the 1-methyl-4-phenyl-1,2,3,6-tetrahydropyridine-treated mouse, *J. Neurochem.* 57:479–482.

Hallman, H., Olson, L., and Jonsson, G., 1984, Neurotoxicity of the meperidine analogue N-methyl-4-phenyl-1,2,3,6-tetrahydropyridine on brain catecholamine neurons in the mouse, *Eur. J. Pharmacol.* 97:133–136.

Hallman, H., Lange, J., Olson, L., Strömberg, I., and Jonsson, G., 1985, Neurochemical and histochemical characterization of neurotoxic effects of 1-methyl-4-phenyl-1,2,3,6-tetrahydropyridine (MPTP) on brain catecholamine neurons in the mouse, *J. Neurochem.* 44:117–127.

Harik, S.I., Schmidley, J.W., Iacofano, L.A., Blue, P., Arora, P.K. and Sayre, L.M., 1987, On the mechanisms underlying 1-methyl-4-phenyl-1,2,3,6-tetrahydropyridine neurotoxicity: the effect of perinigral infusion fo 1-methyl-4-phenyl-1,2,3,6-tetrahydropyridine, its metabolite and their analogs in the rat, *J. Pharm. Exp. Ther.* 241:669–676.

Hasegawa, E., Takeshige, K., Oishi, T., Murai, Y., and Minikami, S., 1990, 1-methyl-4-phenylpyridinium (MPP^+) induces NADH-dependent superoxide formation, and enhances NADH-dependent lipid peroxidation in bovine heart submitrochondrial particles, *Biochem. Biophys. Res. Commun.* 170:1049–1055.

Heikkila, E., 1985, Differential neurotoxicity of 1-methyl-4-phenyl-1,2,3,6-tetrahydropyridine (MPTP) in Swiss-Webster mice from different sources, *Eur. J. Pharmacol.* 117:131–133.

Heikkila, R.E., Hess, A., and Duvoisin, R.C., 1984a, Dopaminergic neurotoxicity of 1-methyl-4-phenyl-1,2,3,6-tetrahydropyridine in mice, *Science* 224:1451–1453.

Heikkila, R.E., Manzino, L., Cabbat, F.S., and Duvoisin, R.C., 1984b, Protection against the dopaminergic neurotoxicity of 1-methyl-4-phenyl-1,2,5,6-tetrahydropyridine by monoamine oxidase inhibitors, *Nature* 311:467–469.

Heikkila, R.E., Hwang, J., Ofori, S., Geller, H.M., and Nicklas, W.J., 1990, Potentiation by the tetraphenylboron anion of the effects of 1-methyl-4-phenyl-1,2,3,6-tetrahydropyridine and its pyridinium metabolite, *J. Neurochem.* 54:743–750.

Hotchkiss, A., Eiden, L., Johannessen, J., Chieuh, C., and Markey, S., 1984, Cholinergic actions of the primate parkinsonism agent *N*-methyl-4-phenyl-1,2,3,6-tetrahydropyridine (NMPTP) on cultured adrenal chromaffin cells. *Fed. Proc.* 43:743.

Hovestadt, A., De Jong, G.J., and Meerwaldt, J.D., 1988, Spatial disorientation in Parkinson's disease: No effect of levodopa substitution therapy, *Neurology* 38:1802–1803.

Janson, A.-M., Agnati, L.F., Fuxe, K., Cintra, A., Sundström, E., Zini, I., Toffano, G., and Goldstein, M., 1988, GM1 ganglioside protects against the 1-methyl-4-phenyl-1,2,3,6-tetrahydropyridine-induced degeneration of nigrostriatal dopamine neurons in the black mouse, *Acta Physiol. Scand.* 132:587–588.

Javitch, J.A., D'Amato, R.J., Strittmatter, S.M., and Snyder, S.H., 1985, Parkinsonism-inducing neurotoxin N-methyl-4-phenyl-1,2,3,6-tetrahydropyridine: Uptake of the metabolite N-methyl-4-phenylpyridine by dopamine neurons explains selective toxicity. *Proc. Natl. Acad. Sci. U.S.A.* 82:2173–2177.

Johannessen, J.N., 1991, A model for chronic neurotoxicity: Long-term retention of the neurotoxin 1-methyl-4-phenylpyridinium (MPP^+) within catecholaminergic neurons, *Neurotoxicology* 12:285–302.

Jonsson, G., Nwanze, E., Luthman, J., and Sundström, E., 1986, Effect of MPTP and its pyridinium metabolites on monoamine uptake and on central catecholamine neurons in mice, *Acta Physiol. Scand.* 128:187–194.

Kordower, J.H., Felten, S.Y., Felten, D.L., and Gash, D.M., 1986, Behavioral sequelae following MPTP administration in mice, *in* "MPTP: A Neurotoxin Producing a Parkinsonian Syndrome" (S.P. Markey, N. Castagnoli, Jr., A.J. Trevor, and I.J. Kopin, eds.), pp. 413–417, Academic Press, Orlando.

Kupsch, A., Löschmann, P.A., Sauer, H., Arnold, G., Renner, P., Pufal, D., Burg, M., Wachtel, H., ten Bruggencate, G., and Oertel, W.H., 1992, Do NMDA receptor antagonists protect against MPTP-toxicity? Biochemical and immunocytochemical analyses in black mice, *Brain Res.* 592:74–83.

Langston, J.W., and Irwin, I., 1986, MPTP: Current concepts and controversies, *Clin. Neuropharmacol.* 9:485–507.

Langston, J.W., Ballard, P., Tetrud, J.W., and Irwin, I., 1983, Chronic parkinsonism in humans due to a product of meperidine-analog synthesis, *Science* 219:979–980.

Langston, J.W., Irwin, I., and Langston, E.B., 1984, Pargyline prevents MPTP-induced parkinsonism in primates, *Science* 225:1480–1482.

Marini, A.M., and Paul, S.M., 1992, N-methyl-D-aspartate receptor-mediated neuroprotection in cerebellar granule cells requires new RNA and protein synthesis, *Proc. Natl. Acad. Sci.* 89:6555–6559.

Markey, S.P., Johannessen, J.N., Chiueh, C.C., Burns, R.S., and Herkenham, M.A., 1984, Intraneuronal generation of a pyridinium metabolite may cause drug-induced parkinsonism, *Nature* 311:464–467.

Martin, F.R., Sanchez-Ramos, J., and Rosenthal, M., 1991, Selective and nonselective effects of 1-methyl-4-phenylpyridinium on oxygen consumption in rat striatal and hippocampal slices, *J. Neurochem.* 57:1340–1346.

Melamed, A., Pikarski, E., Goldberg, A., Rosenthal, J., Uzzan, A., and Conforti, N., 1986, Effect of serotonergic, corticostriatal and kainic acid lesions on the dopaminergic neurotoxicity of 1-methyl-4-phenyl-1,2,5,6-tetrahydropyridine (MPTP) in mice, *Brain Res.* 399:178–180.

Michel, P.P., and Agid, Y., 1992, The glutamate antagonist, MK-801, does not prevent dopaminergic cell death induced by the 1-methyl-4-pyridinium ion (MPP^+) in rat dissociated mesencephalic cultures, *Brain Res.* 597:233–240.

Nicklas, W.J., Vyas, I., and Heikkila, R.E., 1985, Inhibition of NADH-linked oxidation in brain mitochondria by 1-methyl-4-phenylpyridine, a metabolite of the neurotoxin 1-methyl-4-phenyl-1,2,5,6-tetrahydropyridine, *Life Sci.* 36:2503–2508.

Ogawa, N., Mizukawa, K., Hirose, Y., Kajita, S., Ohara, S., and Watanabe, Y., 1985, MPTP-induced parkinsonian model in mice: Biochemistry, pharmacology and behavior, *Eur. Neurol.* 26 (suppl) 1:435–441.

Otto, D., and Unsicker, K., 1990, Basic FGF reverses chemical and morphological deficits in the nigrostriatal system of MPTP-treated mice, *J. Neurosci.* 10:1912–1921.

Perry, T.L., Yong, V.W., Clavier, R.M., Jones, K., Wright, J.M., Foulks, J.G., and Wall, R.A., 1985, Partial protection from the dopaminergic neurotoxin N-methyl-4-phenyl-1,2,3,6-tetrahydropyridine by four different antioxidants in the mouse, *Neurosci. Less.* 60:109–114.

Pileblad, E., and Carlsson, A., 1985, Catecholamine-uptake inhibitors prevent the neurotoxicity of 1-methyl-4-phenyl-1,2,3,6-tetrahydropyridine (MPTP) in mouse brain, *Neuropharmacology* 24:689–692.

Pileblad, E., Nissbrandt, H., and Carlsson, A., 1984, Biochemical and functional evidence for a marked dopamine releasing action of N-methyl-4-phenyl-1,2,3,6-tetrahydropyridine (NMPTP) in mouse brain, *J. Neural Transm.* 60:199–203.

Pileblad, E., Fornstedt, B., Clark, D., and Carlsson, A., 1985, Acute effects of 1-methyl-4-phenyl-1,2,3,6-tetrahydropyridine on dopamine metabolism in mouse and rat striatum, *J. Pharm. Pharmacol.* 37:707–712.

Ramsay, R.R., and Singer, T., 1986, Energy-dependent uptake of N-methyl-4-phenylpyridinium, the neurotoxic metabolite of 1-methyl-4-phenyl-1,2,3,6-tetrahydropyridine, by mitochondria, *J. Biol. Chem.* 261:7885–7587.

Ramsay, R.R., Salach, J.I., and Singer, T.P., 1986, Uptake of neurotoxin 1-methyl-4-phenylpyridine (MPP^+) by mitochondria and its relation to the inhibition of the mitochondrial oxidation of NAD^+ linked substrates by MPP^+ *Biochem. Biophys. Res. Commun.* 134:743–748.

Riachi, N.J., Behmand, A., and Harik, S.I., 1991, Correlation of MPTP neurotoxicity *in vivo* with oxidation of MPTP by the brain and blood–brain barrier *in vitro* in five rat strains, *Brain Res.* 555:19–24.

Ricaurte, G.A., Langston, J.W., DeLanney, L.E., Irwin, I., and Brooks, J.D., 1985, Dopamine uptake blockers protect against the dopamine depleting effect of 1-methyl-4-phenyl-1,2,3,6-tetrahydropyridine (MPTP) in the mouse striatum, *Neurosci. Lett.* 59:259–264.

Ricaurte, G.A., Irwin, I., Forno, L.S., DeLanney, L.E., Langston, E., and Langston, J.W., 1987, Aging and 1-methyl-4-phenyl-1,2,3,6-tetrahydropyridine-induced degeneration of dopaminergic neurons in the substantia nigra, *Brain Res.* 403:43–51.

Rollema, H., Damsma, G., Horn, A.S., De Vries, J.B., and Westerink, B.H., 1986, Brain dialysis in conscious rats reveals an instantaneous massive release of striatal dopamine in response to MPP^+, *Eur. J. Pharmacol.* 126:345–346.

Sanchez-Ramos, J.R., Michel, P., Weiner, W.J., and Hefti, F., 1988, Selective destruction of cultured dopaminergic neurons from fetal rat mesencephalon by 1-methyl-4-phenylpyridinium: Cytochemical and morphological evidence, *J. Neurochem.* 50:1934–1944.

Schneider, J.S., and Yuwiler, A., 1989, GM1 ganglioside treatment promotes recovery of striatal dopamine concentrations in the mouse model of MPTP-induced parkinsonism, *Exp. Neurol.* 105:177–183.

Schultz, W., Scarnati, E., Sundström, E., Tsutsumi, T., and Jonsson, G., 1986. The catecholamine uptake blocker nomifensine protects against MPTP-induced parkinsonism in monkeys, *Exp. Brain Res.* 63:216–220.

Scotchner, K.P., Irwin, I., DeLanney, L.E., Langston, J.W., and Di Monte, D., 1990, Effects of 1-methyl-4-phenyl-1,2,3,6-tetrahydropyridine and 1-methyl-4-phenylpyridinium ion on ATP levels of mouse brain synaptosomes, *J. Neurochem.* 54:1295–1301.

Sershen, H., Hashim, A., and Lajhta, A., 1987, Behavioral and biochemical effects of nicotine in an MPTP-induced mouse model of Parkinson's disease, *Pharmacol. Biochem. Behav.* 28:299–303.

Sonsalla, P.K., and Heikkila, R.E., 1986, The influence of dose and dosing interval on MPTP-induced dopaminergic neurotoxicity in mice, *Eur. J. Pharmacol.* 129:339–345.

Sonsalla, P.K., Nicklas, W.J., and Heikkila, R.E., 1989, Rose for excitatory amino acids in methamphetamine-induced nigrostriatal toxicity, *Science* 243:398–400.

Sonsalla, P.K., Zeevalk, G.D., Manzino, L., Giovanni, A., and Nicklas, W.J., 1992, MK-801 fails to protect against the dopaminergic neuropathology produced by systemic 1-methyl-4-phenyl-1,2,3,6-tetrahydropyridine in mice or intranigral 1-methyl-4-pyridinum in rats, *J. Neurochem.* 58:1979–1982.

Sundström, E., 1988, The parkinsonism-inducing neurotoxin MPTP (1-methyl-4-phenyl-1,2,3,6-tetrahydropyridine), *Thesis,* Karolinska Institutet.

Sundström, E., and Jonsson, G., 1985, Pharmacological interference with the neurotoxic action of 1-methyl-4-phenyl-1,2,3,6-tetrahydropyridine (MPTP) on central catecholamine neurons in the mouse, *Eur. J. Pharmacol.* 110:293–299.

Sundström, E., and Jonsson, G., 1986, Differential time course of protection by monoamine oxidase inhibition and uptake inhibition against MPTP neurotoxicity on central catecholamine neurons in mice, *Eur. J. Pharmacol.* 122:275–278.

Sundström, E., Goldstein, M., and Jonsson, G., 1986, Uptake inhibition protects nigro-striatal dopamine neurons from the neurotoxicity of 1-methyl-4-phenylpyridine (MPP^+) in mice, *Eur. J. Pharmacol.* 131:289–292.

Sundström, E., Strömberg, I., Tsutsumi, T., Olson, L., and Jonsson, G., 1987, Studies on the effect of 1-methyl-4-phenyl-1,2,3,6-tetrahydropyridine (MPTP) on central catecholamine neurons in C57 BL/6 mice: Comparison with three other strains of mice, *Brain Res.* 405:26–38.

Sundström, E. Luthman, J., Goldstein, M., and Jonsson, G., 1988, Time course of MPTP-induced degeneration of the nigrostriatal dopamine system in C57 BL/6 mice, *Brain Res. Bull.* 21:257–263.

Sundström, E., Fredriksson, A., and Archer, T., 1990, Chronic neurochemical and behavioral changes in MPTP-lesioned C57BL/6 mice: A model for Parkinson's disease, *Brain Res.* 528:181–188.

Tabatabaei, A., Perry, T.L., Hansen, S., and Krieger, C., 1992, Partial protective effect of MK-801 on MPTP-induced reduction of striatal dopamine in rats, *Neurosci. Lett.* 141:192–194.

Tadano, T., Satoh, N., Sakuma, I., Matsumura, T., Kisara, K., Arai, Y., and Kinemuchi, H., 1987, Behavioral and biochemical changes following acute administration of MPTP and MPP^+, *Life Sci.* 40:1309–1318.

Turski, L., Bressler, K., Rettig, K.-J., Löschmann, P.-A., and Wachtel, H., 1991, Protection of substantia nigra from MPP^+ neurotoxicity by *N*-methyl-D-aspartate antagonists, *Nature* 349:414–418.

Ungerstedt, U., 1971, Postsynaptic supersensitivity after 6-hydroxydopamine induced degeneration of the nigro-striatal dopamine system, *Acta Physiol. Scand. Suppl.* 367:69–93.

Vriezen, E.R., and Moscovitch, M., 1990, Memory for temporal order and conditional associative-learning in patients with Parkinson's disease, *Neuropsychologia,* 28:1283–1293.

Wagner, G.C., and Walsh, S.L., 1987, Increased sensitivity of mice to tremorogenic agents following MPP^+, *Psychopharmacology* 92:470–472.

Wagner, G.C., Jarvis, M.F., and Rubin, J.G., 1986, L-dopa reverses the effects of MPP^+ toxicity, *Psychopharmacology.* 88:401–402.

Weihmuller, F.B., Hadjiconstantinou, M., and Bruno, J.P., 1988, Acute stress or neuroleptics elicit sensorimotor deficits in MPTP-treated mice, *Neurosci. Lett.* 85:137–142.

Whishaw, I.Q., and Dunnett, S.B., 1985, Dopamine depletion, stimulation or blockade in the rat disrupts spatial navigation and locomotion dependent upon beacon or distal cues, *Behav. Brain Res.* 18:11–29.

Willis, G.L., and Donnan, G.A., 19987a, Histochemical, biochemical and behavioural consequences of MPTP treatment in C-57 black mice, *Brain Res.* 402:269–274.

Willis, G.L., and Donnan, G.A., 1987b, Motor function in the 1-methyl-4-phenyl-1,2,3,6-tetrahydropyridine-treated mouse, *J. Neurol. Sci.* 77:185–191.

Youngster, S.K., Duvoisin, R.C., Hess, A., Sonsalla, P.K., Kindt, M.V., and Heikkila, R.E., 1986, 1-methyl-4-(2′methylphenyl)-1,2,3,6-tetrahydropyridine (2′-CH_3-MPTP) is a more potent dopaminergic neurotoxin than MPTP in mice, *Eur. J. Pharmacol.* 122:283–287.

Yurek, D.M., Deutch, A.Y., Roth, R.H., and Sladek, J.R. Jr., 1989, Morphological, neurochemical, and behavioral characterizations associated with the combined treatment of diethyldithiocarbamate and 1-methyl-4-phenyl-1,2,3,6-tetrahydropyridine in mice, *Brain Res.* 497:250–259.

Zuddas, A., Corsini, G.U., Schinelli, S., Barker, J.L., Kopin, I.J., and di Porzio, U., 1989, Acetaldehyde directly enhances MPP^+ neurotoxicity and delays its elimination from the striatum, *Brain Res.* 501:11–22.

Zuddas, A., Oberto, G., Vaglini, F., Fascetti, F., Fornai, F. and Corsini, G.U., 1992, MK-801 prevents 1-methyl-4-phenyl-1,2,3,6-tetrahydropyridine-induced parkinsonism in primates, *J. Neurochem.* 59:733–739.

6

Behavioral Effects of MPTP Administration in the Vervet Monkey

A Primate Model of Parkinson's Disease

J.R. TAYLOR, J.D. ELSWORTH, R.H. ROTH, J.R. SLADEK, Jr., and D.E. REDMOND, Jr.

The neurotoxin 1-methyl-4-phenyl-1,2,3,6-tetrahydropyridine (MPTP) produces parkinsonian neurochemical and functional deficits in primates, which may be useful for evaluating the effects of possible antiparkinsonian treatments. Sixty-six adult male vervet monkeys (*Cercopithecus aethiops sabaeus*) were treated with MPTP (0.3–0.4 mg/kg., given 4–5 doses over a 5-day period, cumulative dose 1.6–2.0 mg), and 11 were saline-treated controls. Trained observers recorded spontaneous behaviors twice daily (5 min per observation, 5 × week) for varying periods after MPTP administration. Some behaviors were recorded if they occurred any time during a 5-sec period (e.g., climb, shift, chew) or lasted for a full 5-sec duration (e.g., freeze, immobility). Behaviors were also rated on a scale of 0–5 (e.g., poverty of movement, limb and head tremor) during spontaneous behavior and after "challenges" with food or threats. Individual behaviors were analyzed, and summary factors were calculated reflecting parkinsonian, intention tremor, anxiety, arousal, and healthy behaviors. MPTP administration resulted in behavior not observed in control subjects (i.e., tremor, freezing, immobility, eating problems, delayed and poverty of movement) and reduced healthy, arousal, and anxiety-related behavior. The parkinsonian score, based on factor analysis from the month after treatment, was used to determine and cate-

J.R. TAYLOR, J.D. ELSWORTH, R.H. ROTH, and D.E. REDMOND, Jr. • Neurobehavior Laboratory and Departments of Psychiatry and Pharmacology, Yale University School of Medicine, 333 Cedar Street, New Haven, Connecticut 06510. **J.R. SLADEK, Jr.** • Department of Neuroscience, University of Health Sciences, The Chicago Medical School, 3333 Green Bay Road, North Chicago, Illinois 60064.

Toxin-Induced Models of Neurological Disorders, edited by Michael L. Woodruff and Arthur J. Nonneman. Plenum Press, New York, 1994.

gorize the severity of MPTP deficits and was hypothesized to predict the outcome or stability of the deficit over time. We found that similar doses of MPTP and schedules of administration led to behavioral deficits ranging from extremely mild to severe. The stability of the deficit was related to severity of parkinsonism 1 month after MPTP treatment, with mildly parkinsonian subjects showing recovery but the most severely affected remaining impaired. These long-term outcomes were paralleled by the dopamine concentrations in the striatum, with improvement of dopamine concentrations occurring in the least but not the most severely affected subjects. Similar medical problems resulted from MPTP administration as are seen in Parkinson's disease (PD). MPTP treatment in the African green monkey, using these methods, appears to model PD phenomenologically and biochemically during the period studied.

1. Usefulness of the MPTP Model of Parkinson's Disease

Several years ago a group of drug addicts developed the classical signs of PD following use of a "designer drug" that was contaminated with 1-methyl-4-phenyl-1,2,3,6-tetrahydropyridine (MPTP) (Davis *et al.*, 1979). The similarities between MPTP-induced parkinsonism and PD have been reported to include virtually all of the classical behavioral, motoric, cognitive, biochemical, and histologic changes that occur in the disease (Burns *et al.*, 1983, 1985; Langston *et al.*, 1983, 1984; Mitchel *et al.*, 1986; Bankiewicz *et al.*, 1986; Chiueh *et al.*, 1984, 1985; Ballard *et al.*, 1985; Stern and Langston, 1985; Doudet *et al.*, 1985, 1986; Schneider *et al.*, 1988; Schultz *et al.*, 1985; Taylor *et al.*, 1990a,b). Biochemical and histological data have shown that MPTP administration leads to a lesion of the dopamine (DA) nigrostriatal pathway, as occurs in PD. Furthermore, both in the MPTP-treated monkey and in PD there is damage to extrastriatal DA and norepinephrine systems (Elsworth *et al.*, 1990b).

The mechanism of toxicity appears to involve the oxidation of MPTP in the brain by monoamine oxidase type B to $MPDP^+$, which is subsequently converted to the toxic metabolite MPP^+. As a substrate for DA uptake, MPP^+ is accumulated within DA neurons, where it is thought to act as an inhibitor of NAD(H)-linked electron transport at the level of Complex 1, which is lethal to the cell (Gerlach *et al.*, 1991).

1.1. Limitations of the Model

Despite the similarities between MPTP-induced parkinsonism and PD, some disparities have been suggested. First, essential elements of PD may not be reproduced by MPTP administration. Second, "spontaneous" recovery of function sometimes reported is not consistent with the usual progressive deterioration of the disease (Eidelberg *et al.*, 1986; Nomoto *et al.*, 1986; Schneider *et al.*, 1986, 1987). Finally, there may be variation in the syndromes that occur after

MPTP administration in different species of monkeys (Eidelberg *et al.*, 1986; Di Paolo *et al.*, 1986; Jenner and Marsden, 1986; Elsworth *et al.*, 1987a, 1989). No study, however, has assessed the full range of variability, course, and duration of the MPTP syndrome in a monkey species that shows most of the characteristics of Parkinson's disease.

We have used the MPTP-treated African green monkey to study the efficacy of experimental treatments for PD (Collier *et al.*, 1987; Elsworth *et al.*, 1987b, 1990a; Lawrence *et al.*, 1991; Redmond *et al.*, 1986, 1988; Sladek *et al.*, 1986, 1987a,b, 1988; Taylor *et al.*, 1990a, 1991a,b). Over a 3-year period 66 monkeys were given MPTP intramuscularly, and 11 randomly selected controls were studied in parallel after receiving saline injections. Behavior of the sample was examined before treatment with MPTP and after the course of treatment, before some of the monkeys were included in experiments examining the functional effects of fetal tissue grafts and novel pharmacological treatments on the signs of MPTP-induced parkinsonism (Taylor *et al.*, 1991a,b). Behavioral studies were blinded as to the nature of the treatments, and the study group reported here was limited to animals only up to the point at which other treatments, such as transplantation, were initiated. The resulting behavioral, neurochemical, and morphological data may help to understand the similarities and limitations of MPTP-induced toxicity in monkeys as a model for PD.

1.2. Objectives for Assessment of the Behavioral Effects of MPTP

Another of our aims has been to refine assessment methods for examining the effects of MPTP. Our objectives were to (1) examine the duration of the observable behavioral effects, (2) determine the severity of the impairments, (3) document the range of symptoms, (4) assess the medical consequences of the MPTP syndrome, and (5) determine the biochemical correlates of MPTP treatment, by assessment of DA function in subregions of the basal ganglia. We hypothesized that the initial level of severity of parkinsonism would predict the occurrence of spontaneous recovery, and that the entire range of parkinsonian signs, which appears over years in a typical patient with PD, would be seen within weeks of MPTP administration in the monkey. These data will provide the basis for a critical evaluation of the behavioral and biochemical similarities between MPTP-induced parkinsonism and idiopathic PD.

2. Methods for Assessment of the Behavioral Effects of MPTP in the Monkey

2.1. Subjects and Drug Administration

Adult male monkeys (*Cercopithecus aethiops sabaeus*) from St. Kitts, West Indies, were injected intramuscularly with 0.3 to 0.4 mg/kg MPTP, four or five times over a 5-day period (cumulative doses of 1.6–2.0 mg/kg, N = 66). Control subjects were treated similarly with saline injections (N = 11). Data were collected over a period of 3 years during which subjects were examined for periods up to 19 months after MPTP or control treatment. Animals were housed individually in standard primate cages in natural daylight. Access to food and water was unlimited. Care and treatment of these monkeys were in strict compliance with the U.S. Public Health Service *Guide for the Care and Use of Animals* (1985), and the protocols were approved by the relevant institutional review committees. Antiparkinsonian medications were not administered so as to avoid confounding our experimental results. Variable numbers of doses were given in an attempt to prevent "overdosing" of animals, which began to show serious signs of drug toxicity during the treatment period. Dose differences were analyzed to determine whether variable outcomes could be accounted for by total doses administered.

2.2. Assessment of Medical Consequences and Procedures for Care of MPTP-Treated Monkeys

All monkeys were observed several times each day for general level of function and motor coordination. Medical consequences of severe parkinsonism were evaluated and treated to minimize the effect of stress, discomfort, or secondary disease or disability on function. Twice daily temperatures and respiratory rates and daily food and fluid intakes were recorded. All monkeys that were sufficiently debilitated to require handling for feeding received physiotherapy consisting of regular passive range-of-motion exercises.

Any monkey failing to take established minimums of fluid or food was assisted as necessary to ensure adequate nutrition and hydration. Feeding assistance included providing a special diet of water-softened commercial monkey chow (Purina®) and fruit by spoon, large syringe, feeding bag, or nasogastric tube. Cimetidine was administered twice daily with tube-feeding solutions if gastric retention was detected at the next scheduled feeding.

Extremely hypokinetic monkeys were maintained on special padded bedding to minimize contact with cage surfaces. Rice-cloth food sacks filled with Styrofoam "peanuts" were ideal for bedding, since they are durable and provide a dry surface and a broad weight distribution for skin contact. Clean beds were provided twice daily. Other medical complications were treated with standard methods.

2.3. Behavioral Analyses

Observers, unaware of the MPTP treatment, cared for and behaviorally assessed the monkeys. These trained observers achieved a coefficient of concordance (Kendall's) greater than 0.95 in their ratings of all behaviors before beginning actual data collection. Interrater reliability was then evaluated at least once a week. Any differences in rating were discussed regularly by the raters to eliminate observer "drift" or idiosyncracy.

Behavior was examined twice daily, for 5 min each time, 5 days each week during the experiment. The first animal to be examined was rotated in the observation sequence such that each animal's observations were evenly spread throughout the observation sessions between 9 and 12 a.m. The observer sat in plain view a few feet in front of an animal's home cage. Subjects were observed before treatment for approximately 1 month (BASE) and for 1-month time periods after MPTP or saline treatment (i.e., MO1, 2, 3, 4, etc.), for as long as 19 months.

2.3.1. Categories of Behavior Examined to Determine the Range of Symptoms

Observers "scored" and "rated" the behavior and motor movements of each monkey individually, as defined below. All behaviors scored had been identified with a reliability > 0.95 in previous experiments and were selected from an ethogram of 200 behaviors representing nearly 100% of the previously reported spontaneous behaviors of the species. Spontaneous behaviors were scored as the event occurred during each 5-sec interval during the 5-min observation, providing a good approximation of the amount of time during which that behavior was exhibited by the monkey. Other behaviors were rated on a scale of 0–5 at the end of the observation period.

2.3.1.a. Scored Behavior These events, defined in Table I, reflected two types of behavior. Common natural behavior exhibited by monkeys of this species, age, and sex, and abnormal behaviors that were anticipated or observed in pilot studies were quantified by their presence or absence in each 5-sec period (so-called one-zero scoring; Altman, 1974) or were scored when they occurred for at least a 5-sec duration (maximum score of 60 for a session). The 5-sec intervals and 5-min periods were timed by the playing of an audiocassette recorder with a timing tape. Responses were counted on a categorical checksheet.

2.3.1.b. Rated Behavior These behaviors were quantified on a scale of 0 to 5, with 0 being "normal" and 5 being severely parkinsonian. Although these behaviors are more difficult to assess reliably, they provide data on abnormal function that cannot be easily counted per unit of time. Rated behaviors are detailed in Table I.

All the scored spontaneous behaviors were previously determined (see Baulu and Redmond, 1978) to be either frequent or stable within the observation time, and were therefore adequately sampled by the observation frequencies utilized.

Additional behaviors were examined that were elicited by "challenges" to

Table I. Definitions of Scored and Rated Behaviors for the Quantitative Behavioral Assessment

A. Scored Behavior	
(one-zero scored)	
Bipedal lookout	Standing on legs with hands off floor, gazing outside cage
Cage pick	Manual exploration/manipulation of cage region
Chew/bruxism	Alternative mandibular-maxillary apposition without food/objects
Masturbation	Rubbing/manipulation of genital area (not self-grooming)
Penile erection	Protrusion/raising of penis
Scratch	Rapid and repeated rubbing of body with fingers/hand/foot
Self-groom	Gentle manipulation of hair/skin with hands/mouth
Shift	Pacing/walking about the cage on all four feet with at least two steps
Tail flag	Tail sticking out perpendicular to subject while standing/walking
Threaten outside	Prolonged stare (with facial expression) at observer or monkeys
Vertical climb	Locomotion on sides/top of cage without feet touching cage floor
Vocalization	Pharyngeal/laryngeal sounds
Yawn	Mouth open so that vertical separation of lips > width of mouth
(scored per 5-sec duration)	
Drinking	Contact between mouth and drinking spout
Eating	Gathering/manipulating, putting in mouth, chewing/swallowing food
Freeze	Remaining motionless but not asleep, prior movement interrupted
Eyes closed	"Freeze" with eyelids closed/motionless in sleep posture
Face down	Lying with ventral/lateral surface (trunk/abdomen) on floor/perch
B. Rated Behavior	(Rated 0–5, and at the end of observation period after presentation of food and vigorous postural and vocal threat from humans; 0 = normal)
Food response	Speed that subject reaches for, handles, and eats presented fruit
Delayed movement	Degree that movement appears delayed but eventually carried out
Difficulty eating	Physical problem in handling/biting/chewing/swallowing food/liquid
Appearance	General condition of grooming and demeanor
Poverty of movement	Slowness, decreased complexity, and small quantity of movement
Threat response	Motor/facial/vocal responses to human postural and vocal threats
Head tremor	Oscillating movement of head at rest/voluntary movement
Limb tremor	Oscillating movement of limbs at rest/voluntary movement
Effect of "intention"	Tremor increases (+)/decreases (−) with "voluntary" or purposeful movement; i.e., resting tremor
Spontaneous freeze	Interruption of any type of ongoing movement pattern (>5 sec)

Note: Behaviors that were assessed after challenges to the subjects are also shown.

the subject by the observer. The presentation of food was used to observe the speed and quality of the subject's response, the effect of voluntary or intentional movement on tremor, and any difficulties in eating, chewing, or swallowing the food. Subjects were also challenged by threat gestures from a human to elicit fear or aggressive responses and to observe the effect on tremor, motor speed, or spontaneous freezing. These behaviors are also shown in Table I. Each one of the rated and scored behaviors was analyzed individually. It should be noted that our previous name for a natural behavior, "freeze," describes merely the absence of movement for 5 sec. It is seen in normal monkeys under various conditions and after a number of drugs. Clinical neurologists use the term "freezing" to describe the behavior we label as "spontaneous freeze," which requires the interruption of an ongoing movement.

2.3.2. *Severity and Pattern of Behavior Modified by MPTP*

We hoped to find a few factors or summary scores that might provide a more accurate measure of disability than individual behaviors or ratings. Because some of the categories of behavior can "compete" with or displace other behaviors, a false impression of the status of the subject might be given if the individual behaviors are only examined separately. For example, severely impaired animals that cannot initiate voluntary movement would necessarily show less "delayed initiation of movement." However, this should not be considered an improvement since one behavior may be competing with the expression of, or displacing, the other behavior. A summary/factor score would add other items to this now missing one, which together would more adequately reflect the increased severity. Other individual items might reflect only a part of the disability, such as tremor. Many treated monkeys show tremor in the early phases of their disability, which diminishes as they are progressively immobile—certainly not reflecting improvement in severity even though their tremor scores might return to 0.

Previously, we employed a *parkinsonian summary score,* based on pilot data, to quantitate the severity of each animal's response to MPTP. We have now performed a principal component factor analysis (utilizing varimax rotation and correlation coefficients) on the data for all subjects in the present study during the month after treatment to develop a reliable and objective measure of parkinsonism and to evaluate empirically the statistical basis of our previous parkinsonian summary score.

2.3.3. *Persistence and Duration of Behavioral Effects of MPTP*

The new parkinsonian summary score was used to classify subjects into five severity groups, based on the ranking of their average parkinsonian factor score at 1 month after treatment. This severity class then was used to examine the effects of MPTP over time in subjects as a group and individually. Both the length of time the subjects survived and the degree of MPTP impairment were examined in terms of this initial severity of the deficit. The cumulative dose of MPTP given to each subject also was examined to determine whether there was a relationship with the initial severity of the impairments. We hypothesized that the initial severity of the MPTP deficit would predict parkinsonism and its permanency, and the more mild the deficit the more likely the subject would be to show some recovery. The basis for this hypothesis was the observation that various compensatory changes are well known to occur after experimental destruction of dopamine (DA) systems, such as increases in turnover rates and postsynaptic sensitivity. Furthermore, neurotoxic effects in some DA neurons might not be lethal, while impairing function for a period. However, if DA neuronal death occurred and all compensatory mechanisms were exhausted, the resulting deficit was felt to mimic the symptomatic parkinsonian patient most accurately and to be unlikely to be spontaneously remitting.

2.4. Biochemical Analyses

Sacrifice, microdissection of the brain, and analysis of tissue for DA content were performed as described by Elsworth *et al.* (1989). Data presented are from tissue micropunched from the dorsolateral caudate nucleus. The sample was removed from the "mid" level (Elsworth *et al.*, 1989), defined by the crossing of the anterior commissure on the caudal face of a 5-mm coronal brain slice.

2.5. Statistical Analyses

Data were entered into a computer at the observation site by the observers and subsequently analyzed using SAS (Statistical Analysis System). Statistical evaluations were made using analyses of variance (Winer, 1974), factor analysis, and linear regression. Analysis of variance was used to determine significant effects of all behaviors during the treatment time periods (BASE, MO1, 2, 3, 4, etc.) and for MPTP and control group subjects separately. Significant interactions were decomposed using Duncan's Multiple Range test (e.g., group × treatment time).

3. Results of Behavioral Analyses

3.1. Initial Time-Course Effects of MPTP Administration

After the first one or two treatments (i.e., daily injections) little effect was seen. Monkeys were slower and appeared generally less responsive than before the treatment. A few showed slight resting tremors. After several injections, some monkeys exhibited hyperarousal, jumping, running in their cages, and hyper-oral behavior such as biting the cage bars or sides (similar to syndromes produced by increased DA via several drug mechanisms). A few animals died shortly after the second or later treatments. Responses to early injections did not necessarily seem predictive of the long-term effects. All MPTP-treated animals decreased eating and drinking after five treatments. Thereafter, some were progressively unable to feed themselves properly. The deficits ranged from manual difficulty of handling or chewing usual foods to uncoordinated efforts to swallow food or liquid. This was dealt with by encouragement and hand feeding or, if necessary, by tube feeding.* Within 3 weeks of the final treatment, the entire usual course of PD was observed: starting with bradykinesia, tremor, and then progressive incoordination. However, animals often were observed to become worse owing to medical complications after prolonged periods of debilitation (see Section 5).

3.2. Validity of Rating Methods and Individual Behavior Categories Examined after MPTP

We examined whether differences existed between observers, despite the rigorous training of observers and the weekly interrater reliability sessions. No significant differences were found, among 20 observers who contributed data to the study, with observer as the dependent variable and each behavior examined during the observation session as the independent variable.

Because we had previously examined behavior during four time periods each day, we examined these earlier studies to determine whether there were any significant differences between two and four daily observation sessions or effects of the time of day of the four observation periods. An analysis of variance was performed on a random selection of observation sessions from previous studies. No differences between the four sessions for any of the behavioral variables were found, with the exception of the variable "chew" ($F(1, 3) = 3.90, p < .01$). This "chew" difference occurred between the first session (just after morning "feeding" and the most disturbance by caretakers, and the last daily session when monkeys eat the least and caretakers have left). Here, we used the second daily session, which showed no differences for any other of the behavioral variables examined.

The mean scored and rated categories of each behavior observed during baseline and the month after treatment for MPTP-treated (N = 66) and control-treated (N = 11) subjects are shown in Figure 1. Generally, no significant differences existed between subjects during the baseline observation sessions. Baseline differences appeared in a few "normal" behaviors (e.g., cagepick, chew,

*It should be noted that ketamine cannot be used for tube feeding as it appeared to be more toxic in combination with MPTP and is known to diminish appetite for a few hours after administration.

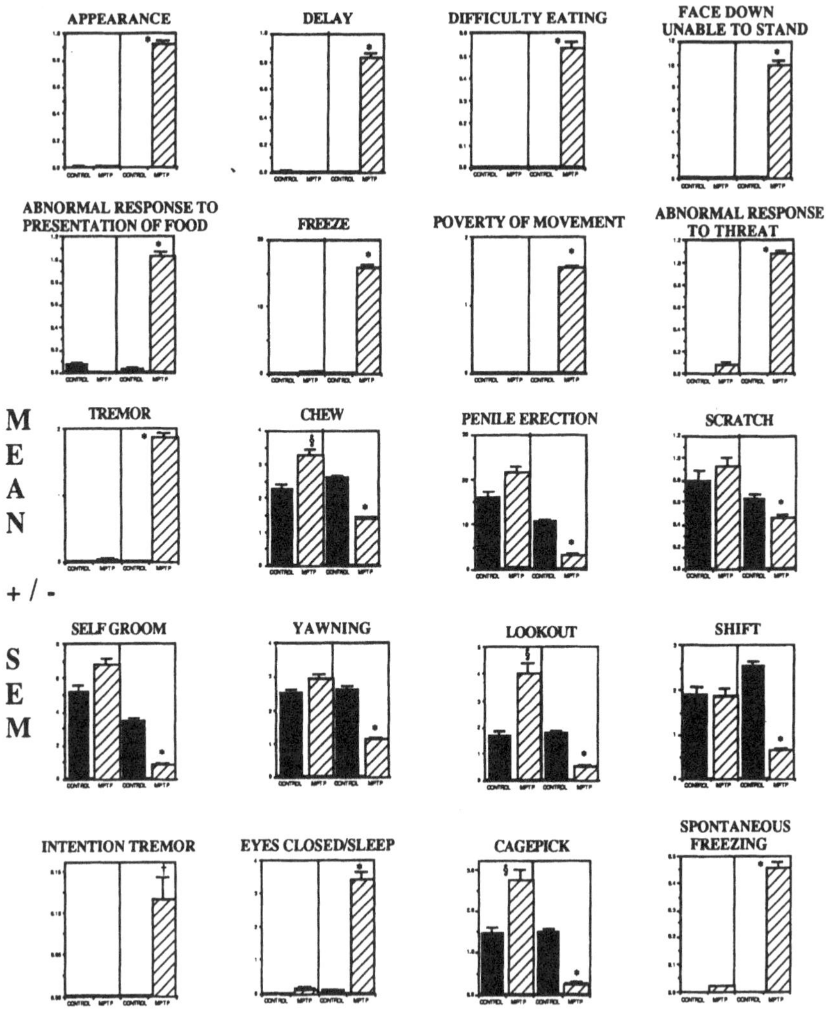

Figure 1 The mean score (± standard error, SEM) is shown for each scored and rated behavior during baseline and the month after treatment for subjects in the MPTP-treated (striped) group (N = 66) and the control-treated (solid) group (N-11). Significant interactions between group (MPTP or Control) and treatment time (baseline or month 1) for the behaviors were as follows (reading from left to right, $F(1,75)$ = 1.87*, 12.00*, 6.05†, 6.07†, 7.51*, 22.90*, 15.96*, 9.36*, 21.08*, 11.08*, 4.99*, 7.43*, 4.32†, 5.07†, 10.10*, 4.10†, 3.97†, 6.59†, 6.31†, 10.14, *$p<.01$ or †$p<.05$). Only significant main effects of group were observed for masturbation, $F(1,75) = 6.79$, $p<.05$. No significant interactions or main effects of group were observed for drink, tail flag, vocalize, eating, vertical climb, and threaten outside (not shown). Significant differences between groups the month after treatment are shown (*$p < .01$ and †$p<.05$) and during baseline (§$p < .05$).

lookout). Abnormal behaviors were increased after MPTP, while other commonly observed normal behaviors were reduced (i.e., normal healthy behaviors).

MPTP administration resulted in significant increases in several categories of behavior, such as difficulties in eating, facedown (or inability to stand), intentional tremor, freeze, head and limb tremor, abnormal response to threat, eyes closed, poverty of movement, spontaneous freeze, delayed initiation of movement, poor appearance, and abnormal response to presentation of food. Significant decreases in normal healthy behaviors were also observed, such as shift, penile erection, drink, tail flagging, lookout, cagepick, yawning, self-groom, chew, eating, and scratch. Shift, for example, is a measure of locomotion, and therefore gives an indication of gross motor activity. Shift was reduced by about 60% the month after MPTP treatment. In some cases, control subjects also showed some changes, often increases, in these behaviors (see Figure 1).

3.3. Identification of Summary Factors

The principal component factor analysis revealed five components with large eigenvalues (all greater than 1) that accounted for approximately 50% of the standardized variance. The remaining factors, or components, had eigenvalues of 1 or less and accounted for 4% or less of the variance each. The five factors with the highest loadings are shown in Table II. These factors were used to analyze the data based on summary scores derived from this analysis.

The factor analysis was in agreement with previous summary factors (Redmond *et al.*, 1986) that were based on small numbers of MPTP-treated monkeys, with minor exceptions. Factor pattern 1 revealed behaviors that were increased by MPTP treatment with high positive correlations (0.6–0.9). This factor had the highest eigenvalue (6.2) and accounted for 23% of the variance. The final communality estimate also showed these behaviors were well accounted for by the components, with estimates averaging 0.7. These behaviors were thus identified as the parkinsonian summary score or factor (i.e., difficulty eating, facedown, freeze, head and limb tremor, abnormal response to threat, poverty of movement, delayed initiation of movement, poor appearance, abnormal response to presentation of food). This factor was largely identical to our previous parkinsonian summary score except "eyes closed" was replaced by "poor appearance" because eyes closed had a very small loading (0.01) and appearance had a high loading (0.85). Although the behaviors were summed to make up the parkinsonian score, freezing was given a diminished weight (divided by 5 as done previously) since it is a part of the subject's normal behavioral repertoire and consequently should not be given equal weight. It should be noted that this freezing behavior is defined as remaining motionless for at least 5 sec and occurs to some extent in normal monkeys. This is in contrast to "spontaneous freezing," which is a clearly pathological behavior and seldom, if ever, is seen in normal monkeys but was not identified as part of the factor.

Table II. Behaviors Identified from the Factor Analysis That Were Used to Determine Summary Scores for Parkinsonian, Anxiety, Arousal, Intention Tremor, and Quiet OK Behavior Scores

FACTOR 1 NAME Behaviors-Loading		FACTOR 2		FACTOR 3		FACTOR 4		FACTOR 5	
PARKINSONIAN SCORE		ANXIETY		AROUSAL		INTENTIONAL TREMOR		QUIET OK	
Appearance	0.85	Chew/bruxism	0.75	Bipedal lookout	0.53	Intention tremor	0.81	Cage pick	0.38
Delayed movement	0.72	Penile erection	0.51	Shift	0.72	Delayed movement	0.43	Self-groom	0.34
Difficulty eating	0.61	Scratch	0.22	Tail flag	0.37	Difficulty eating	0.66	Drinking	0.22
Facedown	0.81	Self-groom	0.39	Threaten outside	0.34			Eating	0.41
Food response	0.91	Yawn	0.53	Vertical climb	0.25				
Freeze	0.57			Vocalization	0.42				
Poverty of movement	0.83								
Threat response	0.84								
Tremor (head and limb)	0.66								
Variance accounted	23%		7%		6%		6%		5%

Note: Loadings for each behavior and % variance accounted for are shown.

The behaviors with negative loadings in factor 1 were the same behaviors identified to have high positive loadings for factors 2, 3, and 5. These behaviors were combined to form the healthy behavior summary score. Individually these components were termed *anxiety, arousal,* and *quiet OK* behavior because they are found to be displayed under normal conditions of anxiety, arousal, or quiet normal activity, respectively. These components also corresponded to previous factors used in earlier studies using large numbers of monkeys. Factor 4 had not been previously identified. This factor had high positive loadings for behavior associated with goal-directed movement (tremor with intentional movement, eating problems, and delayed movement). Because this component had an especially high correlation with tremor that increased with intentional movement (0.8), it was termed the *intention tremor summary score.*

3.3.1. Mean Summary Scores before and after MPTP or Control Treatment

Mean parkinsonian summary scores based on the factor analysis of all data from the first month for MPTP-treated and control-treated subjects during baseline and the first month after treatment are shown in Figure 2. The groups differed in terms of parkinsonian summary score over time (group = MPTP vs control × treatment time interaction = baseline vs MO1, $F(1, 75) = 11.11, p < .01$). Mean parkinsonian summary scores were significantly increased compared with baseline after MPTP treatment ($F(1, 75) = 66.74, p < .01$), and compared with control subjects ($F(1, 75) = 12.83, p < .01$). The posttreatment mean was 20 the month after MPTP compared with means of less than 1 for sessions before MPTP treatment and for control subjects both before and after sham treatment. The incidence of mild movement deficits (intention tremor summary score) was also significant (group × treatment time, $F(1, 75) = 8.49, p < .01$). This factor was markedly increased after MPTP treatment compared with control injections ($F(1, 75) = 8.32, p < .01$) and was increased compared with these subjects' baseline values ($F(1, 75) = 42.66, p < .01$), shown also in Figure 2. No significant differences existed between the groups during baseline for parkinsonian or intention tremor summary scores, and there were no differences during baseline compared with the month after sham treatment for control objects.

Healthy behavior scores of the groups overall were different ($F(1, 75) = 19.16, p < .01$). For MPTP-treated subjects, healthy behaviors were significantly reduced compared with controls ($F(1, 75) = 116.40, p < .01$) and compared with their pretreatment values ($F(1, 75) = 114.08, p < .01$), shown in Figure 3. Incidence of these behaviors was decreased on average by 75%. The arousal, anxiety, and quiet OK factors all showed a similar pattern to the changes found for the healthy behavior summary score, to which they contributed (Figure 3). Significant interactions occurred between group and treatment for each of these summary factors ($F(1, 75) = 9.99, 8.76, 6.62, p < .01$, respectively). These differences were found to be due to significant decreases for arousal,

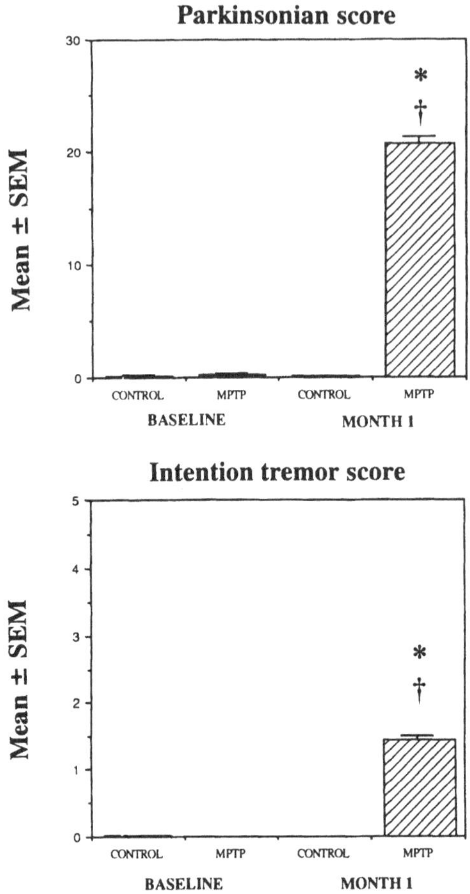

Figure 2 The mean score (± SEM) is shown for parkinsonian summary score and intention tremor summary score during baseline and the month after treatment for subjects in the MPTP-treated (striped) group (N = 66) and the control-treated (solid) group (N = 11). Significant differences between groups the month after treatment are shown (*$p < .001$) and for each group between baseline and the month after (†$p < .001$).

anxiety, and quiet OK scores after MPTP compared with controls [$F(1, 75) = 64.53, 20.59, 48,09$ $p < .01$, respectively] and to decreases after MPTP compared with their mean values during baseline ($F(1, 75) = 54.58, 44.79, 67.15$, $p < .01$, respectively). There were no significant differences between the groups during baseline, or between baseline and the month after sham treatment for control subjects, for healthy, anxiety, arousal, and quiet OK scores.

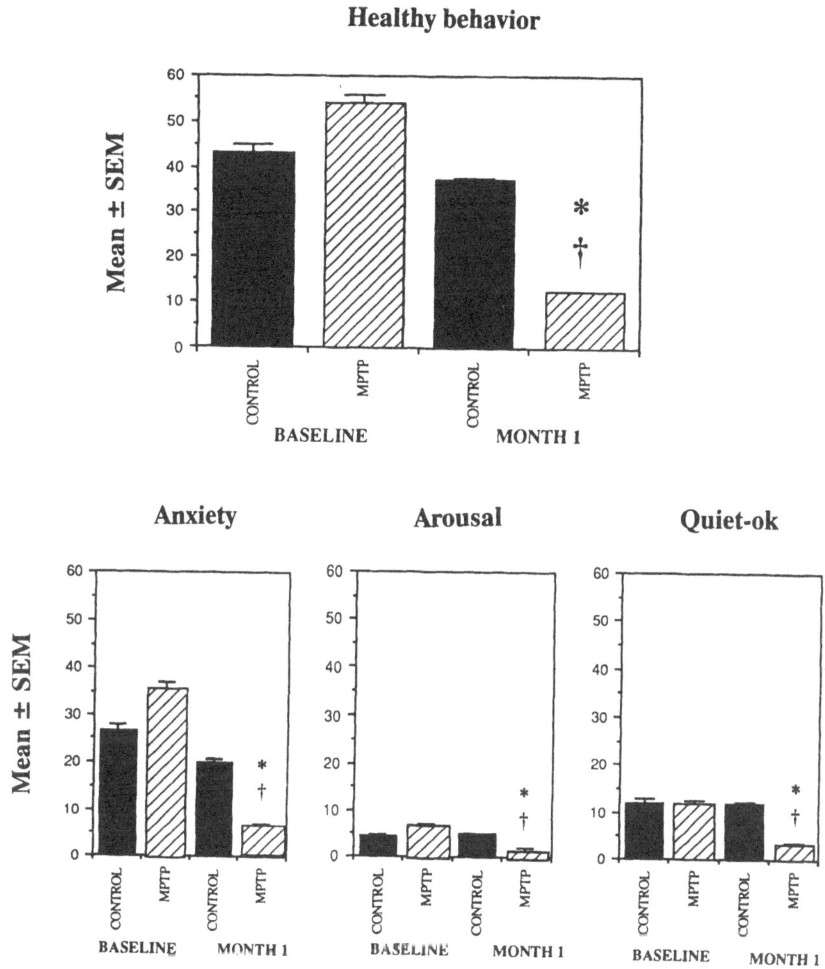

Figure 3 The mean score (± SEM) healthy behavior and the summary factors anxiety, arousal, and quiet-OK (which make up healthy behavior score) are shown during baseline and the month after treatment for subjects in the MPTP-treated (striped) group (N = 66) and the control-treated (solid) group (N = 11). Significant differences between groups the month after treatment are shown (*$p < .001$) and for each group between baseline and the month after (†$p < .001$).

3.4. Analyses of MPTP-Induced Behavioral Deficits by Severity

Each subject's parkinsonian summary score was ranked and then grouped into quintiles. The five resulting severity categories (0 = lowest) were used to examine the pattern of behaviors associated with the deficit, the range of severity,

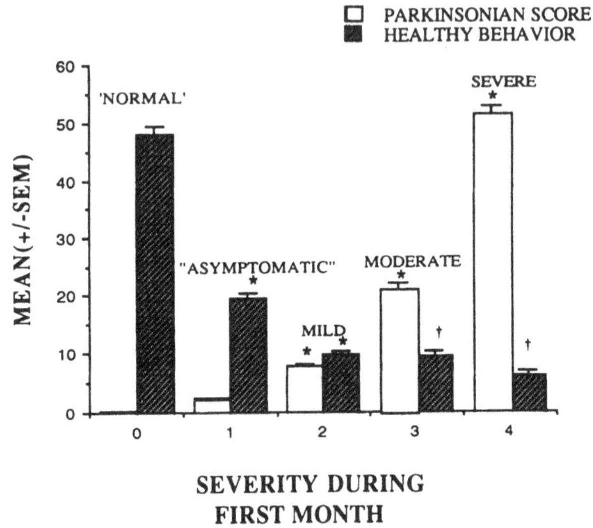

Figure 4 The mean (± SEM) parkinsonian summary score (open) and healthy behavior score (striped) are shown during baseline and the month after treatment for all subjects categorized by severity (0,1,2,3,4). Significant differences from previous severity category (*$p < .001$) and significant differences from only category 1 or 0 (†$p < .001$) are shown.

the relationship between level of impairment and recovery over time, and to examine whether the dose of MPTP correlated with degree of impairment or with medical complications.

All control subjects (N = 11) were in the first quintile (category 0). Figure 4 shows the mean parkinsonian summary scores and healthy behavior scores for all subjects grouped into each of the five severity categories. There was a significant difference between the five severity groupings the month after treatment for mean parkinsonian scores and healthy behavior scores ($F(4, 72) = 349.69$ and 30.83, $p < .0001$, respectively). Subjects in the lowest quintile (category 0) had high healthy behavior scores and low parkinsonian scores (there were 4 MPTP-treated subjects in this group with scores < 1 and all the control subjects). There were differences in healthy behavior, but not parkinsonian, summary scores for those in the next quintile (category 1) compared with subjects in category 0. With higher severity groups, both healthy behavior scores decreased and parkinsonian summary scores increased. In most cases, there were significant differences with the previous severity score grouping as well as with severity group 0. These five categories based on mean parkinsonian score therefore represented real divisions in terms of the range of impairment induced by MPTP. Subjects in the highest quintile (category 4) had mean parkinsonian scores of 50 and low healthy behavior scores, accurately reflecting their functional impairment. The nature and the

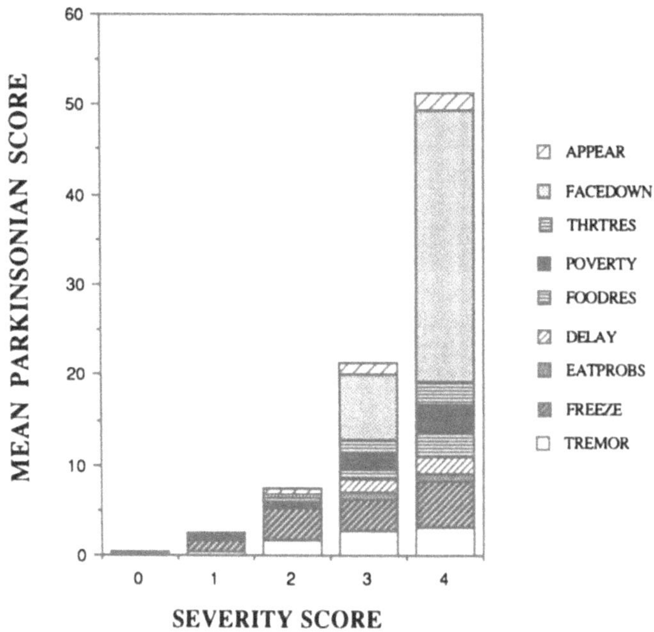

Figure 5 Profile of the individual behaviors that make up the parkinsonian summary score for subjects in each of the severity categories the month after treatment. The mean of each behavior is shown as a different pattern in the histogram (indicated by the key).

range of the parkinsonian deficits found here corresponded very well to, and thus confirmed, the classification used in our previous studies. Severity category 0, 1, 2, 3, and 4 correspond to "normal," "asymptomatic," "mild," "moderate," and "severe," respectively. *Normal* refers to subjects observed not to be different from untreated control subjects. *Asymptomatic* (category 1) corresponds to subjects that do not have observable motor deficits, using these observational methods, compared with untreated subjects, but that have previously been found to have subtle motor deficits and cognitive impairments that can be elicited by special testing procedures (Taylor *et al.*, 1990b,c). *Mild, moderate,* and *severely* parkinsonian subjects (categories 2, 3, and 4) have been found in our previous studies to have increasingly severe and observable motor deficits without treatments and to show motor improvements after fetal tissue transplantation (see Taylor *et al.*, 1990a, 1991a).

Figure 5 depicts the profile of different variables that make up the parkinsonian score for the severity classifications. Freezing was the most prevalent behavior in the more mildly affected subjects (i.e., categories 1 and 2). In the upper quintiles, there was a similar amount of freezing, but with increasing debilitation, facedown became the most frequent behavior in the highest quintile.

Table III. Mean (±SEM) for Each of the Nine Behaviors Comprising the Parkinsonian Summary Derived from the Factor Analysis Are Shown as Subjects Assigned by Severity Group (0–4) for Baseline and the First 4 Months after Treatment (MO1–4)

Parkinsonian summary score	Severity score	Baseline mean (SEM)	MO1	MO2	MO3	MO4
Appearance	0	0 (0.00)	0.02 (0.01)	0.02 (0.01)	0.01 (0.01)	0.04 (0.02)
	1	0 (0.00)	0.21 (0.02)	0.05 (0.02)	0.02 (0.01)	0.04 (0.02)
	2	0.01 (0.00)	0.46 (0.03)	0.24 (0.03)	1.20 (0.09)	1.67 (0.14)
	3	0 (0.00)	1.08 (0.05)	1.90 (0.12)	3.64 (0.14)	4.51 (0.09)
	4	1.95 (0.06)	3.34 (0.10)	3.32 (0.11)	3.52 (0.12)	2.20 (0.07)
Delayed movement	0	0.01 (0.00)	0.01 (0.00)	0.02 (0.01)	0.01 (0.01)	0.02 (0.02)
	1	0 (0.00)	0.09 (0.09)	0.01 (0.01)	0.01 (0.01)	0.05 (0.02)
	2	0.01 (0.01)	0.28 (0.03)	0.18 (0.03)	0.95 (0.09)	0.96 (0.12)
	3	0.01 (0.00)	1.31 (0.07)	1.98 (0.15)	3.17 (0.16)	4.05 (0.17)
	4	0 (0.00)	1.70 (0.08)	3.07 (0.16)	3.16 (0.14)	3.29 (0.15)
Difficulty eating	0	0 (0.00)	0 (0.00)	0 (0.00)	0.01 (0.01)	0.03 (0.02)
	1	0 (0.00)	0.04 (0.01)	0.01 (0.01)	0.01 (0.01)	0.03 (0.02)
	2	0 (0.00)	0.19 (0.03)	0.06 (0.02)	0.29 (0.05)	0.46 (0.09)
	3	0 (0.00)	0.96 (0.07)	1.71 (0.14)	3.24 (0.18)	4.12 (0.24)
	4	0 (0.00)	1.00 (0.07)	3.47 (0.14)	3.35 (0.13)	2.72 (0.18)
Facedown	0	0.02 (0.01)	0.02 (0.02)	0 (0.00)	0.17 (0.12)	0.16 (0.16)
	1	0.01 (0.00)	0.18 (0.07)	0.26 (0.18)	0.28 (0.18)	0.23 (0.23)
	2	0.02 (0.01)	0.42 (0.16)	0.02 (0.02)	0.01 (0.01)	0 (0.00)
	3	0.03 (0.02)	7.36 (0.79)	24.64 (2.63)	53.44 (2.20)	55.60 (2.36)
	4	0 (0.00)	30.04 (1.17)	57.93 (1.02)	53.91 (2.20)	34.07 (3.86)

Food response	0	0.08 (0.02)	0 (0.00)	0 (0.00)	0.02 (0.01)	0.07 (0.03)
	1	0.03 (0.01)	0.02 (0.01)	0.02 (0.01)	0.02 (0.01)	0.04 (0.02)
	2	0.01 (0.00)	0.06 (0.02)	0.03 (0.01)	0.81 (0.09)	0.42 (0.11)
	3	0.09 (0.02)	1.22 (0.08)	1.81 (0.17)	4.26 (0.15)	4.50 (0.22)
	4	0.03 (0.01)	2.73 (0.09)	3.22 (0.15)	2.78 (0.16)	3.21 (0.19)
Freeze	0	0.15 (0.08)	0.55 (0.11)	0.46 (0.14)	0.53 (0.21)	0.80 (0.28)
	1	0.46 (0.15)	5.79 (0.46)	0.83 (0.17)	0.25 (0.11)	0.43 (0.11)
	2	0.05 (0.01)	17.81 (0.86)	11.96 (1.33)	8.40 (1.48)	13.97 (1.96)
	3	0.07 (0.02)	17.50 (0.96)	18.91 (2.27)	7.23 (2.24)	37.14 (3.92)
	4	0.11 (0.03)	25.14 (1.05)	15.44 (2.41)	42.33 (3.27)	52.41 (2.12)
Poverty of movement	0	0 (0.00)	0.03 (0.01)	0.01 (0.01)	0.01 (0.01)	0.05 (0.02)
	1	0 (0.00)	0.28 (0.03)	0.03 (0.01)	0.03 (0.01)	0.09 (0.02)
	2	0 (0.00)	0.86 (0.05)	0.31 (0.03)	1.31 (0.10)	1.94 (0.15)
	3	0 (0.00)	1.85 (0.07)	2.31 (0.17)	3.89 (0.09)	4.53 (0.10)
	4	0 (0.00)	3.22 (0.07)	3.93 (0.09)	4.03 (0.10)	4.19 (0.08)
Threat response	0	0.01 (0.01)	0.08 (0.02)	0.04 (0.02)	0.01 (0.00)	0.02 (0.01)
	1	0.01 (0.01)	0.13 (0.02)	0.05 (0.02)	0.03 (0.01)	0.06 (0.02)
	2	0 (0.00)	0.40 (0.04)	0.08 (0.02)	0.37 (0.06)	0.70 (0.11)
	3	0.01 (0.01)	1.30 (0.07)	1.45 (0.13)	2.83 (0.22)	3.05 (0.21)
	4	0 (0.00)	2.48 (0.08)	1.76 (0.12)	1.20 (0.12)	1.60 (0.18)
Tremor (head and limb)	0	0 (0.00)	0.05 (0.01)	0.02 (0.01)	0.01 (0.01)	0.01 (0.01)
	1	0 (0.00)	0.41 (0.03)	0.12 (0.02)	0.02 (0.01)	0.02 (0.01)
	2	0 (0.00)	1.59 (0.07)	2.07 (0.10)	2.54 (0.17)	3.14 (0.18)
	3	0 (0.00)	2.63 (0.10)	2.99 (0.19)	3.28 (0.25)	3.67 (0.39)
	4	0 (0.00)	3.15 (0.10)	3.00 (0.20)	3.62 (0.21)	3.53 (0.21)

The means for the behaviors that make up the parkinsonian score in Figure 5 are shown in Table III for up to 4 months after MPTP for subjects in each of the severity scores.

Although the number of subjects in each severity ranking was equal (due to the division of subjects into quintiles), there was a marked difference in the range of mean scores within the five categories. The greater the parkinsonian score, the greater the range of scores in each severity category. For example, the range of individual parkinsonian scores for subjects in the five rankings were as follows: 0 or normal = 0–0.79; 1 or asymptomatic = 1.00–4.59; 2 or mild = 5.00–11.59; 3 or moderate = 13.00–36.99; and 4 or severe = 40.43–61.22

The differences in response to MPTP administration shown by these parkinsonian summary score rankings could not, however, be accounted for by differences in the total dose of MPTP administered (or other factors such as batch, numbers of injections, baseline effects, etc.) No significant differences existed between the severity groups for the total dose of MPTP given ($F < 1$). There was also no correlation between dose and parkinsonian score ($r = -0.092$, df 64, p = NS). Average doses of MPTP for the 4 MPTP-treated subjects in group 0 were 1.87 and for the other severity groups 1, 2, 3, and 4 they were, 1.89, 1.90, 1.84, and 1.87 mg/kg MPTP, respectively.

The severity of parkinsonism based on the factor averaged from the month after treatment significantly predicted whether the subject died or was sacrificed prior to the end of the experiment ($\chi^2(4) = 27.54$, $p < .01$). Fifty percent of subjects in category 4 (severe) and 25% in category 3 (moderate) died or were sacrificed for humane reasons within 4 months of MPTP treatment despite constant nursing care and medical treatments (see Section 5). No deaths occurred in the other severity categories.

3.5. Duration and Persistence of MPTP-Induced Behavioral Deficits

The severity categories from 1 month after MPTP treatment were used to examine the duration of the behavioral deficits over time. Unfortunately, it was not possible to examine all 77 subjects for 19 months after treatment. Many subjects in the moderate and severely affected groups became a part of other experiments, and most of the severely affected subjects that remained in this study died or were sacrificed as previously described. Thus, long-term data (up to 19 months) were only obtained for subjects in severity category 0 or 1.

The onset of MPTP-induced deficits and the stability of the deficits were examined over successive days for all subjects during the first 2 months. Mean parkinsonian summary scores in the five severity categories were examined using a regression analysis to determine whether subjects were becoming more or less impaired during this time (see Figure 6). Subjects with severity ratings of 2, 3, and 4 (mildly, moderately, and severely affected) showed significant, positive

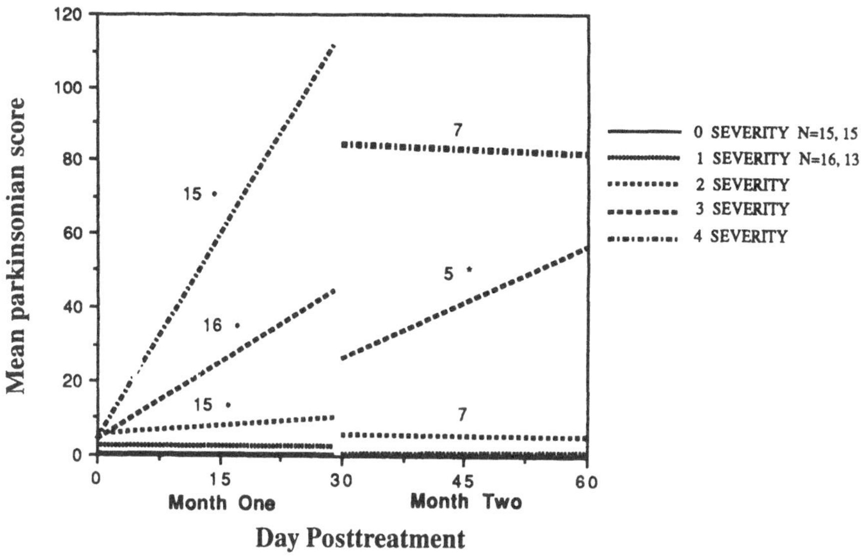

Figure 6 The mean parkinsonian summary scores for subjects in the severity categories are shown over days for the first 2 months. Significant slopes and number of subjects in each severity group are shown (•p < .0001 and *p < .01).

slopes for the first month, as the onset of the behavioral deficits became more pronounced. Conversely, subjects in severity category 0 (comprised mostly of untreated controls) and 1 (which are asymptomatic for gross motor impairments) were found to have no changes in parkinsonian score over days during the first month. In the second month, the severity-3 group was the only group to show a significant positive slope. Of note is the finding that the category-4 (severe) animals remained stable during the second month after MPTP treatment, even though many of the subjects in this group died or were sacrificed at this time.

The course of behavioral deficits was also examined for the first 4 months, as shown in Figure 7. Subjects in the severity groupings differed during the first 4 months after MPTP ($F(4, 16) = 13.62$, $p < .001$). As expected, differences existed between each severity category at months 1, 2, 3, and 4 ($F(4, 72) = 349.69$; $F(4, 42) = 105.95$; $F(4, 33) = 290.22$; $F(4, 29) = 81.31$, $p < .001$, respectively), but no differences existed between the severity groups during baseline. There were no differences between subjects as a group in category 0 and 1 at each of the 4 months examined. Subjects in the quintile category 2 did show increased parkinsonian summary scores during the days of the first month (as detailed above), and they showed increased parkinsonian scores during months 1 and 3 only. Parkinsonian summary scores for subjects in severity

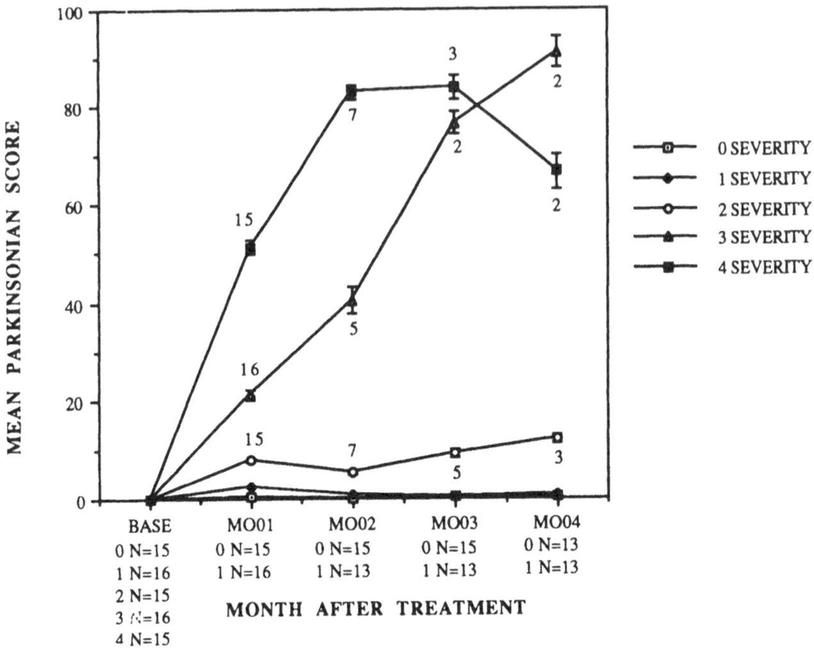

Figure 7 The mean parkinsonian summary scores (± SEM) for subjects in the severity categories are shown for the first 4 months after treatment. The numbers of subjects examined at each month for each severity group are shown.

category 3 and category 4 were markedly increased at all time points compared with the other subjects, and scores for subjects in category 4 were significantly increased compared with subjects in category 3, except for month 4 where parkinsonian scores were higher for subjects in category 3 than in category 4. Also, significant differences occurred in scores for subjects in severity categories 1, 2, 3, and 4, over time ($F(4, 66) = 21.17$; $F(4, 39) = 14.09$; $F(4, 37) = 61.99$; $F(4, 37) = 140.69$, $p < .001$, respectively), but not for subjects in severity category 0. Subjects in category 1 had increased scores during the first month compared with baseline. Parkinsonian summary scores for subjects ranked in category 3 or category 4 increased significantly in the first month compared with baseline and were further increased during the second month. By the third month, subjects in category 3 continued to have increased scores, after which no further increases were observed. At this time, there were also no further increases in mean scores in category-4 subjects. Parkinsonian scores continued to be significantly increased (compared with the first month); however, there appeared to be a decrease by month 4 compared with month 3. This was probably due to decreasing numbers of subjects, since a further increase in parkinsonism

Table IV. Mean (±SEM) for the Other Summary Scores Are Shown by Subjects Assigned by Severity Group (0–4) for Baseline and the First 4 Months after Treatment (MO1–4)

Summary score	Severity	Baseline mean (SEM)	MO1	MO2	MO3	MO4
Intention tremor	0	0.01 (0.00)	0.00 (0.00)	0.02 (0.01)	0.02 (0.02)	0.06 (0.04)
	1	0.00 (0.00)	0.15 (0.03)	0.04 (0.02)	0.02 (0.02)	0.08 (0.04)
	2	0.01 (0.00)	0.85 (0.07)	1.16 (0.08)	1.35 (0.18)	1.95 (0.29)
	3	0.01 (0.00)	2.41 (0.15)	4.76 (0.30)	7.72 (0.32)	8.90 (0.43)
	4	0.00 (0.00)	2.56 (0.17)	7.79 (0.34)	8.44 (0.26)	7.89 (0.27)
Anxiety	0	25.54 (1.12)	31.96 (1.24)	25.33 (1.27)	29.28 (1.37)	33.67 (1.69)
	1	18.33 (0.93)	10.88 (0.69)	17.17 (1.28)	21.08 (1.21)	25.47 (1.64)
	2	19.42 (0.095)	5.34 (0.41)	8.99 (1.06)	12.95 (1.64)	13.05 (2.24)
	3	18.63 (0.98)	5.01 (0.57)	1.96 (0.79)	1.71 (1.25)	0.26 (0.15)
	4	24.39 (1.20)	3.78 (0.52)	0.27 (0.13)	0.06 (0.05)	0.44 (0.19)
Arousal	0	4.86 (0.34)	5.97 (0.48)	5.88 (0.52)	3.25 (0.26)	2.29 (0.21)
	1	4.58 (0.31)	2.45 (0.19)	3.11 (0.25)	3.08 (0.27)	3.03 (0.35)
	2	5.53 (0.29)	1.26 (0.18)	2.45 (0.30)	1.94 (0.30)	0.53 (0.12)
	3	4.94 (0.32)	1.40 (0.24)	0.69 (0.39)	0.31 (0.23)	0.14 (0.11)
	4	4.11 (0.25)	0.75 (0.16)	0.03 (0.02)	0.10 (0.08)	0.20 (0.14)
Quiet OK	0	11.44 (0.69)	10.23 (0.56)	10.45 (0.65)	9.72 (0.27)	10.45 (0.78)
	1	12.81 (0.74)	6.12 (0.03)	7.61 (0.64)	10.14 (0.82)	8.78 (0.85)
	2	9.53 (0.66)	3.09 (0.39)	1.84 (0.41)	3.73 (0.79)	4.14 (1.24)
	3	10.30 (0.67)	3.02 (0.38)	0.92 (0.55)	0.57 (0.56)	0.21 (0.21)
	4	15.93 (0.89)	1.73 (0.31)	0.08 (0.08)	0.10 (0.08)	0.00 (0.00)
Healthy	0	41.84 (1.48)	48.17 (1.48)	41.67 (1.54)	42.55 (1.74)	46.62 (2.14)
	1	35.73 (1.32)	19.46 (0.97)	27.90 (1.41)	34.31 (1.60)	37.3 (2.13)
	2	34.94 (1.18)	9.70 (0.83)	13.29 (1.24)	18.62 (2.06)	17.72 (2.96)
	3	33.88 (1.31)	9.44 (0.83)	3.57 (1.17)	2.61 (1.78)	0.60 (0.37)
	4	44.44 (1.52)	6.28 (0.71)	0.37 (0.20)	0.16 (0.10)	0.65 (0.24)

Note: Data for Parkinsonian Summary Score Are in Figure 7. Data for tremor are shown in Table III.

would have and did result in death. The subjects that were still alive at this time were also in the less affected range within this severity group.

A similar pattern of behavioral changes was also observed for the other summary scores shown in Table IV. For example, the intention tremor score ($F(4, 16) = 9.83, p < .01$) paralleled the effects of the parkinsonian score over time for severity categories 1, 2, 3, and 4 ($F(4, 66) = 4.49$; $F(4, 39) = 8.66$; $F(4, 37) = 36.34$; $F(4, 37) = 23.40, p < .01$, respectively), but not for subjects in severity category 0.

3.6. Individual Changes in Severity over Time

The initial severity of parkinsonian signs in the first month determined both the rapidity of onset and the stability of the syndrome over time. Because of possible individual subject variation, which might be obscured by group analysis, and the decreasing numbers of subjects in the five severity categories (subjects "dropped out" because of death or inclusion in other experiments), we examined the stability of subjects' severity category by comparing their severity during the first month with the third to fifth month (depending on subject) after the administration of MPTP, as depicted in Figure 8. These results confirmed the group results for subjects in categories 3, 4, and 0. For example, severely (category 4) affected subjects are stable over time, moderately (category 3) affected subjects either remain stable or show an increase in parkinsonism, and normal (category 0) subjects are also stable. In contrast to the group analysis, the individual data for asymptomatic (category 1) subjects showed consistent improvements to control levels, and mild (category 2) subjects showed a large variation in behavioral stability—that is, subjects displayed increases, decreases, and no changes in severity category over time.

3.7. Summary of Behavioral Effects

In the earliest months after MPTP, there was a dramatic increase in parkinsonian scores in the more severely affected subjects, which then stabilized. In subjects with more mild deficits (severity category 2), parkinsonian summary scores appeared either not to change much over the first 4 months or variable changes were observed. Although some increases in the first month after MPTP administration appeared to occur for subjects in severity category 1 (asymptomatic subjects), within a few months after MPTP administration these subjects were not different from subjects in category 0 (mostly untreated control subjects). Subjects in severity category 0 also remained stable.

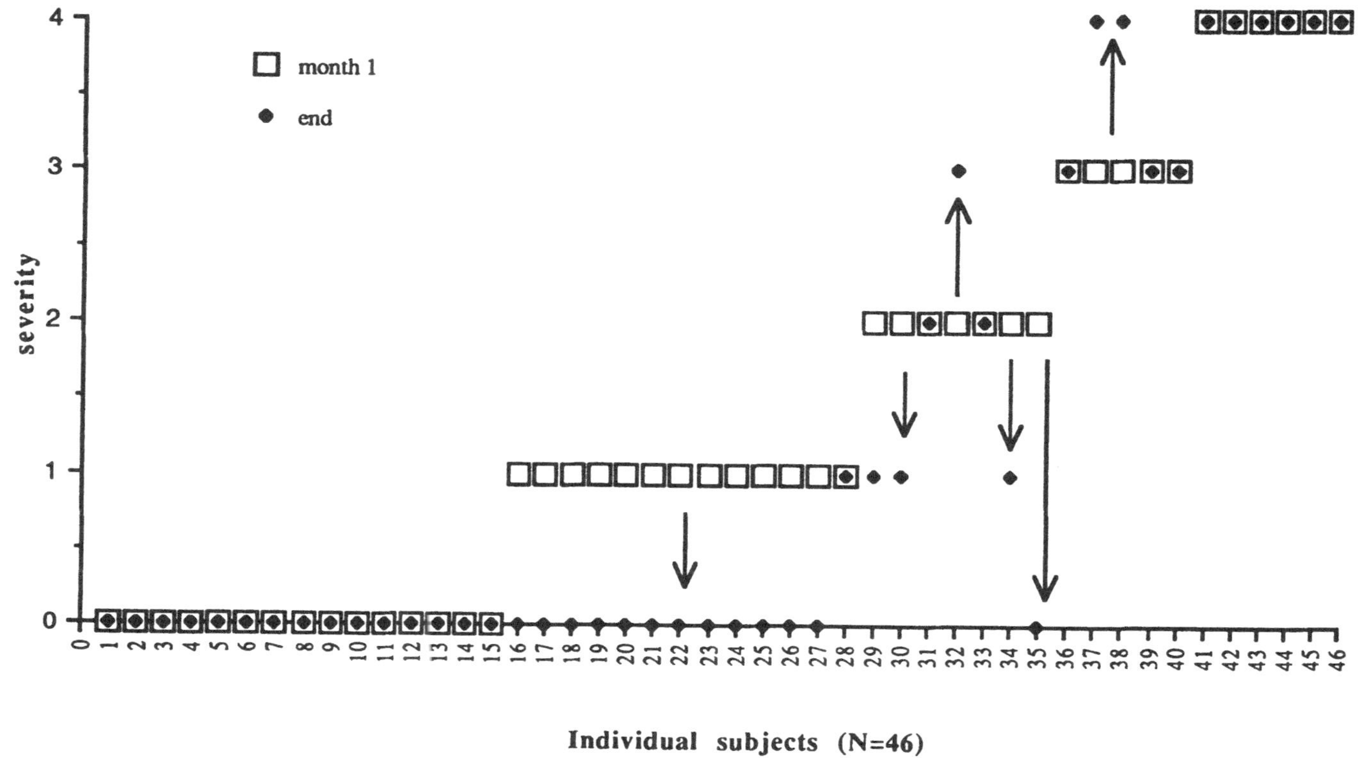

Figure 8 Individual subject's severity over time is shown. The severity category for each subject (0,1,2,3,4) is shown the month after MPTP or control treatment as a square. At the end of the comparison period (3–5 months after MPTP) each subject's severity, assessed by mean parkinsonian score at that time, is shown as a diamond. The direction of the arrow indicates whether there is an increase or decrease in severity. No change is shown by the diamond and square being superimposed.

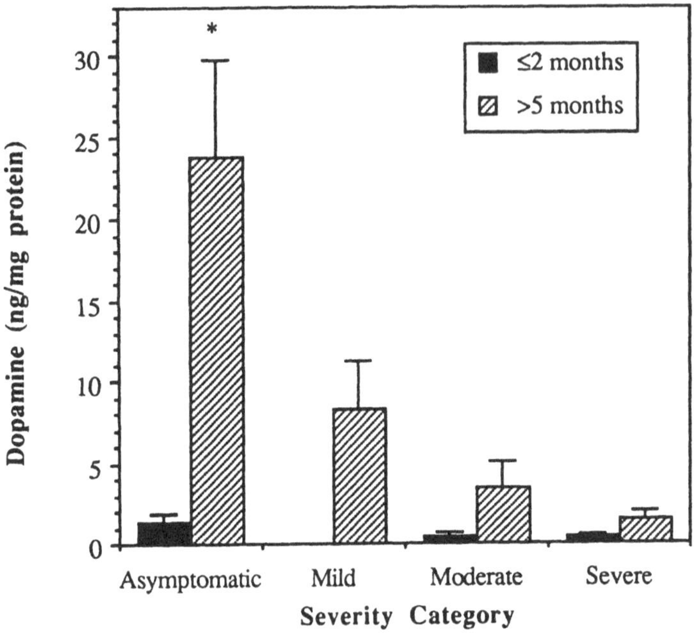

Figure 9 Mean (± SEM) DA concentration in the dorsolateral caudate nucleus of monkeys grouped by severity category and time of sacrifice after MPTP treatment is shown. The severity category refers to the status of the monkeys during the first month of treatment with MPTP. The monkeys were sacrificed at or less than 2 months following MPTP treatment (n = 4 for severe, 3 for asymptomatic, 2 for moderate), or at least 5 months following MPTP treatment (n = 21 for mild, 12 for asymptomatic, 8 for moderate, 8 for severe). The asterisk indicates difference from all other groups (ANOVA, followed by Duncan's multiple range test). The DA concentration in this brain region of untreated monkeys is 136 ± 10.5 ng/mg protein (n = 21).

4. Biochemical Correlations with Individual Changes in Parkinsonian Severity

Biochemical values were analyzed to compare the effect of severity category of animals during the first month of MPTP treatment and sacrifice time after MPTP on DA levels in the caudate nucleus. As we have previously documented (Elsworth *et al.*, 1989), dramatic reductions in striatal regions were found in all MPTP-treated monkeys. However, it was observed here that only in the asymptomatic group was there a significant difference between the DA concentrations in monkeys sacrificed ≤ 2 months or > 5 months after MPTP treatment (see Figure 9).

5. Medical Consequences Associated with MPTP Administration

Medical consequences of MPTP treatment were most prevalent in the severely affected subjects (Table III). Feeding with a special diet or assisted feeding was significantly different for control compared with MPTP-treated subjects. There were also differences between subjects when examined by severity category for the need for nasogastric feeding. Even with nasogastric feeding of theoretically "adequate" calories (often several times the required daily calories for normal monkeys), many severely parkinsonian monkeys lost weight and muscle mass (as do end-stage Parkinson's patients) and eventually appeared cachectic. Severely debilitated monkeys sometimes got aspiration pneumonia probably from a combination of poor ventilatory efforts and inhalation of their own oral secretions. The incidence of pneumonia was significantly different in the severity groups (see Table V).

Several severely affected monkeys developed gastric or duodenal ulcers. The earliest sign of this problem was gastric retention. Retention was carefully monitored in all monkeys receiving tube feeding. When retention was found, we adjusted the dietary intake and, after some experience, began to treat it prophylactically with cimetidine (20 mg/kg). Because of this preventative intervention, the number of subjects that actually developed ulcers was probably an underestimation of the incidence potentially resulting from MPTP treatment. The incidence of ulcers, however, was significant (see Table V.)

Table V. Medical Consequences of MPTP Treatment Are Shown by Severity Category and by Whether Subjects Were MPTP Treated

Severity	Special diet	Assisted feeding	NG tube	Pre-ulcer/ ulcer	Pneumonia
0	2	1	0	0	0
1	8	4	0	0	0
2	8	2	2	0	1
3	6	5	3	0	0
4	4	6	8†	3†	9†
CONTROL	1	0	0	0	0
MPTP	27*	18*	13	3	10

Note: The numbers of subjects in each severity category 0–4 (top) and in each group, control or MPTP (bottom), are shown for having received a special diet (e.g., softened biscuits with added sugar and fruit), assisted feeding (e.g., syringe or spoon feeding of similar preparations), feeding by nasogastric tube (NG tube), evidence of an ulcer or determination of an ulcer at postmortem examination, and having had evidence of pneumonia. Significant differences existed between severity classification for NG tube, pre-ulcer/ulcer and pneumonia ($\chi^2(4) = 20.67$, 11.32, and 27.6, †$p < .01$, respectively). Significant differences occurred between subjects that were control or MPTP-treated for special diet and assisted feeding ($\chi^2(1) = 4.12$ and 3.9, *$p < .05$).

6. Phenomenological Similarities with Idiopathic Parkinson's Disease

A number of other similarities with idiopathic PD were recorded. It was noted that fear or stress induced by human threats eliminated tremor and spontaneous freezing episodes for short periods, in the same way that a wheelchair-ridden parkinsonian patient may be able to flee a burning building on foot (Schwab, 1972; Schwab and Zeiper, 1976). Exercise appeared also to diminish symptoms. Severe subjects were often noted to have "masked facies"; however, this observation was extremely difficult to quantify reliably between observers and was dropped from our rating system. When subjects were physically examined during passive exercise, they were noted to show rigidity as well as cogwheeling. Some monkeys showed urinary retention and constipation (off all drugs), suggestive of the autonomic disturbances found in PD. Finally, three monkeys suffered pathological fractures during normal movement in a social cage at a time when their general function was otherwise greatly improved.

7. Discussion of the Usefulness of MPTP-Induced Parkinsonism as an Animal Model of Parkinson's Disease

7.1. Summary of the Results of MPTP in the Vervet Monkey

This report describes a detailed characterization of the behavioral response to MPTP in an Old World macaque, the African green monkey. It is clear from the data presented that stability of MPTP-induced parkinsonism can be achieved and can be predicted reliably by the initial degree of impairment.

Factor analysis revealed clusters of individual behaviors associated with MPTP treatment and with healthy "normal" behavior. Categories of behavior found to be significantly associated with MPTP administration were eating problems, delayed movement, decreased response to the presentation of food, freeze, poverty of movement, decreased response to threat, head and limb tremor, immobility, and poor appearance. These behavioral signs made up the parkinsonian summary score. Several of these behaviors have not been studied or reported in other nonhuman primate species (e.g., tremor); some may correspond to "activities of daily living" items used to assess the functioning of patients with idiopathic PD (Fahn *et al.*, 1987; Schwab and England, 1969). As the severity of the MPTP-induced behavior was increased some of these behaviors were masked by others and behavior associated with normal function was decreased. For example, MPTP-induced deficits, such as delayed movement, difficulties in eating, and poverty of movement, were no longer observed in severely debilitated animals that had difficulty sitting or standing. Because of this competition for behavioral expression, the parkinsonian summary score and other summary measures are critical for the accurate assessment of the overall status of the subject. We validated the reliability of these measures with quantitative analysis and found a striking similarity to the signs and behavioral effects of PD. Even though the initial severity of MPTP-induced deficits was found to vary widely between animals, no significant evidence of "spontaneous" recovery for the time period studied was found in the more severe cases. In this respect, the St. Kitts African green monkey is similar to young adult humans after unintentional MPTP ingestion (David *et al.*, 1979; Langston *et al.*, 1983) and may be different from other species of monkeys. We found that the level of the behavioral deficit remained fairly stable over time in the most severely impaired animals. It persisted at least up to 4 months. It is also of note that other similarly severe subjects were part of experiments in which they received fetal substantia nigra grafts into the caudate nucleus. They subsequently showed marked behavioral improvements, whereas other severe subjects that received various control grafts did not improve (Taylor *et al.*, 1991a). Such severely affected subjects, like patients with end-stage PD, may be difficult to maintain for periods that match the survival times of similarly affected transplanted monkeys, because, as we have concluded, successfully grafted monkeys improve. Nevertheless, evidence of "spontaneous" recovery was found in the less severe (categories 1 and 2) subjects. Asymptomatic subjects (severity category 1) showed some small deficits the first month, which were not present in subsequent months. Subjects in category 2 showed inconsistent responses over time. It is also possible that there are short-term toxic effects of MPTP or MPP^+, not related to permanent striatal DA losses (Irwin *et al.*, 1990), that rapidly disappear and are masked in more severely parkinsonian subjects by more pronounced impairments. In another group of asymptomatic monkeys that were tested over a year after MPTP administration, and that showed evidence of a dopaminergic lesion, we found deficits that were characterized by subtle motor and cognitive impairments (Taylor *et al.*, 1990b,c).

7.2. Range of Severity of Behavioral Deficits

Because a similar dose of MPTP resulted in a wide range of parkinsonian deficits, and the level of disability predicts the stability of parkinsonism, it is useful to control experimentally for the range of observed deficits. It is unclear why such a wide range of behavioral deficits exists in response to identical doses of MPTP administration in African green monkeys and other species (Eidelberg *et al.*, 1986; Nomoto *et al.*, 1986; Di Paolo *et al.*, 1986; Elsworth *et al.*, 1989; Schneider *et al.*, 1987). First, there may be differences in the metabolism of MPTP between subjects within a species or other factors may affect susceptibility to the toxicity of MPTP and/or compensatory changes (Rose *et al.*, 1989).

Second, it is also possible that there is significant species variation in response to MPTP that critically determines the severity since some species (i.e., rodents) are comparatively insensitive to MPTP toxicity (Johannessen *et al.*, 1985; Rose *et al.*, 1990). For example, two studies using MPTP-treated New World monkeys (marmosets) have reported some variability in response to MPTP within a number of weeks, but have not reported the range of severity observed in the MPTP-treated African green monkey (see Jenner and Marsden, 1986; Nomoto *et al.*, 1986). Moreover, these studies have not reported a severe parkinsonian-like syndrome. The MPTP syndrome achieved in the marmoset appears to be relatively mild and has been characterized by dystonia, action myoclonus, and exaggerated startle (Jenner and Marsden, 1990). These effects may not correspond closely to the symptomatology of idiopathic or MPTP-induced parkinsonism or to the features of MPTP-induced parkinsonism in Old World primates such as the African green monkey and the human.

7.3. Stable Deficits Depend on Severity of Parkinsonism

Some studies using Old World monkeys described spontaneous recovery (Eidelbert *et al.*, 1986), whereas others have not (Degryse and Colpaert, 1986; Imai *et al.*, 1988); however, many studies have used relatively small numbers of animals. Often these subjects appeared to be only mildly parkinsonian, since improvement in performance on gross measures of motor function such as lever pulling have been described; one problem is that lever pulling would not have been possible in our severe animals. The most widely cited report on recovery from MPTP toxicity in Old World monkeys (*Macaca fascicularis*) is that of Eidelbert *et al.* (1986), who, although not using blinded comparisons with untreated controls, reported that recovery was not observed in approximately half of the subjects. In the marmoset, recovery often has been based on rotation measures or increases in spontaneous activity that may not correlate with, or directly relate, to PD. In our study, measures of spontaneous activity (e.g., vertical climbing and shifting) either were not significantly reduced by MPTP treatment or not reliably identified as an MPTP-associated factor (i.e., part of the parkinsonian summary score). Thus, apparent differences in the literature regarding recovery from MPTP toxicity in monkeys possibly can be understood when it is considered that there are species differences in response to MPTP and that stability of deficits may only be achieved when the experimental parkinsonism is sufficiently severe to include all features of the idiopathic disease.

7.4. Possible Factors Determining the Severity of the Behavioral Deficit

It has been suggested that the most important factor in determining the severity of the behavioral deficit induced by 6-OHDA in the rat and monkey is the extent of the DA depletion (i.e., Zigmond and Stricker, 1989). In addition, the persistence of the deficit and occurrence of recovery have been linked to the intensity of the dopaminergic alterations. In MPTP-treated monkeys, some small biochemical differences may account for large behavioral differences (Elsworth *et al.*, 1989). With our standardized behavioral analysis, which characterized several levels of deficit, we can more precisely examine the correlation between regional DA depletion in the basal ganglia and the level of severity at different times after MPTP treatment. Intramuscular administration of MPTP produces an uneven loss of DA in the striatum of the African green monkey, with the dorsolateral caudate nucleus being one of the most susceptible regions (Elsworth *et al.*, 1989). When sacrificed at ≤ 2 months after MPTP treatment, DA concentration in the dorsolateral caudate nucleus was reduced by 99%, even in the asymptomatic group. However, when animals that were originally in the asymptomatic category were examined 7–24 months later, the level of DA depletion was only 82%. This level of depletion seen at long time periods after MPTP administration is more comparable to the DA losses measured in the striatum of patients with Parkinson's disease than the depletions measured soon after MPTP treatment in the monkey. This restoration of DA levels corresponds well with the behavioral recovery seen in the asymptomatic group over time and suggests that the improvement is DA-related. The severe group of MPTP-treated monkeys remained behaviorally stable over time, and, in concert with the behavioral data, DA levels in the caudate nucleus of the severe group were not time-dependent, at least during the periods studied. The basis of this biochemical recovery is not known, but one possible explanation is widespread sprouting of surviving fibers in mildly affected animals with substantial numbers of surviving nigral DA neurons.

7.5. Further Restrictions of the Model

Although the behavioral effects and neuropathology that are produced by systemic injection of MPTP in nonhuman primates mimic PD extremely well, the model has been criticized on the basis of a lack of progression of behavioral abnormalities. There may, however, be some advantage to this feature. Furthermore, no other model of PD (i.e., the unilateral MPTP and 6-OHDA or bilateral 6-OHDA) in nonhuman primates mimics the full behavioral syndrome or neuropathology (cf. Annett *et al.*, 1990; Jenner and Marsden, 1990; Bankiewicz *et al.*, 1990; Guttman *et al.*, 1990). Specifically, the different severity categories associated with the bilateral administration of MPTP could be used to parallel the stages of severity associated with PD. For example, by determining the functional effects of novel treatments in monkeys over a range of severity, one may be able to predict whether putative treatments would be limited to patients within a particular severity level or "stage" of parkinsonism. Nevertheless, despite an accurate assessment of the behavioral effects of MPTP treatment and the many similarities between the symptoms of Parkinson's disease and MPTP toxicity, the progressive pathological changes should be considered when novel treatments for PD are based on predictions using the MPTP model.

7.6. Summary of Medical Consequences of MPTP Administration

The medical consequences of MPTP administration in monkeys also closely mimicked the clinical complications associated with idiopathic PD (e.g., Hornykiewicz, 1982). In most cases, as in PD, pneumonia was the cause of death. Difficulties in eating and drinking resulted in weight losses that also required special nursing care. In addition, gastric and duodenal ulcers and pathological fractures were observed. Although clinical phenomena associated with idiopathic PD, such as spasticity, rigidity, and cogwheeling, were observed in the more symptomatic subjects, these behaviors could not be quantified reliably. They may contribute nonetheless to many of the behavioral impairments induced by MPTP in the monkey and further underscore the similarities with PD and the animal model.

7.7. Final Considerations

We urge investigators preparing to utilize the MPTP model to consider the species of monkey employed and the method used to document behavioral deficits. These decisions will affect the range of deficits that can be obtained, their relevance to PD, and the stability of deficits that can be achieved. By validating and standardizing our analyses of the behavioral effects of MPTP in the monkey, we can use the model more effectively to confirm stable parkinsonism in severely affected subjects reported here and to determine the effects of pharmacological and other novel treatment strategies, such as fetal neural grafting, on MPTP-induced parkinsonism. The persistence of the behavioral impairments or possible deterioration of subjects for periods longer than 5 months will be of great interest if longer-term studies can be completed.

ACKNOWLEDGMENTS. We thank the staff of the St. Kitts Biomedical Research Foundation for care of the monkeys and behavioral observation. In addition, we thank several students who assisted with various aspects of the study over a 3-year period: C.A. Walts, D.P. Kennedy, R. Block, R.S. Watson, C. Lewis, L.H. Siegel, D.L. Hulsizer, L.M. Woodhouse, J.R. Redmond, J.D. Lory, D.H. Bilder, K.J. Guleserian, W.G. Urban, S. Gardner, S. Bures, P. Chen, M. Gillespie, R. Jimenez, J. Schaffer, A. Ung, K. Whitney, M.S. Lawrence, J. Swet, and R. Guilette. We thank M.S. Crouse-Artus and J. Swet for assistance with data analyses. Funding came from NINDS PO1-NS24032, Axion Research Foundation, and RSA MH00643 to D.E.R.

8. References

Altman, J., 1974, Observational study of behavior: Sampling methods, *Behavior* 49:222-267.

Annett, L.E., Rogers, D.C., and Dunnett, S.B., 1990, Unilateral 6-OHDA lesions in marmosets: A primate model for neural transplantation in Parkinson's disease, *in* "Function and Dysfunction of the Basal Ganglia" (A.J. Franks, J.W. Ironside, R.H.S. Mindham, R.J. Smith, E.G.S. Spokes, and W. Winlow, eds.), pp. 181–197, Manchester, Manchester University Press.

Ballard, P.A., Tetrud, J.W., and Langston, J.W., 1985, Permanent human parkinsonism due to 1-methyl-4-phenyl-1,2,3,6-tetrahydropyridine (MPTP): Seven cases, *Neurology* 35:949–956.

Bankiewicz, K.S., Oldfield, E.H., Chiueh, C.C., Doppman, J.L., Jacobowitz, D.M., and Kopin, I.J., 1986, Hemiparkinsonism in monkeys after unilateral internal carotid artery infusions of 1-methyl-4-phenyl-1,2,3,6-tetrahydropyridine (MPTP), *Life Sci.* 39:7–16.

Bankiewicz, K.S., Plunkett, R.J., Jacobowitz, D.M., Porrino, L., Di Porzio, U., London, W.T., Kopin, I.J., and Oldfield, E.H., 1990, The effect of fetal mesencephalon implants on primate MPTP-induced parkinsonism: Histochemical and behavioral studies, *J. Neurosurg.* 72:231–244.

Baulu, J., and Redmond, D.E., Jr., 1978, Some sampling considerations in the quantitation of monkey behavior under field and captive conditions, *Primates* 19:391–399.

Burns, R.S., Chiueh, C.C., Markey, S.P., Ebert, M.H., Jacobowitz, D.M., and Kopin, I.J., 1983, A primate model of parkinsonism: Selective destruction of dopaminergic neurons in the pars compacta of the substantia nigra by N-methyl-4-phenyl-1,2,3,6-tetrahydropyridine, *Proc. Natl. Acad. Sci. USA*, 80:4546–4550.

Burns, R.S., LeWitt, P.A., Ebert, M.H., Pakkenberg, H., and Kopin, I.J., 1985, The clinical syndrome of striatal dopamine deficiency Parkinsonism induced by 1-methyl-4-phenyl-1,2,3,6-tetrahydropyridine (MPTP), *N. Engl. J. Med.* 312(22):1418–1421.

Chiueh, C.C., Markey, S.P., Burns, R.S., Johannessen, J.N., Jacobowitz, D.M., and Kopin, I.J., 1984, Neurochemical and behavioral effects of 1-methyl-4-phenyl-1,2,3,6-tetrahydropyridine (MPTP) in rat guinea pig and monkey, *Psychopharmacol. Bull.* 20(3):548–553.

Chiueh, C.C., Burns, R.S., Markey, S.P., Jacobowitz, D.M., and Kopin, I.J., 1985, Primate model of Parkinsonism: Selective lesion of nigrostriatal neurons by 1-methyl-4-phenyl-1,2,3,6-tetrahydropyridine produces an extrapyramidal syndrome in rhesus monkeys, *Life Sci.* 36:213–218.

Collier, T.J., Redmond, D.E., Jr., Sladek, C.D., Gallagher, M.J., Roth, R.H., and Sladek, J.R., Jr., 1987, Intracerebral grafting and culture of cryopreserved primate dopamine neurons, *Brain Res.* 436:363–366.

Davis, G.C., Williams, A.C., Markey, S.P., Ebert, M.H., Caine, E.D., Reichert, C.M., and Kopin, I.J., 1979, Chronic parkinsonism secondary to intravenous injection of meperidine analogues, *Psychiatry Res.* 1:249–254.

Degryse, A.-D., and Colpaert, F.C., 1986, Symptoms and behavioral features induced by 1-methyl-4-phenyl-1,2,3,6-tetrahydropyridine (MPTP) in an old Java monkey [*Macaca cynamolgus fascicularis* (Raffles)]. *Brain Res. Bull.* 16:561–571.

Di Paolo, T., Bedard, P., Daigle, M., and Boucher, R., 1986, Long-term effects of MPTP on central and peripheral catecholamine and indoleamine concentrations in monkeys, *Brain Res.* 379:286–293.

Doudet, D., Gross, C., Lebrun-Grandie, P., and Bioulac, B., 1985, MPTP primate model of Parkinson's disease: A mechanographic and electromyographic study, *Brain Res.* 335:194–199.

Doudet, D., Gross, C., Lebrun-Grandie, P., and Bioulac, B., 1986, Effect of increasing regimens of levodopa on chronic MPTP-induced parkinsonism in monkeys: Mechanographic and electromyographic data, *Electromyog. Clin. Neurophysiol.* 26:711–727.

Eidelberg, E., Brooks, B.A., Morgan, W.W., Walden, J.G., and Kokemoor, R.H., 1986, Variability and functional recovery in the *N*-methyl-4-phenyl-1,2,3,6-tetrahydropyridine model of parkinsonism in monkeys, *Neuroscience* 18(4):817–822.

Elsworth, J.D., Deutch, A.Y., Redmond, D.E., Jr., Sladek, J.R., Jr., and Roth, R.H., 1987a, Effects of 1-methyl-4-phenyl-1,2,3,6-tetrahydropyridine (MPTP) on catecholamines and metabolites in primate brain and CSF, *Brain Res.* 415:293–299.

Elsworth, J.D., Deutch, A.Y., Redmond, D.E., Jr., Sladek, J.R., Jr., and Roth, R.H., 1987b, Differential responsiveness to 1-methyl-4-phenyl-1,2,3,6-tetrahydropyridine toxicity in subregions of the primate substantia nigra and striatum, *Life Sci.* 40:193–202.

Elsworth, J.D., Deutch, A.Y., Redmond, D.E., Jr., Taylor, J.R., Sladek, J.R., Jr., and Roth, R.H., 1989, Symptomatic and asymptomatic 1-methyl-4-phenyl-1,2,3,6-tetrahydropyridine (MPTP)-treated primates: Biochemical changes in striatal regions, *Neuroscience* 33:323–331.

Elsworth, J.D., Redmond, D.E., Jr., Sladek, J.R., Jr., Deutch, A.Y., Collier, T.J., and Roth, R.H., 1990a, Reversal of MPTP-induced parkinsonism in primates by fetal dopamine cell transplants, *in* "Function and Dysfunction of the Basal Ganglia" (A.J. Franks, J.W. Ironside, R.H.S. Mindham, R.J. Smith, E.G.S. Spokes, and W. Winlow, eds.), pp. 161–181, Manchester, Manchester University Press.

Elsworth, J.D., Deutch, A.Y., Redmond, D.E., Jr., Sladek, J.R., Jr. and Roth, R.H., 1990b, MPTP reduces dopamine and norepinephrine concentrations in the supplementary motor area and cingulate cortex of the primate, *Neurosci. Lett.* 114:316–322.

Fahn, S., Elton, R., and members of the UPDRS Development Committee, 1987, Unified Parkinson's disease rating scale, *in* "Recent Advances in Parkinson's Disease" (S. Fahn, C.D. Marsden, M. Goldstein, and C. Calne, eds.), pp. 153–163, Macmillan, Florham Park, NJ.

Gerlach, M., Riederer, P., Przuntek, H., and Youdim, M.B.H., 1991, MPTP mechanisms of neurotoxicity and their implications for Parkinson's disease, *Eur. J. Pharmacol.* 208:273–286.

Guttman, M., Fibiger, H.C., Jakubovic, A., and Calne, D.B., 1990, Intracarotid 1-methyl-4-phenyl-1,2,3,6-tetrahydropyridine administration: Biochemical and behavioral observations in a primate model of hemiparkinsonism, *J. Neurochem.* 54:1329–1334.

Hornykiewicz, O., 1982, Brain neurotransmitter changes in Parkinson's disease, *in* "Movement Disorders" (D. Marsden and S. Fahn, eds.), pp. 41–58, Butterworth, London.

Imai, H., Nakamura, T., Endo, K., and Narabayashi, H., 1988, Hemiparkinsonism in monkey after unilateral caudate nucleus infusions of 1-methyl-4-phenyl-1,2,3,6-tetrahydropyridine: Behavior and histology, *Brain Res.* 474:327–332.

Irwin, I., DeLanney, L.E., Forno, L.S., Finnegan, K.T., Di Monte, D.A., and Langston, J.W., 1990, The evolution of nigrostriatal neurochemical changes in the MPTP-treated squirrel monkey, *Brain Res.* 531:242–252.

Jenner, P., and Marsden, C.D., 1986, The actions of 1-methyl-4-phenyl-1,2,3,6-tetrahydropyridine in animals as a model of Parkinson's disease, *J. Neural Transm.* Suppl XX:11–39.

Jenner, P., and Marsden, C.D., 1990, 1-Methyl-4-phenyl-1,2,3,6-tetrahydropyridine (MPTP), an update on its relevance to the cause and treatment of Parkinson's disease, *in* "Function and Dysfunction of the Basal Ganglia" (A.J. Franks, J.W. Ironside, R.H.S. Mindham, R.J. Smith, E.G.S. Spokes, and W. Winlow, eds.), pp. 140–160, Manchester, Manchester University Press.

Johannessen, J.N., Chiueh, C.C., Burns, R.S., and Markey, S.P., 1985, Differences in the metabolism of MPTP in the rodent and primate parallel differences in the sensitivity to the neurotoxic effects, *Life Sci.* 36:219–224.

Langston, J.W., Ballard, P., Tetrud, J.W., and Irwin, I., 1983, Chronic parkinsonism in humans due to a product of meperidine-analog synthesis, *Science* 219:979–980.

Langston, J.W., Forno, L.S., Rebert, C.S., and Irwin I., 1984, Selective nigral toxicity after systemic administration of 1-methyl-4-phenyl-1,2,3,6-tetrahydropyridine (MPTP) in the squirrel monkey, *Brain Res.* 292:390–394.

Lawrence, M.S., Redmond, D.E., Jr., Elsworth, J.D., Taylor, J.R., and Roth, R.H., 1991, The D1 receptor antagonist, SCH 23390, induces signs of parkinsonism in African green monkeys, *Life Sci. Pharmacol. Lett.* 49:PL-229–234.

Mitchel, I.J., Cross, A.J., Sambrook, M.A., and Crossman, A.R., 1986, N-methyl-4-phenyl-1,2,3,6-tetrahydropyridine-induced parkinsonsim in the monkey: Neurochemical pathology and regional brain metabolism, *J. Neural Transm.* (Suppl. XX):41–46.

Nomoto, M., Jenner, P., and Marsden, C.D., 1986, Alterations in motor behavior produced by the isomers of 3-PPP in the MPTP-treated marmoset, *Eur. J. Pharmacol.* 121:123–128.

Redmond, D.E., Jr., Sladek, J.R., Jr., Roth, R.H., Collier, T.J., Elsworth, J.D., Deutch, A.Y., and Haber, S.N., 1986, Fetal neuronal grafts in monkeys given methylphenyltetra-hydropyridine, *Lancet* (8490):1125–1127.

Redmond, D.E., Jr., Sladek, J.R. Jr., Roth, R.H., Collier, T.J., Elsworth, J.D., Taylor, J.R., Deutch, A.Y., and Haber, S.N., 1988, Transplanted fetal dopamine neurons: A future treatment for Parkinson's disease?, *in* "Progress in Catecholamine Research. Part B: Central Aspects" (M. Sandler, A. Dahlstöm, and R.H. Belmaker, eds.), pp. 117–123, Alan R. Liss, New York.

Rose, S., Nomoto, M., Kelly, E., Kilpatrick, G., Jenner, P., and Marsden, C.D., 1989, Increased caudate dopamine turnover may contribute to the recovery of motor function in marmosets treated with the dopaminergic neurotoxin MPTP, *Neurosci. Lett.* 101:305–310.

Rose, S., Nomoto, M., Jackson, E.A., Gibb, W.R.G., Jenner, P., and Marsden, C.D., 1990, 1-Methyl-4-(2′-methylphenyl)-1,2,3,6-tetrahydropyridine (2′-methyl-MPTP) is less neurotoxic than MPTP in the common marmoset, *Eur. J. Pharmacol.* 181:97–103.

Schneider, J.S., Yuwiler, A., and Markham, C.H., 1986, Production of a Parkinson-like syndrome in cat with *N*-methyl-4-phenyl-1,2,3,6-tetrahydropyridine (MPTP): Behavior, histology, and biochemistry, *Exp. Neurol.* 91:293–307.

Schneider, J.S., Yuwiler, A., and Markham, C.H., 1987, Selective loss of subpopulations of ventral mesencephalic dopamine neurons in the monkey following exposure to MPTP, *Brain Res.* 411:144–150.

Schneider, J.S., Unguez, G., Yuwiler, A., Berg, S.C., and Markham, C.H., 1988, Deficits in operant behaviour in monkeys treated with *N*-methyl-4-phenyl-1,2,3,6-tetrahydropyridine (MPTP), *Brain* 111:1265–1285.

Schultz, W., Studer, A., Jonsson, G., Sundström, E., and Mefford, I., 1985, Deficits in behavioral initiation and execution processes in monkeys with 1-methyl-4-phenyl-1,2,3,6-tetrahydropyridine-induced parkinsonism, *Neurosci. Lett.* 59:225–232.

Schwab, R.S., 1972, Akinesia paradoxica, *Electroencephalogr Clin. Neurophysiol.* 31:87–92.

Schwab, R.S., and England, A.C., Jr., 1969, Projection techniques for evaluating surgery in Parkinson's disease, *in* "Third Symposium on Parkinson's Disease" (F. Gillingham and I. Donaldson, eds.), pp. 152–157, E. and S. Livingstone, Edinburgh, Scotland.

Schwab, R.S., and Zeiper, I., 1965, Effects of mood, motivation, stress, and alterness on the performance in Parkinson's disease, *Psychiatr. Neurol. (Basel)* 150:345–357.

Sladek, J.R., J., Collier, T.J., Haber, S.N., Roth, R.H., and Redmond, D.E., Jr., 1986, Survival and growth for fetal catecholamine neurons transplanted into primate brain, *Brain Res. Bull.* 17(6):809–818.

Sladek, J.R., Jr., Redmond, D.E., Jr., Collier, T.J., Haber, S.N., Elsworth, J.D., Deutch, A.Y., and Roth, R.H., 1987a, Transplantation of fetal dopamine neurons in primate brain reverses MPTP-induced parkinsonism, *Prog. Brain Res.* 71:309–322.

Sladek, J.R., Jr., Collier, T.J., Haber, S.N., Deutch, A.Y., Elsworth, J.D., Roth, R.H., and Redmond, D.E., Jr., 1987b, Reversal of parkinsonism by fetal nerve cell transplants in primate brain, *Ann. NY Acad. Sci.* 495:641–657.

Sladek, J.R., Jr., Redmond, D.E., Jr., Collier, T.J., Blount, J.P., Elsworth, J.D., Taylor, J.R., and Roth, R.H., 1988, Fetal dopamine neural grafts: Extended reversal of methylphenyltetrahydrapyridine-induced parkinsonism in monkeys, *Prog. Brain Res.* 78:497–506.

Stern, Y., and Langston, W., 1985, Intellectual changes in patients with MPTP-induced parkinsonism, *Neurology* 35:1506–1509.

Taylor, J.R., Elsworth, J.D., Roth, R.H., Collier, T.J., Sladek, J.R., Jr., and Redmond, DE, Jr., 1990a, Improvements in MPTP-induced object retrieval deficits and behavioral deficits after fetal nigral grafting in monkeys, *Prog. Brain Res.* 82:543–559.

Taylor, J.R., Elsworth, J.D., Roth, R.H., Sladek, J.R., Jr., and Redmond, D.E., Jr., 1990b, Cognitive and motor deficits in the acquisition of an object retrieval/detour task in MPTP-treated monkeys, *Brain* 113(3):617–637.

Taylor, J.R., Roth, R.H., Sladek, J.R., Jr., Redmond, D.E., Jr., 1990c. Cognitive and motor deficits in the performance of an object retrieval task with barrier/detour in monkeys treated with MPTP: Long-term performance and effect of transparency of the barrier, *Behav. Neurosci.* 104(4):564–576.

Taylor, J.R., Elsworth, J.D., Roth, R.H., Sladek, J.R., Jr., Collier, T.J., and Redmond, D.E., Jr., 1991a, Grafting of fetal substantia nigra to striatum reverses behavioral deficits induced by MPTP in primates: A comparison with other types of grafts as controls, *Exp. Brain Res.* 85:335–348.

Taylor, J.R., Lawrence, M.S., Redmond, D.E. Jr., Elsworth, J.D., Roth, R.H., Nichols, D.E., and Mailman, R.B., 1991b, Dihydrexidine, a full D1 agonist, reduces MPTP-induced parkinsonism in African green monkeys, *Eur. J. Pharmacol.* 199(III):387–388.

U.S. Public Health Service, 1985, "Guide for the Care and Use of Animals." U.S. Government Printing Office, Washington, DC.

Winer, B.J., 1974 "Statistical Principles in Experimental Design" (2nd ed.), McGraw-Hill, New York.

Zigmond, M.J., and Stricker, E.M., 1989, Animal models of parkinsonism using selective neurotoxins: Clinical and basic implications, *Int. Rev. Neurobiol.* 31:1–79.

7

Intranigral Iron Infusion as a Model for Parkinson's Disease

G.W. ARENDASH, G.J. SENGSTOCK, C.W. OLANOW, S. BARONE, Jr., and A.J. DUNN

Parkinson's disease (PD) is characterized by a progressive degeneration of monoaminergic neurons in the brainstem, particularly nigrostriatal dopaminergic (DA) neurons originating in zona compacta of the substantia nigra (SNc) (Forno, 1982; Hirsch *et al.*, 1988). Although the neuropathology and neurochemistry of the disease are reasonably well understood, the pathophysiology and etiologic factor(s) remain unknown. In the absence of direct support for an infectious or genetic etiology, recent interest has centered on possible involvement of exogenous (environmental) or endogenous toxins and the role played by free radicals generated from oxidation reactions (Halliwell, 1989; Youdim *et al.*, 1989; Olanow, 1990; Olanow and Cohen, 1992). Support for possible involvement of environmental toxins has been derived from the discovery that a meperidine analog, 1-methyl-4-phenyl-1,2,3,6-tetrahydropyridine (MPTP), selectively destroys nigrostriatal DA neurons to induce parkinsonism in humans and experimental parkinsonism in animals (Langston *et al.*, 1983; Stern, 1990) and from a series of epidemiological studies (for review, see Tanner and Langston, 1990). In recent years, however, a substantial body of both clinical and experimental evidence has accumulated that supports the concept of an "endogenous neurotoxic mechanism" in PD, wherein increased oxidative stress within the substantia nigra (SN) resulting from excessive oxygen radical species formation is directly

---

G.W. ARENDASH and G.J. SENGSTOCK • Department of Biology and Institute for Biomolecular Science, University of South Florida, Tampa, Florida 33620. **C.W. OLANOW** • Departments of Neurology, Pharmacology, and Psychiatry, University of South Florida College of Medicine, Tampa, Florida 33620. **S. BARONE, Jr.** • Man Tech Environmental Technology, P.O. Box 12313, 2 Triangle Drive, Research Triangle Park, North Carolina 27709. **A.J. DUNN** • Department of Pharmacology and Therapeutics, Louisiana State University Medical Center, Shreveport, Louisiana 71130.

Toxin-Induced Models of Neurological Disorders, edited by Michael L. Woodruff and Arthur J. Nonneman. Plenum Press, New York, 1994.

responsible for the death of dopaminergic neurons (Youdim *et al.*, 1989, 1991; Olanow, 1990; Jenner, 1991, 1992, Olanow and Cohen, 1992).

This chapter describes an "intranigral iron infusion" model for PD that was developed because: (1) there is excess iron in the SN of PD brains, (2) iron is well known to catalyze the formation of cytotoxic free radical species, and (3) the consequent hypothesis that oxidation reactions/free radical formation are involved in the pathogenesis of PD. Iron-induced degeneration of nigrostriatal DA neurons may involve oxidative damage to the SN through mechanisms that could cause the progressive loss of nigrostriatal DA neurons in PD. In this chapter we first review the evidence supporting a role of free radicals/oxidation reactions in PD, followed by a characterization of our nigral iron infusion model from the prospectus of neuropathological, neurochemical, and behavioral effects. We then compare and contrast this model with two currently utilized models for PD—animals treated with either MPTP or 6-hydroxydopamine.

1. Involvement of Free Radicals/Oxidation Reactions in Parkinson's Disease

Free radicals are atoms or molecules that contain an unpaired electron in one of its orbitals. Molecular oxygen qualifies as a di-radical because it contains two orbitals with unpaired electrons that have parallel spin in the ground state. Oxygen can be reduced by the addition of one, two, or three electrons in a stepwise fashion to generate superoxide anion (O_2^-), hydrogen peroxide (H_2O_2), and hydroxyl free radical ($OH^{\cdot}$). These reactive oxygen species (ROS) and other free radicals are normal by-products of oxidative metabolism and are generally removed by antioxidant defenses (Halliwell, 1991). Under normal circumstances, there is a balance between free radical production and defense mechanisms. A shift in this equilibrium favoring the formation of ROS is known as *oxidant stress*. Increasing evidence indicates that oxidant stress can damage essential biological molecules such as proteins, lipids, and DNA (Halliwell and Gutteridge, 1989). Concern that nigrostriatal DA neurons may be vulnerable to oxidant stress has been raised because of the manner in which dopamine is metabolized (Olanow, 1990; Olanow and Cohen, 1992). Dopamine is oxidized to form a pool of hydrogen peroxide in a reaction catalyzed by monoamine oxidase (MAO). Hydrogen peroxide is normally cleared from the brain by glutathione, through the enzyme glutathione peroxidase. However, the glutathione pool is saturable. Excess H_2O_2 in the presence of ferrous iron (Fe^{2+}) can proceed through the Fenton reaction and be reduced to form $OH^{\cdot}$, which appears to be the primary mediator of oxidative damage. Dopamine can also be nonenzymatically "auto-oxidized" to form H_2O_2, as well as potentially toxic quinones, semiquinones, and superoxide anions (Carlsson and Fornstedt, 1991).

Increased free-radical-induced damage to nigrostriatal DA neurons and consequent PD could result from (1) increased dopamine metabolism (turnover) resulting in increased formation of hydrogen peroxide, (2) decreased hydrogen peroxide clearance due to reductions in glutathione or glutathione peroxidase, and/or (3) excessive levels of reactive iron in the substantia nigra, resulting in increased $OH^{\cdot}$ formation. Regarding the last of these possibilities, a direct relationship exists between local iron concentrations and oxidant-induced damage (Halliwell and Gutteridge, 1984). It is noteworthy that iron must usually be in a "free or low-molecular weight" state (i.e., unbound to proteins such as ferritin or transferrin) for it to participate in redox reactions and catalyze the formation of cytotoxic hydroxyl free radicals.

Several lines of evidence are consistent with the involvement of free radicals/oxidation reactions in the pathogenesis of PD. First, a specific increase in iron concentration within the SN of postmortem PD brains has been reported by several laboratories (Earle, 1968; Dexter *et al.*, 1987,; 1989a; Sofic *et al.*, 1988; Hirsch *et al.*, 1991), particularly for the ferric (Fe^{3+}) form (Sofic *et al.*, 1988; Riederer *et al.*, 1989). One recent report indicates that this increased nigral iron occurs primarily in SNc (Sofic *et al.*, 1991). Although a histochemical study by Jellinger *et al.* (1990) concluded that most of the increased SNc iron is associated with reactive microglia, astrocytes, macrophages, and nonpigmented neurons, more recent and sophisticated studies involving X-ray microanalysis and laser microprobe mass analysis (LAMMA) show that in PD iron is significantly increased in neuromelanin granules of pigmented SNc neurons (Jellinger *et al.*, 1992; Good *et al.*, 1992).

Why iron accumulates in the SN of PD brains is not presently known. Its accumulation may be secondary to degeneration of SNc neurons, but one study of patients with Progressive Supranuclear Palsy who have substantial neuronal degeneration in the SNc notes no increase in SNc iron content (Hirsch *et al.*, 1991). These results are not, however, consistent with a study by Dexter *et al.* (1991) in which increased levels of nigral iron were found in patients with either Progressive Supranuclear Palsy or multiple system atrophy. One possible explanation for nigral iron accumulation in PD involves the capacity of neuromelanin to bind ferric iron (Fe^{3+}) and reduce it to Fe^{2+}, resulting in enhanced $OH^{\cdot}$ production (Youdim *et al.*, 1989). Along this line, neuromelanin-containing dopaminergic neurons within the midbrain appear to be selectively vulnerable to degeneration in PD (Hirsch *et al.*, 1988; Kastner *et al.*, 1992; Jellinger *et al.*, 1992). Because iron bound to ferritin is in a relatively unreactive state and therefore less capable of promoting free radical formation, it is noteworthy that nigral ferritin levels have been found to be significantly decreased in PD (Dexter *et al.*, 1990, 1991). Although this finding requires confirmation, the combination of increased iron and decreased ferritin suggests the presence of excess "free or low-molecular weight" iron within the SN of PD patients—iron that is capable of catalyzing formation of cytotoxic free radicals. Therefore, nigral iron excess may function as an endogenous neurotoxic factor and, consequently, a primary etiological factor in the pathogenesis of PD (Youdim *et al.*, 1989).

Additional evidence consistent with oxidative damage in PD is provided by studies showing that striatal dopamine turnover is dramatically increased in PD, suggesting enhanced dopamine synthesis and release within surviving nigrostriatal DA neurons (Zigmond *et al.*, 1990). Increased dopamine turnover has the potential to induce oxidative stress in nigrostriatal DA neurons through a greater production of hydrogen peroxide. This increased hydrogen peroxide must be detoxified by the glutathione/glutathione peroxidase system to prevent an increased conversion of hydrogen peroxide to cytotoxic $OH^{\cdot}$. In this context, it is noteworthy that in PD levels of glutathione and in some cases glutathione peroxidase are reduced (Kish *et al.*, 1985; Perry and Yong, 1986; Riederer *et al.*, 1989). Decreased protection against oxidant damage may, therefore, characterize the Parkinsonian brain, making free radical formation and resultant damage more likely.

A major cytotoxic action of free radicals in the brain involves their ability to react with polyunsaturated fatty acids in phospholipid membranes, thus initiating and promoting the process of lipid peroxidation (Halliwell and Gutteridge, 1989). Studies by Dexter *et al.* (1989b) indicate that levels of malondialdehyde (MDA), a breakdown product of lipid peroxidation, are selectively increased in the SN of PD brains, while polyunsaturated lipids are reduced. These data suggest that increased lipid peroxidation is occurring in the SN of PD brains. The role of iron in initiating lipid peroxidation is well recognized, with either the ferrous or ferric form capable of initiating peroxidation damage providing there is a mechanism for interconversion of these two redox states (Zaleska and Floyd, 1985; Braughler *et al.*, 1986; Minotti and Aust, 1987). Furthermore, iron may actually be bound to or solubilized by membrane lipid (Ernster and Nordenbran, 1982). It is important to note that the rate of lipid peroxidation depends on the Fe^{2+}/Fe^{3+} ratio and that maximal rates are at a ratio near 1:1. In this context, the Fe^{2+}/Fe^{3+} ratio in the SN of normal human brains has been reported to be 2:1, whereas the ratio in PD brains appears to be approximately 1:1 (Riederer *et al.*, 1989; Sofic *et al.*, 1991)—this reduced ratio may contribute to the elevation in nigral lipid peroxidation reported in PD brains (Dexter *et al.*, 1989b).

If free radical formation and oxidation reactions contribute to the neuronal damage that underlies PD, strategies designed to interfere with these reactions may have therapeutic value. Recent clinical studies have, in fact, demonstrated that in patients with untreated PD, the MAO-B inhibitor deprenyl delays the development of disability requiring symptomatic therapy with L-dopa (Parkinson's Study Group, 1989; Tetrud and Langston, 1989). It is unknown whether deprenyl acts by (1) decreasing PD symptomatology by increasing dopamine availability/transmission, (2) providing a protective (antioxidant) effect by limiting the oxidation of dopamine and thus reducing free radical formation (Olanow and Calne, 1991), or (3) some other unknown mechanism. At the present time, these encouraging clinical data are at least consistent with involvement of free radical/oxidative damage in the pathogenesis of PD.

In summary, a significant body of evidence supports a role for free radicals/oxidation reactions in the development of PD. A formidable array of free radical-promoting conditions appear to characterize the SN of the Parkinsonian brain, including increased iron content, decreased protective mechanisms (glutathione, glutathione peroxidase), and increased dopamine turnover. Separately or in combination, these characteristics could provide an "oxidative stress," with consequent nigral neuronal degeneration in PD. An animal model for PD based on oxidative damage within the SN may be extremely valuable, particularly for assessing potential antioxidant regimens. We believe that iron infusions into the SN of experimental animals could provide such a model and have, thus, developed a model of intranigral iron infusion in the rat.

2. Development of an Iron Infusion Model for Parkinson's Disease

Given that nigral iron content is increased in the SN of PD brains (primarily in the SNc) and that this excess nigral iron may contribute to oxidative damage to NS-DA neurons, a logical animal model for PD would involve experimental elevation of nigral iron levels. In considering the development of such a model, several points are worthy of consideration. First, although iron is normally present throughout the mammalian brain, it is concentrated in only a few brain regions, most notably the globus pallidus, neostriatum, and substantia nigra (Hill and Switzer, 1984; Dwork *et al.*, 1988). How and why iron normally accumulates in these areas is currently unknown. Second, controversy exists as to the exact mechanism by which iron enters the brain parenchyma. Iron, conjugated to transferrin, is generally thought to bind to transferrin receptors on the luminal side of the blood-brain barrier, after which the entire complex is internalized. Iron then separates from this complex and, in some unknown manner, is secreted on the abluminal side of the blood-brain barrier (Roberts *et al.*, 1992). Additionally, a transferrin-independent process has been described (Basset *et al.*, 1986; Irie and Tavassoli, 1987). Supportive of the latter mechanism are recent studies showing that atransferrinemic mice (<1% transferrin) achieve a normal brain distribution/content of iron (Connor and Benkovic, 1992), suggesting that transferrin is not necessary for iron to cross the blood-brain barrier. Regardless of how iron enters the brain, a regulatory mechanism exists that limits the ability of systemic iron to cross the blood-brain barrier (Banks *et al.*, 1988). In fact, attempts to increase rat brain levels of iron by peripheral administration have not been successful (Dexter *et al.*, 1989c; Ben-Shachar *et al.*, 1991a). Finally, the intracellular localization of iron has yet to be determined, although it is believed that most intracellular iron is stored in association with ferritin in the Fe^{3+} form.

Studies involving the direct "intracortical" infusion of iron chloride have clearly demonstrated the brain's general susceptibility to excess iron since such infusions induce transient superoxide radical formation, lipid peroxidation, tissue necrosis, and macrophage infiltration (Wilmore and Rubin, 1982; Wilmore *et al.*, 1983; Triggs and Wilmore, 1984). However, a major drawback to most of these studies involves the fact that solutions of iron chloride are extremely acidic and cannot be buffered without precipitation of the iron. We have recently devised an iron-citrate solution for intracerebral infusion that is pH balanced and isosmotic to brain tissue (Sengstock *et al.*, 1992a). Any effects observed following infusion of this solution should therefore be directly attributable to iron.

In our initial intranigral infusion studies, 1.0–1.5 μl of a 12.6 mM iron citrate solution (pH 7.4) was infused unilaterally and an identical infusion was performed 1 week later. At sacrifice 1 month after the initial infusion, neostriatal levels of dopamine, DOPAC, and HVA were reduced by more than 90% on the ipsilateral side compared to levels on the contralateral side and to untreated control animals (Sengstock, Menzies, and Arendash, unpublished observations). However, serotonin and 5-HIAA levels were also reduced ipsilaterally (by approximately 60%), perhaps reflecting involvement of serotoninergic fibers ascending medially to the SN. Histologic evaluation of infusion sites indicated that the iron infusate volume was too large to limit iron diffusion to the relatively small SN. In addition, the iron-citrate infusate was determined to be hyperosmotic and thus the effects could not with certainty be attributed to iron. Therefore, utilizing a small infusate volume (0.25 μl) and several iron concentrations that were pH-balanced and isosmotic, we investigated three different infusion protocols to determine the most effective protocol for limiting iron diffusion to the SN and for inducing neuropathologic and neurochemical changes to the nigrostriatal DA system.

Adult male Sprague-Dawley rats received infusion(s) of vehicle or an iron (Fe) solution at the junction between the substantia nigral zona compacta (SNc) and (Fe) zona reticularis (SNr) according to one of the following protocols: *Sequential Infusion:* Animals received a single 0.25 μl infusion of 0.63 nmole Fe or vehicle solution into the SN unilaterally, followed by an identical infusion 7 days later (total Fe infused was 1.25 nmole [Sengstock *et al.*, 1993]); *Side-by-Side Infusion:* Animals received two side-by-side infusions (0.25 μl each) of vehicle, 1.25 nmole, 2.10 nmole, 3.15 nmole, or 6.3 nmole total Fe unilaterally into the SN during the same surgical session (Sengstock *et al.*, 1992a); *Single Infusion:* Animals received a single 0.25 μl infusion of 0.63 nmole Fe into the SN unilaterally (Sengstock *et al.*, 1992b).

2.1. Histochemical and Neuropathological Changes within the SN following Nigral Iron Infusion

Coronal sections through the SN were stained with either thionine or Perls' stain plus diaminobenzidine (DAB) intensification. The basis of the Perls' stain is the release of Fe^{3+} ions from iron-containing compounds by HCl and the ensuing formation of ferric ferrocyanide (Prussian blue). The reaction is usually pale in the rat but can be intensified with the addition of DAB, which is oxidized by ferric ferrocyanide and forms an insoluble brown precipitate (Hill and Switzer, 1984). It should be noted that the rat SNc normally does not stain for iron (Figure 1A). By contrast, the SNr normally does stain for iron, most of which is localized to oligodendrocytes or fibers of the neuropil (Hill and Switzer, 1984; Dwork *et al.*, 1988; Connor *et al.*, 1990). At both 1 and 2 months following sequential Fe infusion (1.25 nmole Fe total), a marked increase in iron-staining intensity was clearly evident and limited to the infused SN. Heaviest iron staining was in the SNc and dorsal-most portions of the SNr. Within this region of heavy iron staining, large numbers of presumed macrophages and reactive astrocytes stained intensely for iron (Figure 1B). Nigral iron infusion enhanced the normal pattern of SNr iron staining in two respects: (1) by increasing fiber staining and (2) by increasing the iron staining of presumptive oligodendrocytes, both in terms of a greater intensity of cellular staining and a larger number of cells (Figure 2). This latter finding is consistent with a proposed "paracrine" role of oligodendrocytes in iron regulation (Connor *et al.*, 1990).

Iron infusion into the SN caused substantial neuronal loss and atrophy, primarily within the SNc (Figure 3); neuronal loss in the SNr was limited to its extreme dorsal region. The area of neurodegeneration correlated closely with an area of reactive gliosis (Figure 3B), as well as with the area of intense iron staining (Figure 1B). Iron-positive neurons were occasionally seen in the SNc and SNr at 1 or 2 months postinfusion, with no apparent change in their staining quality compared to the uninfused SN. However, iron-positive neurons were much more prevalent in the infused SNc at 24 hr following the first iron infusion, suggesting that most iron-positive neurons within the SNc may degenerate after iron uptake. Characteristic of iron-positive SNc neurons was a faint granular appearance to the cytoplasm and an intensely stained nucleus containing a very heavily stained nucleolus and a variable number of large iron granules (Figure 2C).

Although the infusion of 1.25 nmole Fe via the sequential infusion protocol clearly provided a selective neuronal degeneration of the SNc, the total area affected was relatively small (no more than one-third of the SNc). Medial neurons in the SNc, as well as those in its most rostral portions, were not affected. Compared to the sequential infusion protocol, side-by-side infusion of the same

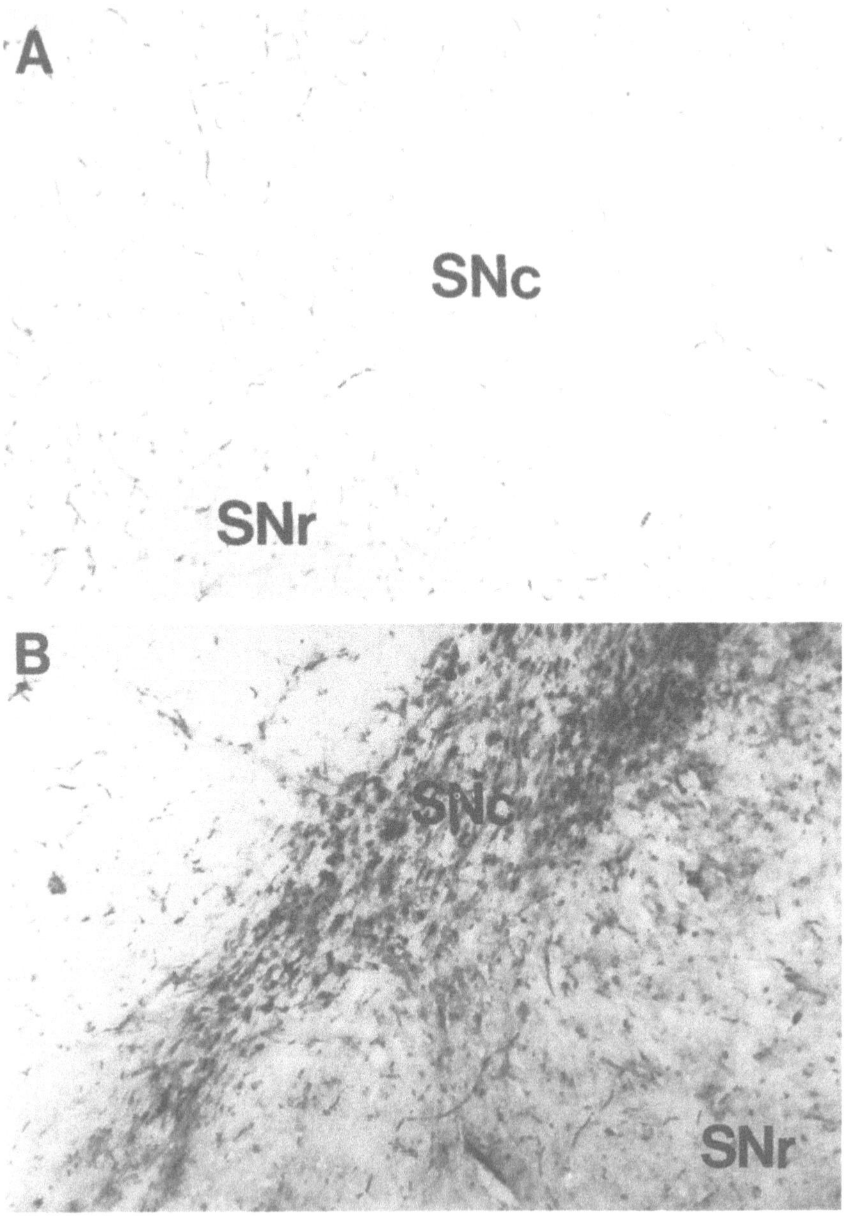

Figure 1 Iron-stained coronal section from an animal that received a sequential infusion of 1.25 nmole Fe at the interface of SNc and SNr 1 month prior to sacrifice. Note the heavy iron staining within the SNC and dorsal-most portions of the SNr on the infused side (**B**) compared to the uninfused side (**A**). Within the infused SNc, most iron-positive cells appeared to be reactive microglia and astrocytes (B). Normally, the SNc does not stain for iron, and the SNr stains only lightly at this age (**A**). ×100. Abbreviations: SNc, zona compacta of the substantia nigra; SNr, zona reticularis of the substantia nigra.

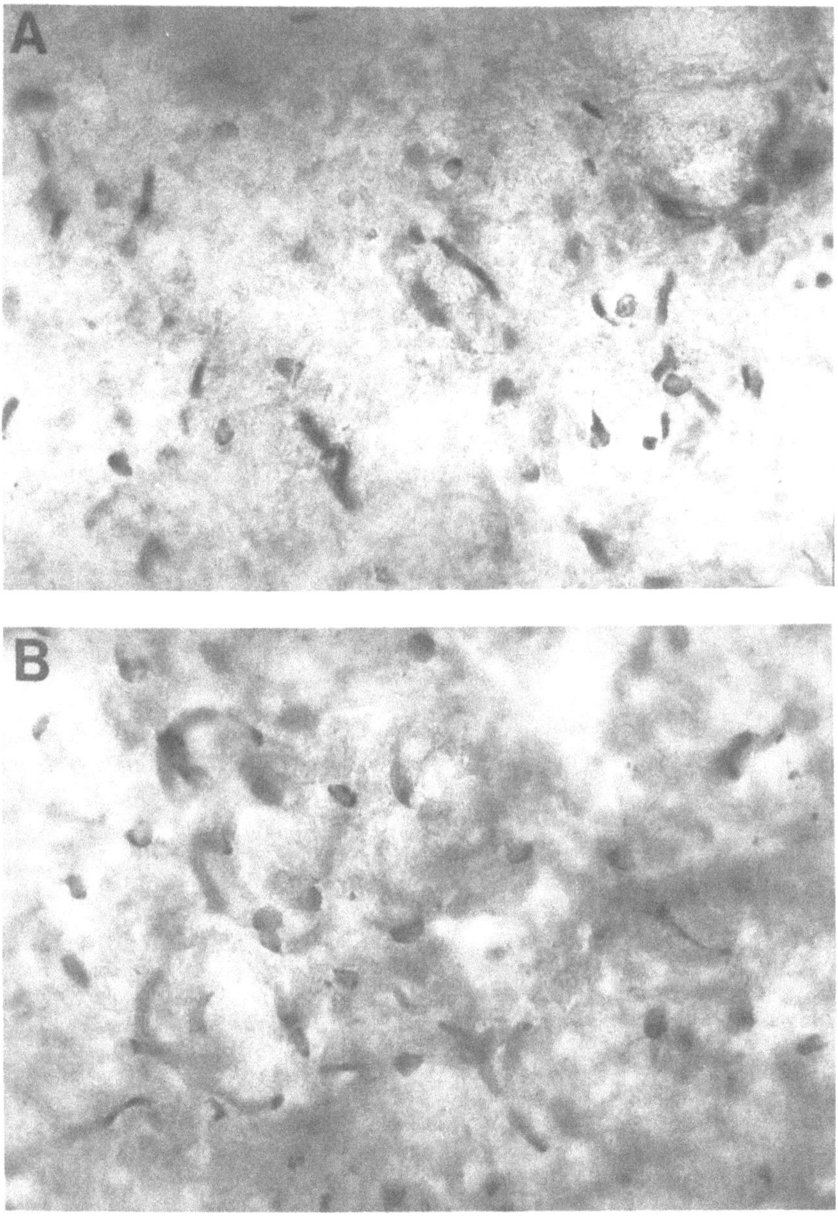

Figure 2 The iron-stained SNr from an animal that received a sequential infusion of 1.25 nmole Fe 1 month before sacrifice. In comparison to the uninfused side (**A**), note the increased number of iron-positive oligodendrocytes and their greater intensity of staining within the infused SNr (**B**). ×400. (**C**) Iron-positive neurons (arrows) within the SNc of an animal that received a single infusion of 0.63 nmole Fe intranigrally and sacrificed 24 hr later. Typically, these neurons have a light staining cytoplasm surrounding a more intensely stained nucleus that contains a very heavily stained nucleolus and several large iron-positive granules.

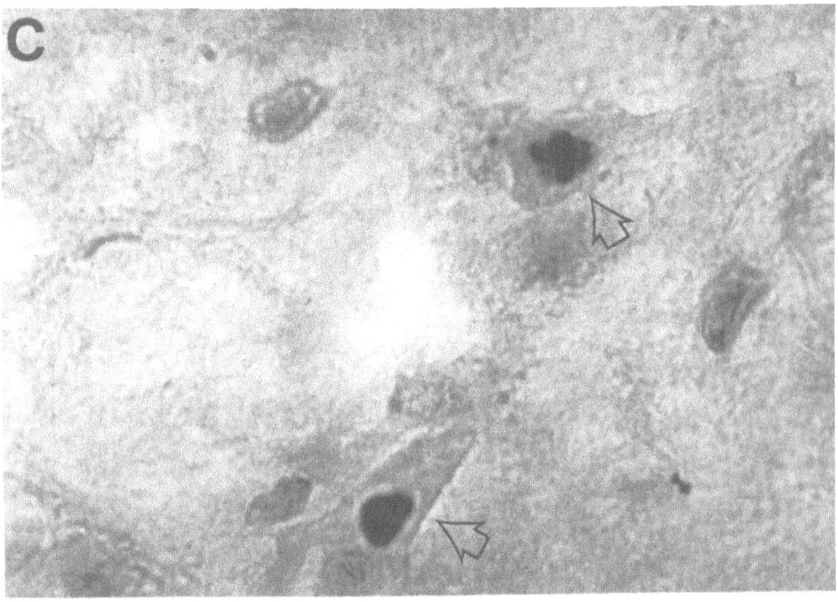

Figure 2 (*Continued*)

total amount of iron (1.25 nmole), resulted in a larger medial to lateral area of the SNc exposed to, and affected by, iron. At 1–2 months postinfusion, increased iron staining was restricted to the infused SN, with the most intense staining located within the SNc and dorsal-most SNr (Figure 4). Cellular staining for iron was consistent with that previously described for the sequential infusion model (compare Figures 1B and 4B). In contrast to the sequential infusion protocol, however, increased iron staining within the SNc extended to its medial-most regions (Figure 4B,C). Consequently, the corresponding gliosis and selective neuronal losses within SNc/dorsal SNr were greater in "medial to lateral" extent. Most SNc neurons within this more extensive gliosis were degenerative (Figure 5), although there still existed a majority of neurons rostral or caudal to the infusion that appeared unaffected.

With increasing amounts of iron (2.10, 3.15, and 6.3 nmole) infused via the side-by-side protocol, neurodegenerative changes were more widespread within the SN at sacrifice 2 months later. Both SNr atrophy and neurodegeneration were particularly evident with higher amounts of iron infused. Infusion of intermediate amounts of iron (2.10 and 3.15 nmole) resulted in destruction of most SNc neurons, whereas infusion of 6.3 nmole Fe induced a near complete loss of SNc neurons. Vehicle infusions had no effect on nigral iron staining and did not result in any determinable neuronal losses in the SN.

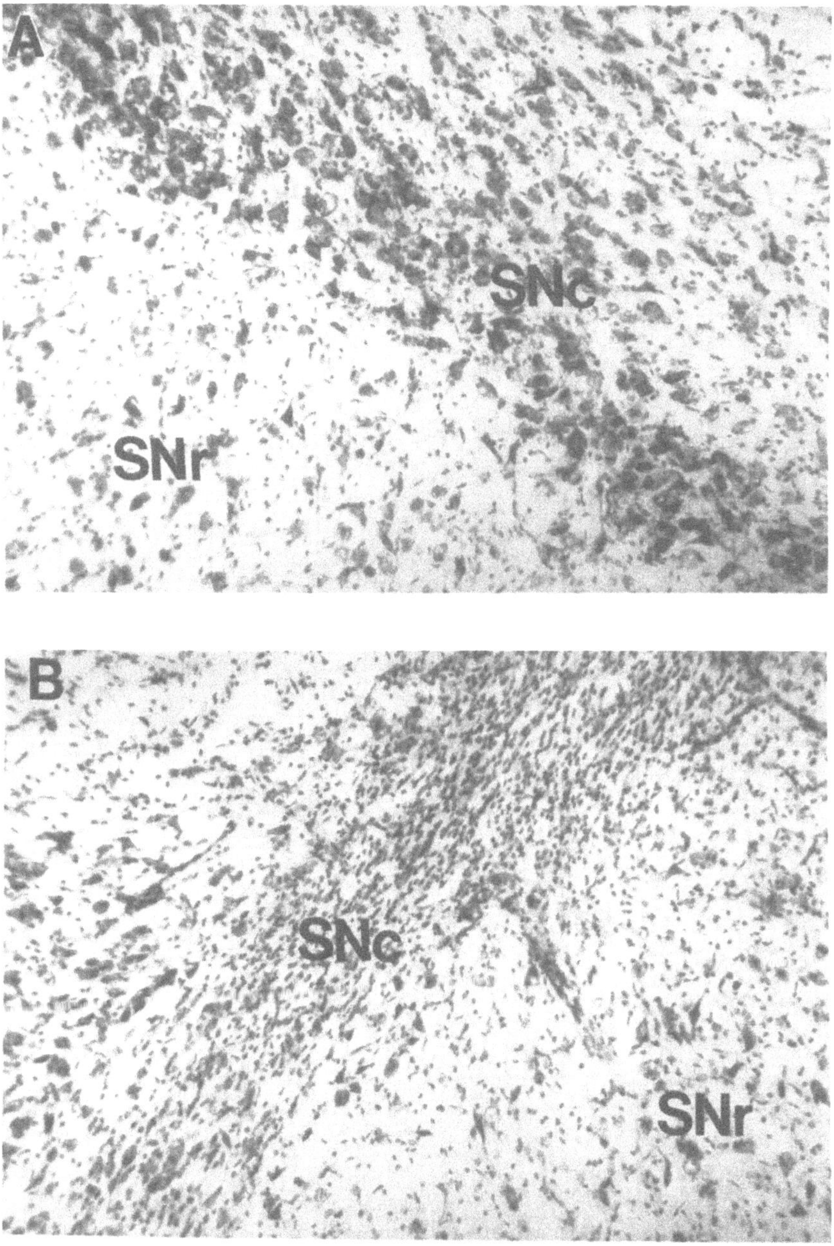

Figure 3 A thionine-stained coronal section from the same rat in Figure 1 that received sequential infusion of 1.25 nmole Fe at the interface between SNc and SNr 1 month prior to sacrifice. This section, which is immediately adjacent to the iron-stained section shown in Figure 1, shows substantial neuronal loss and gliosis within the infused SNc (**B**) in comparison to the uninfused side (**A**). ×100. The area of neuronal degeneration and gliosis corresponds with the area of increased iron staining (compare Figure 1B with Figure 3B).

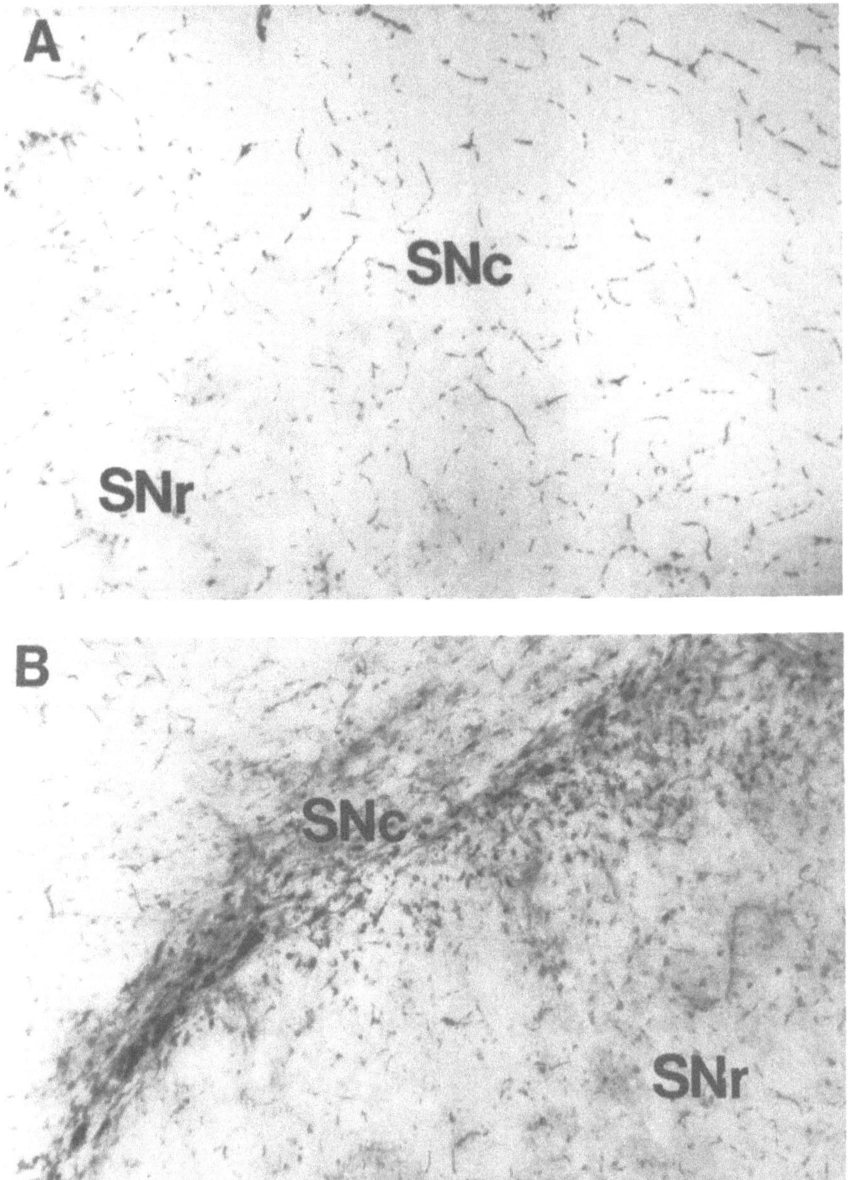

Figure 4 (**A** and **B**) Iron-stained coronal section through the SN at 1 month following side-by-side infusion of 1.25 nmole Fe. Compared to the uninfused side (**A**), intense iron staining is evident within the infused SN (**B**), particularly in the SNc. ×100. (**C**) From the same animal, a computer-enhanced negative image of iron staining within the iron-infused SN; areas of intense iron staining appear white in this image. Note the striking accumulation of infused iron within the SNc; heavy iron staining is seen throughout the entire medial-to-lateral extent of the SNc at this level.

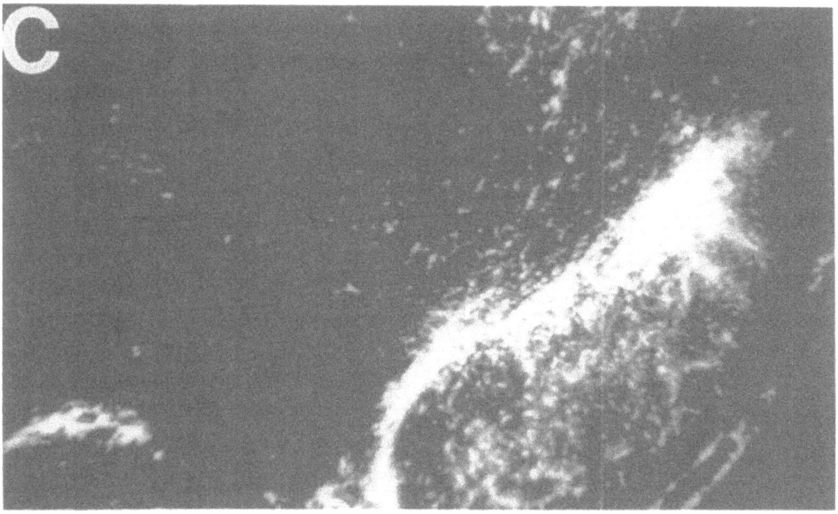

Figure 4 *(Continued)*

2.2. Striatal Neurochemical Changes following Nigral Iron Infusion

Neostriatal tissues were dissected out bilaterally and analyzed for biogenic amines using high-performance liquid chromatography (HPLC) as described in Sengstock *et al.* (1992a). The limited degree of iron-induced degeneration within the SNc following sequential infusion of 1.25 nmole Fe was underscored by the lack of any ipsilateral versus contralateral changes in striatal DA or its catabolites (DOPAC and HVA) at both 1 and 2 months postinfusion. The most likely explanation for these neurochemical results involves a compensatory upregulation of dopamine metabolism in surviving nigrostriatal DA neurons (Zigmond *et al.*, 1990) and/or collateral sprouting of nigrostriatal DA neurons. Sequential infusions of vehicle solution into the SN also had no effect on striatal DA or its catabolites. Similarly, following side-by-side infusion of 1.25 nmole Fe, neostriatal concentrations of DA, DOPAC, and HVA ipsilateral to nigral infusion were unaffected in comparison to contralateral values. Nonetheless, as was the case with sequential infusions of the same total amount of iron, a rather selective degeneration of SNc neurons occurred. This may relate to an increased sensitivity of SNc neurons to iron-induced damage, as we have proposed (Sengstock *et al.*, 1992a).

Consistent with the dose-related destruction of SNc neurons following side-by-side infusions were dose-related decreases in striatal levels of DA, DOPAC, and HVA. Two months after side-by-side infusion of 2.10 nmole Fe, the striatal

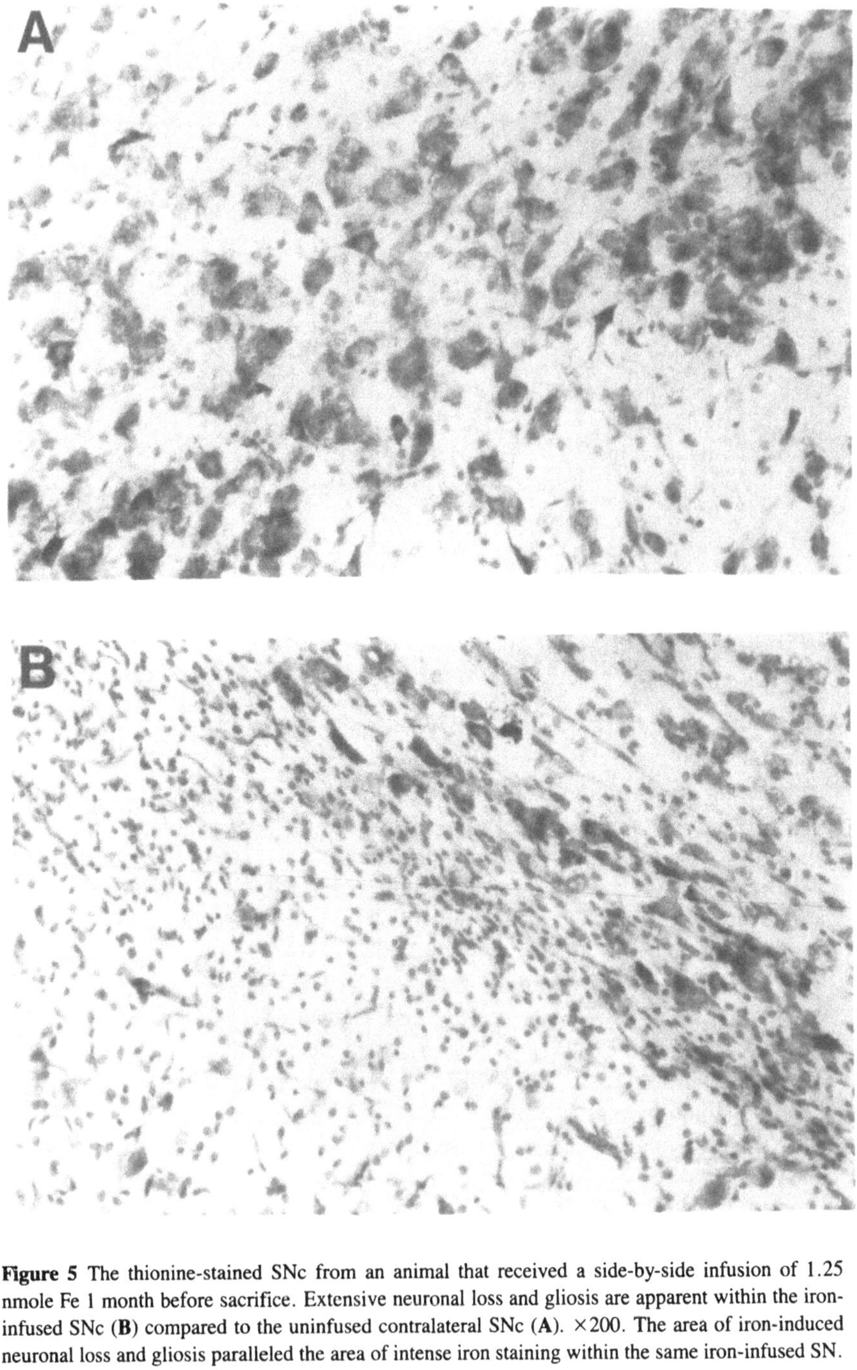

Figure 5 The thionine-stained SNc from an animal that received a side-by-side infusion of 1.25 nmole Fe 1 month before sacrifice. Extensive neuronal loss and gliosis are apparent within the iron-infused SNc (**B**) compared to the uninfused contralateral SNc (**A**). ×200. The area of iron-induced neuronal loss and gliosis paralleled the area of intense iron staining within the same iron-infused SN.

concentrations of these neurochemicals were reduced ipsilaterally by 40%–47% compared to the contralateral uninfused side. Reductions in striatal DA and its catabolites were greatest following infusion of 6.3 nmole Fe (DA −82%; DOPAC −77%; HVA −69%). Moreover, DA turnover (as indicated by [DOPAC+HVA]/DA) was substantially elevated by 50% at 2 months following infusion of 6.3 nmole Fe, suggesting that the small number of surviving SNc dopaminergic neurons had upregulated their metabolism of DA. Nondopaminergic markers in striatum (i.e, NE, 5-HT, 5-HIAA) were essentially unaffected by any amount of iron infused, affirming that iron diffusion and resultant neurodegeneration were restricted to the SN region.

2.3. Induction of Lipid Peroxidation following Nigral Iron Infusion

We have utilized a single infusion protocol, involving a 0.25 μl nigral infusion of 0.63 nmole Fe, to determine whether enhanced lipid peroxidation occurs within the rat SN as a result of the presence of excess iron. As discussed in Section 1, increased lipid peroxidation appears to occur in the SN of PD brains as measured by nigral levels of malondialdehyde (MDA), a reaction product of lipid peroxidation (Dexter, 1989b). Because iron can initiate lipid peroxidation (Braughler *et al.*, 1986; Minotti and Aust, 1987), increased lipid peroxidation in the Parkinsonian SN may be due, at least in part, to the elevated levels of iron found in the SN of PD brains (Youdim *et al.*, 1989; Olanow and Cohen, 1992). For our determinations of lipid peroxidation in the rat SN following iron infusion, only *basal* MDA levels were measured. This was accomplished by incorporating both desferrioxamine and BHT into our spectrophotometric MDA microassay (Sengstock *et al.*, 1992b) to chelate all available iron, thus controlling for the disparity in iron levels between infused and uninfused tissues and preventing further iron-catalyzed or nonspecific MDA formation following decapitation (Zaleska and Floyd, 1985).

The MDA levels in the iron-infused SN were significantly elevated (+72%) at 1 hour postinfusion as compared to values from unoperated control animals and were elevated (+25%) in other iron-infused animals sacrificed 24 hr postinfusion, although the elevation in MDA levels at this later time-point did not reach statistical significance. Nigral MDA levels at 1 week and 1 month following iron infusion were not significantly elevated, and vehicle infusions had no effect on nigral MDA levels at any time-point between 1 hr and 1 month postinfusion. These data indicate that iron infusion into the SN acutely increases lipid peroxidation within the infused SN. The results suggest that iron-induced lipid peroxidation may be involved not only in the degeneration of nigrostriatal DA neurons resulting from nigral iron infusion in rats, but also in the degeneration of nigrostriatal DA neurons in PD. An important caveat is that the aforementioned MDA data do not eliminate the possibility that lipid peroxidation within the iron-infused SN is elevated chronically, but simply at a level insufficient for detection by our spectrophotometric MDA microassay.

2.4. Progressive Changes Induced by Iron Infusion over 6 Months

The aforementioned results from side-by-side iron-infused animals all involved sacrifice at 1–2 months postinfusion. However, to elucidate any iron-induced "progressive changes" that may occur over time, it would be of interest to determine the time-course of neuropathological and neurochemical changes resulting from such infusions. This is an issue of particular importance since an ideal animal model for PD should not only reflect the disease's etiology, but also provide for the progressive decrease in dopaminergic function and associated neurodegenerative changes in PD. Neither of the currently used neurotoxin-based animal models for PD (i.e., MPTP- and 6-hydroxydopamine-treated animals) have been shown to provide a progressive loss of nigrostriatal DA neurons, and neither model appears to provide progressive neurochemical or behavioral changes over a long-term period following neurotoxin administration. It is possible, however, that a single nigral exposure to excess iron could initiate and sustain oxidative reactions that may result in protracted formation of free radicals and stimulation of self-catalytic cycles (such as lipid peroxidation) that eventually become cytotoxic.

Fundamental to the determination of any iron-induced progressive changes is the selection of a dose of iron that creates only a moderate initial loss of nigrostriatal DA neurons in conjunction with modest reductions in striatal dopaminergic markers. Fulfilling this dual criterion is the side-by-side infusion of 1.25 nmole or 2.10 nmole Fe. Both of these doses were therefore used in a time-course study involving sacrifice at 2, 4, or 6 months after nigral iron infusion. Following infusion of either iron amount, a semiquantitative analysis of SNc neuronal numbers in thionine-stained brain sections indicated no greater neuronal loss at 4 or 6 months postinfusion compared to the moderate loss seen at 2 months; vehicle infusions did not result in any discernible SNc neuronal losses at any of these time-points after infusion. However, a computer-assisted determination of SN volumes from iron-infused animals revealed that a progressive atrophy of the iron-infused SN occurred after 2 months and by 4 months postinfusion in comparison to both the contralateral uninfused side and vehicle-infused controls (Figures 6 and 7). Moreover, a highly significant ($p < 0.0001$) correlation was present between the degree of SN atrophy and extent of SNc neuronal losses induced by intranigral iron infusion, irrespective of sacrifice time.

Consistent with the semiquantitative analysis of iron-induced neuronal losses in SNc, tyrosine hydroxylase (TH) immunohistochemical staining at 2, 4, and 6 months following infusion of 1.25 or 2.10 nmole Fe indicated persistent decreases in TH immunoreactivity within the infused SN, particularly in its more

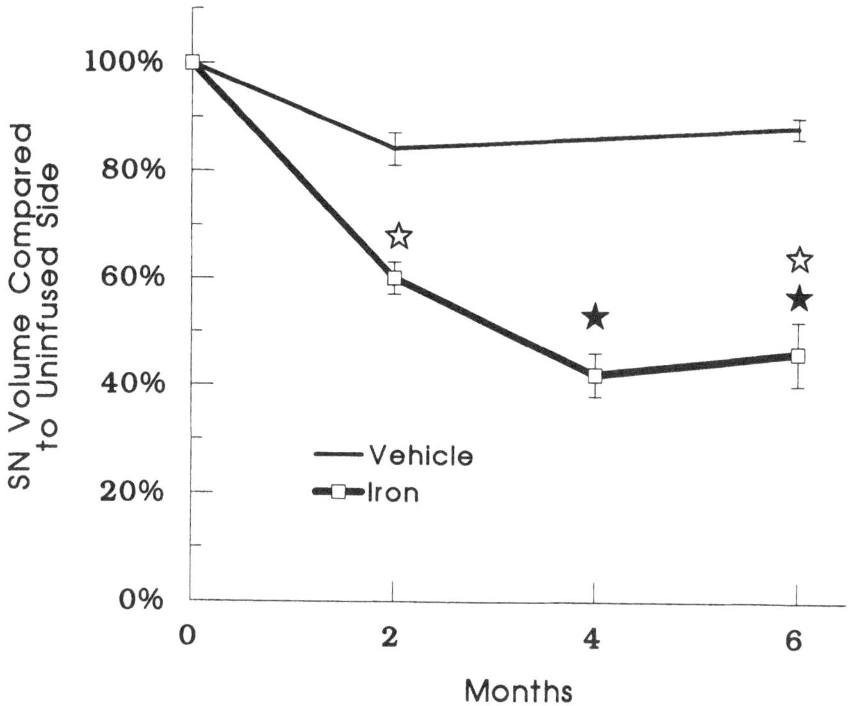

Figure 6 Volume of the SN through 6 months following side-by-side infusion of iron or vehicle solution into the SN. Since no volume differences existed at any time-point between groups infused with 1.25 and 2.10 nmole Fe, data from these two groups were combined into a single iron group for statistical comparisons involving one-way analysis of variance. Solid stars = significantly different from 2 month volume after iron infusion; hollow stars = significantly different from volume of vehicle-infused group at that time-point.

lateral aspects (Figures 7 and 8). This was due to a qualitative decrease in both immunoreactive cells and fibers within the SNc; persistent decreases in immunoreactive fibers within the SNr were evident and dose-related at 6 months following iron infusion. TH immunohistochemical staining further indicated that the more medial aspects of the SNc were less affected by iron infusion, whereas the ventral tegmental area (another dopaminergic cell population with mesolimbic projections) appeared unaffected (Figure 7). Through 6 months, vehicle infusions alone caused a minimal decrease in TH immunoreactivity within the SN (Figures 7A and 8A).

A determination of striatal DA concentrations at 2, 4, and 6 months after infusion of 2.10 nmole Fe revealed that a significant reduction in ipsilateral DA at 2 months postinfusion (−24%) was not significantly intensified at 4 or 6

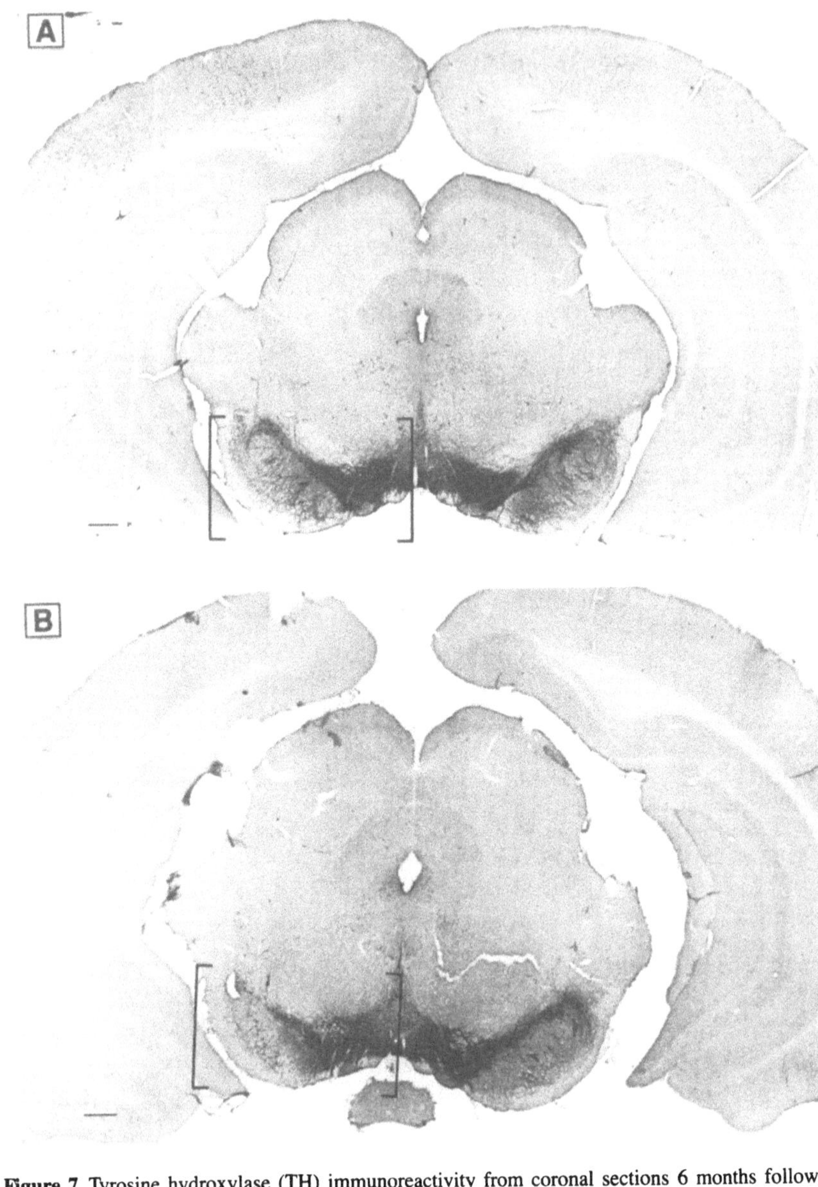

Figure 7 Tyrosine hydroxylase (TH) immunoreactivity from coronal sections 6 months following side-by-side unilateral infusion of vehicle (**A**), 1.25 nmole Fe (**B**), or 2.10 nmole Fe (**C**) into the SN. The boxed regions (A,B,C) on the infusion side are shown at higher magnification in Figure 8. Note the decrease in TH immunoreactivity within the iron-infused SN (B,C). Note also the dose-related shrinkage (B,C) following iron infusion into the SN. Bar = 400 μm. TH immunohistochemistry was performed according to modifications of the method of Haykal-Coates *et al.* (1991), which employs pargyline and butyrolactone enhancement.

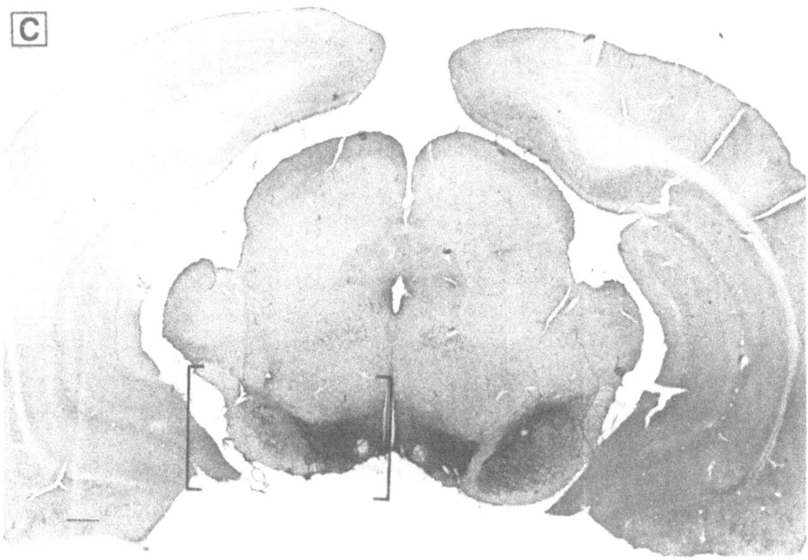

Figure 7 (*Continued*)

months postinfusion (Figure 9). However, a progressive decrease in striatal DA levels occurred between 2 and 4 months following intranigral infusion of 1.25 nmole Fe (Figure 9); animals sacrificed at 2 months postinfusion had a nonsignificant 13% decrease in striatal DA, while those sacrificed at 4 and 6 months postinfusion had significant 22% reductions in striatal DA levels. A similar progressive decrease in striatal HVA levels was also observed with this dose of iron. It is significant to note that, despite only modest iron-induced reductions in striatal DA and HVA concentrations at 2 months following infusion of either amount of iron, no recovery occurred in either striatal DA or HVA levels at later time-points. This lack of neurochemical recovery following intranigral iron infusion could indicate that excess iron within surviving SNc dopaminergic neurons prevented an upregulation in DA metabolism. Without regard to the time of sacrifice, significant correlations existed between (1) the degree of iron-induced SN atrophy and ipsilateral decreases in striatal DA concentrations, and (2) the degree of iron-induced neuronal loss in SNc and ipsilateral decreases in striatal DA concentrations. Thus, the extent of iron-induced nigral damage correlates closely with the magnitude of reductions in striatal DA content.

Because the above results suggest that intranigral iron infusion can induce a progressive SN atrophy, as well as a progressive decrease in striatal DA and HVA

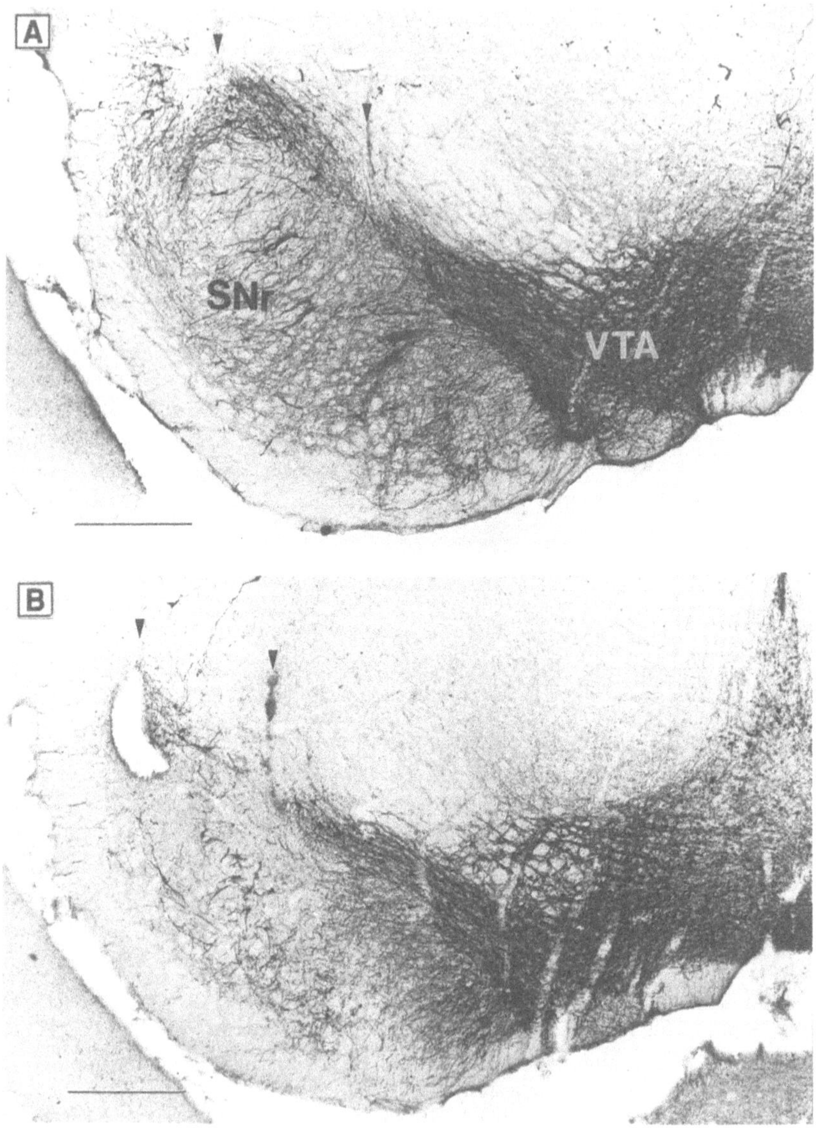

Figure 8 Higher magnification views of the boxed regions in Figure 7 to show tyrosine hydroxylase (TH) immunoreactivity 6 months following infusion of vehicle (**A**), 1.25 nmole Fe (**B**), or 2.10 nmole Fe (**C**) into the SN. Arrowheads denote needle tracks. Note the apparent dose-related decrease in TH immunoreactivity in the SNc and SNr resulting from nigral iron infusion (B,C), as well as iron-induced SN atrophy (B,C). Abbreviations: SNr, zona reticularis of the substantia nigra; VTA, ventral tegmental area. Bar = 400 μm.

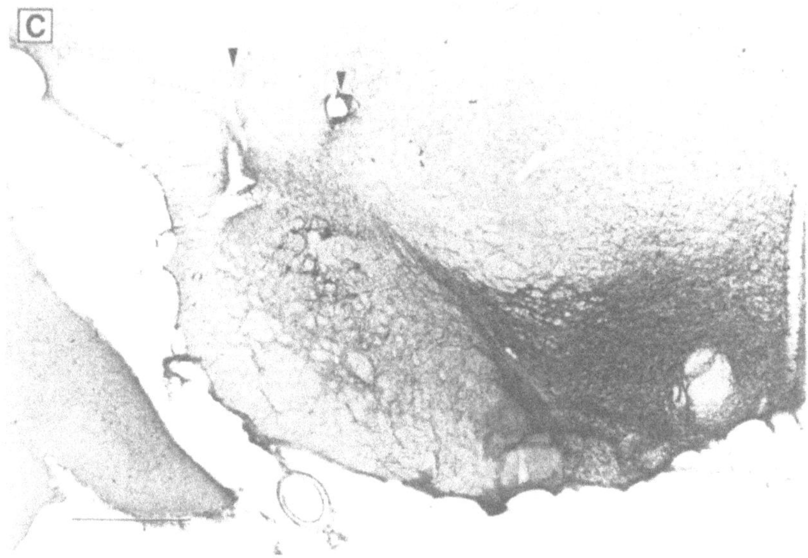

Figure 8 *(Continued)*

levels at least through 4 months, is it possible that apomorphine-induced rotational behavior will occur and increase progressively following nigral iron infusion? Even though iron-induced depletions in striatal DA and HVA levels never exceeded 32% of contralateral values throughout this 6-month study, the answer to that question appears to be yes, with rotation occurring *ipsilaterally*. Figure 10 shows apormorphine-induced rotational behavior through 6 months after side-by-side infusions of iron or vehicle into the SN. A significant progressive increase in net ipsilateral turns is evident for animals in both infusion groups ($p < 0.005$; 2 factor ANOVA for repeated measures) through 6 months.

The direction of apomorphine-induced rotation in iron-infused animals is opposite to that observed in animals with large lesions of the nigrostriatal pathway induced by unilateral infusion of 6-hydroxydopamine. Because depletions in striatal DA exceeding 90% are typically required to achieve the upregulation of striatal DA receptors important for apomorphine-induced *contralateral* rotation, it is not surprising that the modest 13%–32% depletions produced by our nigral infusion of 1.25 or 2.10 nmole Fe did *not* result in contralateral rotation after apomorphine administration. However, the mechanism(s) responsible for the prominent and progressive *ipsilateral* rotation observed in these iron-infused animals after apomorphine is less clear. One likely explanation could involve a

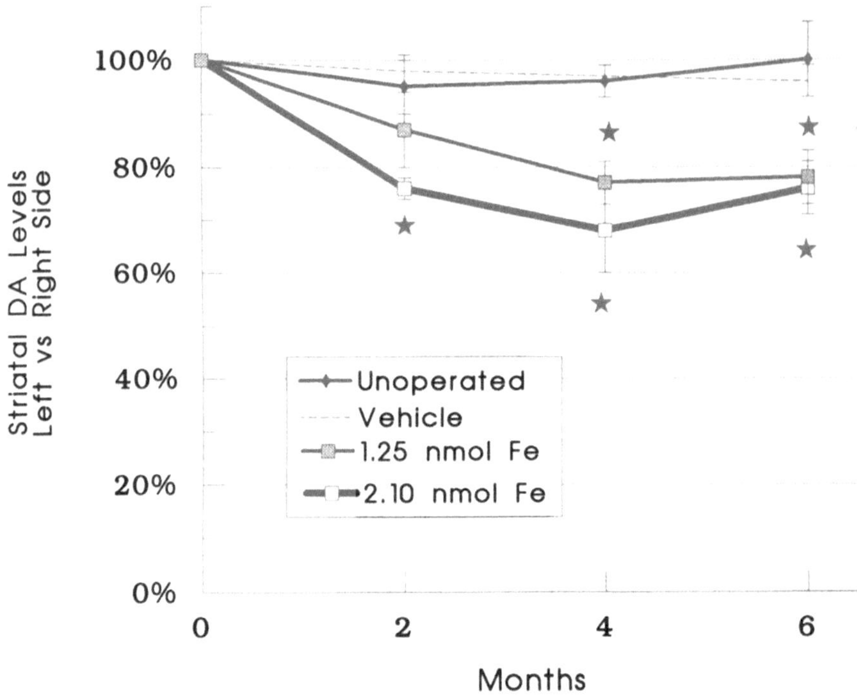

Figure 9 Changes in striatal dopamine concentrations through 6 months following side-by-side unilateral infusions of iron or vehicle solution into the left SN. Solid stars = significantly different from contralateral uninfused side.

degree of iron-induced neuronal dysfunction or loss within the SNr. As mentioned earlier, nigral iron infusion increases iron staining within both the SNr and SNc, with considerable numbers of SNr neurons staining for iron. In this context, several studies have demonstrated that, when both the SNr and SNc are damaged by electrolytic lesioning or neurotoxin infusion, spontaneous and apomorphine-induced rotational behavior occur in a direction *opposite* to that observed in animals with damage only to the SNc via 6-hydroxydopamine lesions (Schwartz *et al.*, 1976; Di Chiara *et al.*, 1977; Levy *et al.*, 1984). Thus, it has been suggested that non-DA neurons originating in the SNr are involved with rotational behavior in a manner opposite to that of nigrostriatal DA neurons. Further studies have even provided evidence that the SNr is an important site of rotational action for apomorphine because apomorphine-induced contralateral rotation following intranigral 6-hydroxydopamine administration is completely blocked by kainate lesioning of the same SN (Herrera-Marschitz and Ungerstedt, 1984, 1985; Robertson, 1992). Increased iron within the SNr resulting from nigral iron infusion may, therefore, induce a neuronal loss or dysfunction within

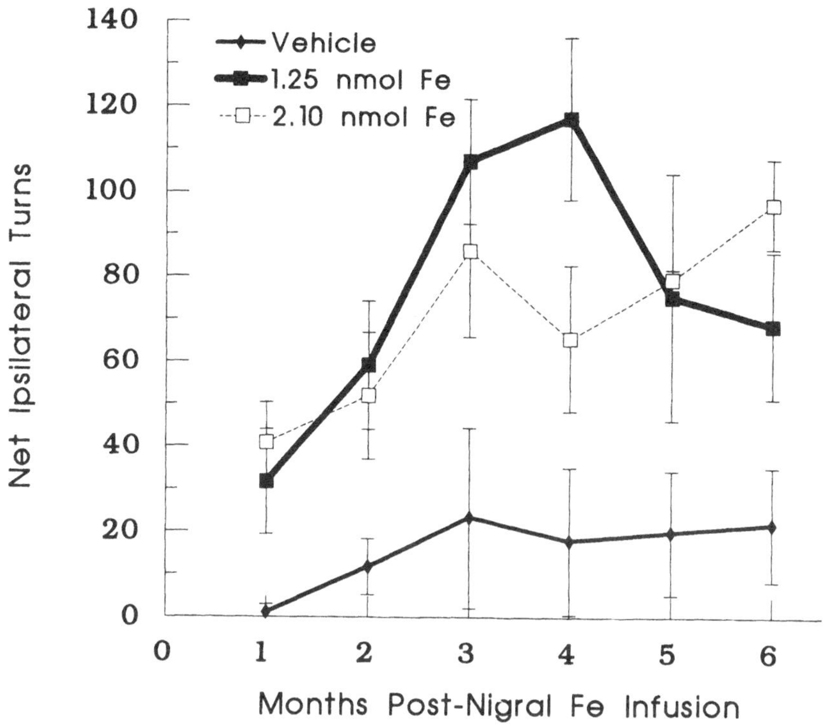

Figure 10 Apomorphine-induced rotational behavior through 6 months following side-by-side unilateral infusions of iron or vehicle solution into the SN. Apomorphine (0.25 mg/kg) was injected subcutaneously; net ipsilateral turns were determined over the ensuing 30-min period. For animals in both infusion groups ipsilateral turns increased significantly ($p < 0.005$; 2 factor ANOVA for repeated measures) over 6 months.

the SNr that could be responsible for the ipsilateral rotation seen following apomorphine administration. Supportive of this notion is the lack of a correlation between the degree of iron-induced *SNc* degeneration and apomorphine-induced ipsilateral rotation, but a strong correlation between the degree of iron-induced SN atrophy and apomorphine-induced rotation.

To summarize, it appears that a single exposure of the SN to a modest amount of iron can induce several progressive changes occurring through a number of months postinfusion. These changes include a progressive atrophy of the SN, a progressive decrease in striatal DA and HVA, and a progressive increase in apomorphine-induced rotational behavior.

2.5. Summary of Effects Induced by Nigral Iron Infusions

Using three different protocols for the intranigral infusion of iron, we have begun to elucidate the neuropathological, neurochemical, and behavioral changes induced by excess iron within the SN and, in so doing, have begun to define a new animal model for Parkinson's disease that may reflect the disease's etiology or pathophysiology (i.e., oxidative damage). Specifically, intranigral iron infusion causes a long-term and dose-related increase in nigral iron content that can be essentially restricted to the SNc, wherein the iron is associated primarily with glial cells. Within the nigral zone of iron diffusion, substantial neuronal losses occur, accompanied by gliosis. Iron-induced SNc neuronal loss and depletions in striatal DA markers are dose-dependent and striatal non-dopaminergic markers are unaffected by nigral iron infusions. The intranigral infusion of iron can provide several long-term "progressive" changes, including (1) a progressive decrease in striatal DA and HVA, (2) a progressive atrophy of the SN, and (3) a progressive increase in apomorphine-induced rotational behavior. Finally, lipid peroxidation within the SN is elevated following nigral iron infusion, suggesting that peroxidative damage may be involved in the neurodegeneration resulting from intranigral iron infusion.

3. Possible Mechanisms of Iron's Intranigral Neurotoxicity

Prior to discussing specific mechanism(s) by which intranigral iron infusions could induce neurodegeneration within the SN, it is appropriate to address the following related questions: (1) Does intranigral infusion of iron act extracellularly and/or intracellularly? (2) If an intracellular site of action appears likely, how does the infused iron enter nigral cells? In answer to the first question, it appears that both extra- and intracellular sites of action are possible. Empirically, intranigrally infused iron is initially located in the extracellular compartment and in a "free or low-molecular weight" state that is capable of catalyzing extracellularly based oxidative damage, such as lipid peroxidation. However, intraneuronal iron staining within the SNc is apparent by 24 hr postinfusion, with neuronal nuclei staining especially intensely for iron (see Figure 2C). These histochemical results are supportive of an intraneuronal uptake of infused iron, although confirmation of this finding requires more precise techniques such as X-ray microanalysis or LAMMA. Assuming, however, that intracellular uptake of infused iron does occur, this uptake could take place in association with the iron transport protein transferrin via transferrin receptors located on cell membranes (Irie and Tavassoli, 1987). Alternatively, infused iron may be taken up via a transferrin-independent mechanism (Basset *et al.*, 1986), particularly in view of evidence that mutant atransferrinemic mice have normal amounts of iron within brain regions that normally concentrate iron (Connor and Benkovic, 1992). This latter finding suggests that transferrin-mediated iron uptake is not necessary to achieve normal iron content/distribution within the brain.

An elucidation of the specific mechanisms through which intranigral iron infusions induce neurodegeneration within the SN must await future study, although these mechanisms are hypothesized to involve damage to membrane lipids and/or DNA incurred as a result of iron-catalyzed oxygen free radical formation. The role of iron in lipid peroxidation is well recognized and appears to involve a catalysis of two processes: (1) the initiation of lipid peroxidation via $OH^{\cdot}$ formation and (2) the decomposition of lipid hydroperoxides (Halliwell and Gutteridge, 1984; Youdim *et al.*, 1989). Consistent with this role, our intranigral infusions of iron increase lipid peroxidation within the SN (see Section 2.3). Another study, involving the infusion of an iron-carbohydrate complex into the neocortex, also demonstrated an increase in lipid peroxidation (Ciuffi *et al.*, 1991). The importance of iron in lipid peroxidation is further underscored by *in vitro* studies showing that dopamine induces lipid peroxidation and neuronal death only in the presence of iron (Tanaka *et al.*, 1991). Lipid peroxidation can produce a number of damaging consequences, such as a decrease in membrane fluidity, resulting in effects on membrane structure/function sufficient to induce neurodegeneration (Slater, 1984). Along this line, CSF infusions of ferrous ammonium sulfate that induce a two-fold increase in iron content of neocortical synapotosomes also result in a decrease in their membrane fluidity (Nagy *et al.*, 1985). Thus, iron-catalyzed lipid peroxidation must be considered as one possible mechanism of iron's neurotoxic action on nigrostriatal DA neurons within the SN. A similar mechanism may be involved in the degeneration of nigrostriatal DA neurons in PD, particularly since increased lipid peroxidation has been reported in the SN of PD brains (Dexter *et al.*, 1989b).

A second possible mechanism of iron-induced neurotoxicity within the SN, also stemming from iron-catalyzed free-radical formation, involves DNA damage. As mentioned in Section 2.1, iron appears to be heavily concentrated within the nucleus of SNc neurons following intranigral iron infusion. If present in a free or low-molecular weight state, this nuclear iron can catalyze formation of $OH^{\cdot}$, resulting in DNA strand breaks and consequent reductions in cellular ATP levels (Halliwell, 1989; Halliwell and Gutteridge, 1989). Intracellular concentrations of free calcium (Ca^{2+}) may than increase, resulting in activation of Ca^{2+}-stimulated proteases, lipases, and endonucleases with consequent neurodegeneration. It is noteworthy that any oxidative damage to DNA resulting from intranigral iron infusion need not involve *nuclear* DNA exclusively, but may involve oxidative damage to *mitochondrial* DNA. In this context, mitochondrial DNA is considerably more vulnerable to the destructive actions of free radicals than is nuclear DNA (Richter *et al.*, 1988). Because mitochondrial DNA encodes seven of the 25 polypeptides comprising Complex I of the mitochondrial electron transport chain (Chomyn *et al.*, 1985), damage to mitochondrial DNA incurred through intranigral iron infusion could result in decreased Complex I activity and consequent depletions in cellular ATP levels that result in neuronal death. Although we have not investigated the effects of intranigral iron infusions on Complex I activity in the rat SN, Hartley *et al.* (1993) have recently found that iron-loaded rat PC12 cells exhibit reductions in both Complex I and IV activities.

Moreover, Schapira *et al.* (1990) have determined that Complex I activity in the SN of human PD brains is significantly reduced. Complex I deficiency in PD could be related to the primary disease process or it may be secondary to mitochondrial damage induced by oxidative damage. Regarding this latter possibility, recent findings of increased iron in neuromelanin granules of pigmented SNc neurons (Jellinger *et al.*, 1992; Good *et al.*, 1992) may be of particular significance because increased intraneuronal iron could increase oxidative damage to mitochondrial DNA through catalysis of free radical formation.

In addition to the possible direct cytotoxic effects of intranigrally infused iron on neurons within the SN, *indirect* neurodegeneration may result from iron-induced damage initially inflicted on surrounding glial cells. Consistent with this premise, intranigrally infused iron becomes associated with oligodendrocyte-like cells in the SNc. Furthermore, increased numbers of intensely iron-positive oligodendrocytes are seen in the SNr following intranigral iron infusion (Sengstock *et al.*, 1992b, 1993). Increased iron in association with SN oligodendrocytes may induce oxidative damage to such cells and, in that fashion, compromise their functionality. Because iron, transferrin, and ferritin are all found primarily in brain oligodendrocytes (Connor *et al.*, 1990), it has been suggested that these glial cells, particularly the "perineuronal satellite" type, regulate the availability of iron to neurons and thus provide a trophic role for neurons (Gerber and Connor, 1989; Connor and Benkovic, 1992). Exposure of oligodendrocytes to excess iron via intranigral infusion could result in dysfunction of iron regulation by oligodendrocytes and consequent neuronal death.

Glial cells may not only participate in, but also respond to, neurodegeneration resulting from intranigral iron infusion. With lower amounts of iron infused, an intense gliotic reaction occurs within the SNc; it appears to be composed of activated microglia and reactive astrocytes that stain intensely for iron. This SNc gliosis is most likely a response to iron-induced degeneration of SNc neurons, with resultant phagocytosis of iron-rich neuronal debris by gliotic cells. It is noteworthy that a similar distribution of iron within the SNc (i.e., iron staining associated primarily with reactive microglia and astrocytes) has been reported to exist within the SNc of PD brains (Jellinger *et al.*, 1990), suggesting that basically the same pathological process may occur within the SN of PD brains as that resulting from intranigral iron infusion. Supporting an active neuropathological process by glial cells in PD is the work by McGeer *et al.* (1988a,b), which demonstrated an ongoing phagocytosis of dopaminergic neurons in the SN by reactive microglia or macrophages.

4. Current Limitations of Intranigral Iron Infusion as an Animal Model for Parkinson's Disease

Parkinson's disease involves multiple neuropathologic and neurochemical alterations that occur in a progressive fashion. Neuropathological changes include not only a massive loss of nigrostriatal DA neurons, but also Lewy body formation in SN neurons and at least some loss of (1) DA neurons in other midbrain groups, (2) noradrenergic neurons in locus ceruleus, and (3) serotoninergic neurons in the raphe nuclei. Accompanying this varied degeneration of brainstem monoaminergic neurons are associated reductions in DA, norepinephrine, and/or serotonin levels within forebrain projection regions of affected monoaminergic neurons. Current animal models for PD, including the intranigral iron infusion model, do not provide this complex array of multisystem neuropathological and neurochemical changes. Rather, these models are used to provide a "hypo-dopaminergic" condition, primarily through destruction of nigrostriatal DA neurons. In that capacity, our current iron infusion model can induce fairly selective neuronal losses within the SNc when appropriately low amounts of iron are infused. However, it is important to recognize that iron is not a specific toxin for dopaminergic neurons; rather, iron can be toxic to all neuronal and glial cell types depending on its local concentration and form. Indeed, the amounts of intranigrally infused iron presently required to induce large (80%–90%) depletions in striatal DA through massive destruction of SNc neurons also result in damage to the SNr and immediately adjacent structures such as the red nucleus. Great care should, therefore, be taken in selecting iron concentrations and infusate volumes for intranigral infusion. In this regard, we have neurochemically determined that the total amount of iron present in the young adult rat SN is 60–70 ng; this compares very favorably with the 70–110 ng (i.e., 1.25–2.10 nmole) Fe that we usually infuse unilaterally into the rat SN.

Because PD is a progressive disease, it is encouraging that acute intranigral infusions of iron in the rat can induce several progressive changes that appear to be relevant to PD, including progressive atrophy of the SN, progressive decreases in striatal DA and HVA, and progressive increases in apomorphine-induced rotational behavior (see Section 2.4). However, an important limitation of the current intranigral iron infusion model is that a single nigral exposure to excess iron does not seem to induce the "progressive" loss of SNc neurons that presumably occurs in PD brains. A more chronic exposure of the rat SN to iron may have the capacity to achieve this important progressive aspect of SNc neuronal loss in PD.

The significance of consistent and progressive *ipsilateral* rotational behavior in iron-infused animals following apomorphine administration is presently unclear, particularly the context in which this behavior relates to PD. Because ipsilateral rotation is observed in iron-infused animals with only moderate damage to nigrostriatal DA neurons, it apparently does not involve an upregulation of DA receptors in the striatum; rather, a destruction/dysfunction of (non-dopaminergic?) neurons in the SN would seem to be the most likely basis for this behavior.

Further characterization of, and refinements to, the intranigral iron infusion model should define more precisely those characteristics of PD that can (and cannot) be provided by intranigral iron infusion. Any number of presently perceived limitations may be eliminated or minimized as a result of future experimentation.

5. Comparison of the Intranigral Iron Infusion Model with Currently Used Animal Models for Parkinson's Disease

Up to the present, two neurotoxins—6-hydroxydopamine (6-OHDA) and MPTP—have been used almost exclusively to induce degeneration of nigrostriatal DA neurons in animals and thus provide models for the loss of similar DA neurons in PD. These two animal models for PD have been extensively characterized in numerous studies—studies that have established each as a viable model, with limitations as well as advantages. Given the limited scope of our present knowledge concerning the intranigral iron infusion model for PD, a direct comparison between it and the aforementioned models may be premature. Nonetheless, some comparisons, both with regards to possible advantage and perceived mechanisms of action, can be made based on currently available data.

First, all three models have the ability to induce a fairly selective degeneration of nigrostriatal DA neurons, but apparently through different mechanisms of action (to be discussed below). Advantages of the intranigral iron infusion model involve its ability to provide an increase in nigral iron content—specifically in the SNc and primarily within reactive glia—a characteristic of the SN in PD brains. This advantage is not, however, unique to the intranigral iron infusion model because we have recently determined that 6-OHDA lesions of the rat NS-DA pathway also increase SNc iron staining, as well as neurochemical iron concentrations within the SN (Oestreicher *et al.*, submitted for publication). Although these latter findings suggest that the increased nigral iron in PD brains may be a consequence of NS-DA neuronal death, a prominent role of increased nigral iron in the pathogenesis of PD remains a distinct possibility. In any event, only the intranigral iron infusion model appears to be capable of inducing an increase in lipid peroxidation within the SN, such as has been observed in PD brains (Dexter *et al.*, 1989b). Neither 6-OHDA nor MPTP have been reported to provide an increase in nigral lipid peroxidation (Corongiu *et al.*, 1987; Kopin and Schoenberg, 1988). By increasing levels of SN iron that enhances SN lipid peroxidation, intranigral iron infusions may induce SNc neuronal loss by the same mechanisms that occur in the Parkinsonian brain (i.e., via iron-induced oxidative/free radical damage).

Although no animal model for PD has yet been demonstrated to induce *progressive* loss of nigrostriatal DA neurons, the presently described intranigral iron infusion model is capable of inducing progressive depletions in striatal DA and HVA, progressive SN atrophy, and progressive increases in apomorphine-induced rotation behavior. To our knowledge, no such progressive effects have been demonstrated to occur chronically following 6-OHDA or MPTP treatment of animals (Riederer and Youdim, 1987). In fact, a certain degree of neurochemical and/or behavioral recovery can occur in animals with 6-OHDA- or MPTP-induced lesions of the nigrostriatal DA pathway, even in animals that incur extensive neurotoxic damage to this pathway (Ricaurte *et al.*, 1986, 1987; Sonsalla and Heikkila, 1986; Zigmond *et al.*, 1990; Pasinetti *et al.*, 1992); parenthetically, both collateral sprouting and metabolic upregulation of surviving nigrostriatal DA neurons have been implicated in this recovery. By contrast, our data indicate that no recovery occurs from the depletion in neostriatal DA and its catabolites induced by intranigral iron infusions.

The neurotoxic effects of intranigral iron infusions can be limited to the infused SN without affecting immediately adjacent mesolimbic DA neurons in the ventral tegmental area or noradrenergic/serotoninergic neurons in nearby brainstem regions. A rather selective and permanent "hypodopaminergic" condition can therefore be achieved in the neostriatum after intranigral iron infusions. This restricted DA hypofunction induced by intranigral iron may be interpreted as an advantage or a disadvantage depending on whether the investigator wishes to evaluate functions of the nigrostriatal DA system in relative isolation from other biogenic amine systems in the midbrain.

It is important to emphasize again that PD is a multisystem disease, associated with at least some loss of mesolimbic DA neurons, as well as loss of noradrenergic and serotonergic neurons within the locus ceruleus and raphe nuclei, respectively. In animals, this multisystem degeneration would appear to be partially mimicked by mesencephalic infusions of 6-OHDA, which induce degeneration of mesolimbic and nigrostriatal DA neurons in addition to some loss of locus cerulear noradrenergic neurons (Breese and Traylor, 1970). For investigators wishing to induce a selective neurotoxic effect of 6-OHDA on dopaminergic neurons alone, noradrenergic neurons can be protected from the neurotoxic actions of 6-OHDA through desmethylimipramine pretreatment (Breese and Traylor, 1970; Zigmond *et al.*, 1990). MPTP treatment of animals also causes degeneration to nigrostriatal and mesolimbic DA neurons, as well as a less robust loss of brainstem noradrenergic and serotoninergic neurons (Hallman *et al.*, 1984; Mitchell *et al.*, 1985; Jarvis and Wagner, 1985, 1990). At long times following MPTP treatment, however, only DA neurons appear to remain affected because persistent neurochemical depletions are usually only detected in forebrain concentrations of DA and its catabolites (Burns *et al.*, 1983; Jarvis and Wagner, 1990). It would seem, then, that all three models have the potential to provide a selective dopaminergic hypofunction in neostriatum. However, a present disadvantage of the intranigral iron infusion model is that intranigral amounts of iron necessary to induce massive depletions in striatal DA also result in damage to the SNr. By contrast, 6-OHDA or MPTP treatment can induce massive depletions in striatal DA concentrations while limiting intranigral damage to the SNc (i.e., to the nigrostriatal DA system).

The likely mechanisms of neurotoxicity provided by intranigral iron, 6-OHDA, and MPTP seem to be basically different, although some evidence (described below) suggests that they may share a common feature—induction of oxygen free radical formation/damage. Neurotoxic mechanisms of intranigral iron infusion would appear to involve iron-catalyzed oxygen free radical formation and resultant membrane lipid peroxidation and/or DNA damage, as discussed in Section 3 above. By contrast, the neurotoxic mechanism of intraventricular or intraparenchymal 6-OHDA administration is thought to involve its selective transport into nerve terminals of catecholaminergic neurons via high-affinity catecholaminergic transport carriers (Zigmond *et al.*, 1990). Once inside DA or noradrenergic neurons, 6-OHDA is rapidly oxidized to form cytotoxic semiquinones and quinones, with concurrent production of superoxide and hydroxyl free radicals (Sachs and Jonsson, 1975; Halliwell and Gutteridge, 1989; Zigmond *et al.*, 1990); together, these products of 6-OHDA metabolism are thought to be responsible for subsequent degeneration of catecholamine neurons. The formation of cytotoxic oxygen free radicals through oxidation of 6-OHDA may involve iron because pretreatment of rats with the iron chelator desferrioxamine has been shown to attenuate the depletions in striatal DA and its catabolites resulting from intracerebroventricular 6-OHDA infusions (Ben-Shachar *et al.*, 1991b). The chain-breaking antioxidant alpha-tocopherol (vitamin E) has also been reported to protect partially against 6-OHDA-induced depletions in striatal DA (Cadet *et al.*, 1989), providing further evidence that the neurotoxicity of 6-OHDA may involve oxidative damage.

The neurotoxicity of MPTP requires its oxidation to MPP^+ (1-methyl-4-phenyl-pyridine) by monoamine oxidase (Singer and Ramsey, 1990), an event that most likely occurs in glial cells (Brooks *et al.*, 1989; McCrodden *et al.*, 1900). The toxic metabolite MPP^+ is then taken up by DA nerve terminals through high-affinity DA transport carriers and concentrated by mitochondria (Javitch and Snyder, 1984; McCrodden *et al.*, 1990). Although some evidence supports involvement of free radicals/oxidation reactions in the mechanism of neurotoxicity of MPP^+, oxidative damage may not be the *primary* cause of MPP^+-induced neurotoxicity (Halliwell, 1989; Halliwell and Gutteridge, 1989). Rather, the neurotoxicity of MPP^+ appears to involve primarily an inhibition of mitochondrial electron transport within DA neurons. More specifically, MPP^+ inhibits Complex I activity in the electron transport chain, resulting in depletions in cellular ATP and consequent neuronal death (Nicklas *et al.*, 1986; Langston *et al.*, 1987; Singer *et al.*, 1987). A reduction in mitochondrial Complex I activity has also been reported in the SN of PD brains (Schapira *et al.*, 1990), suggesting that some neurotoxin may be taken up into nigral mitochondria of PD patients in a manner similar to that of MPP^+ and induce Complex I deficiency. Alternatively, the reduction in nigral Complex I activity in PD may reflect free radical/oxidative damage within the SN, possibly induced by increased concentrations of reactive-free iron.

The rat intranigral iron infusion model has, as its primary focus, the potential involvement of increased nigral iron (and presumed free radical/oxidative damage) in the degeneration of nigrostriatal DA neurons. By contrast, MPTP-induced degeneration of nigrostriatal DA neurons in mice and monkeys has not been shown to alter iron content within the SN. Moreover, conflicting evidence exists regarding the extent to which free radical formation is involved in MPTP's neurotoxicity. Several studies have provided data against oxidative processes in the mechanism of MPTP's neurotoxicity (Martinovits *et al.*, 1986; Frank *et al.*, 1987). Consistent with these reports are others indicating that MPTP treatment in normal animals does not result in increased brain lipid peroxidation (Corongiu *et al.*, 1987; Kopin and Schoenberg, 1988; Adams *et al.*, 1990)—a finding that may relate to the inability of MPTP to alter brain iron content. However, failure to find MPTP-induced lipid peroxidation does not exclude the possibility of oxidant damage following MPTP treatment because other molecular targets for oxidant damage exist, most notably DNA (Halliwell, 1989).

Potential involvement of oxidative damage in MPTP's mechanism of neurotoxicity may result from MPTP's extraneuronal oxidation to MPP^+, a metabolic conversion that appears to involve formation of oxygen free radicals (Cohen and Mytilineou, 1985; Rossetti *et al.*, 1988). In support of this premise are studies reporting that pretreatment with the antioxidants alpha-tocopherol or ascorbic acid can partially prevent the neurochemical and cytotoxic effects of MPTP in rats and mice (Perry *et al.*, 1985; Wagner *et al.*, 1985, 1986; Jarvis and Wagner, 1990). Additionally, the superoxide dismutase inhibitors diethyl-dithiocarbamate (DDC) potentiates the neurotoxicity of MPTP in mice (Corsini *et al.*, 1985). Because glutathione (which is the primary mechanism for removal of hydrogen peroxide) is reduced in PD brains (Riederer *et al.*, 1989), it is noteworthy that MPTP causes a depletion in brainstem glutathione levels in mice and that this effect is prevented by antioxidant pretreatment (Yong *et al.*, 1986). Another study has reported the formation of superoxide radicals in mouse brain mitochondrial preparations as a result of a redox reaction between MPTP metabolites (Rossetti *et al.*, 1988). Finally, iron contributes to the oxidation of MPTP *in vitro*, perhaps resulting in the formation of a new MPTP reactive species that accelerates the autoxidation of DA (Poirier *et al.*, 1985). Together, the aforementioned studies support the premise that formation of reactive oxygen radicals contributes to the neurotoxicity of MPTP (Adams and Odunze, 1991). Nonetheless, an MPTP-induced generation of free radicals/oxidative stress has not been shown to be the *primary* cause of MPTP-induced neurotoxicity.

As an animal model for idiopathic PD, MPTP treatment would appear to have both advantages and limitations. Advantages include (1) selective degeneration of nigrostriatal DA neurons resulting in a selective hypodopaminergic condition in striatum; (2) in primates, motor dysfunctions that closely resemble those of PD; and (3) reduction in Complex I within the SN. However, acute exposure of animals to MPTP induces massive degeneration of DA neurons and, thus, does not mimic the slowly progressive neurochemical/neuropathological changes that characterize idiopathic PD. Furthermore, no studies have shown that acute toxicity of MPTP results in accelerated DA degeneration (Riederer and Youdim, 1987). Whether chronic administration of low amounts of MPTP to animals could result in the progressive nature of PD has not yet been established. An additional limitation is that the neurotoxicity of MPTP is species-specific, with primates being much more sensitive than rodents (Jenner and Marsden, 1986). Moreover, MPTP has some neurochemical actions that may be undesirable for a PD model, such as its inhibition of tyrosine hydroxylase, its binding to muscarinic receptors, and its ability to act as a cholinergic agonist (McCrodden *et al.*, 1990).

To summarize, none of the three animal models for idiopathic PD provide the full spectrum of progressive changes associated with the disease. All three models have associated advantages and limitations that should be considered in their use. As the newest and consequently least well characterized of these animal models for PD, the intranigral iron infusion model may provide several key aspects of PD not apparent in the other two models, including enhanced nigral lipid peroxidation, certain progressive changes, and a mechanism of action that may more closely model the pathogenesis of PD.

6. Future Directions and Concluding Remarks

Whenever a potential new animal model for a human disease is presented, the model invariably raises a number of questions concerning exactly how closely it mimics the actual disease—both in terms of the disease's characteristics and pathogenetic mechanisms. In the case of PD, the etiologic factors are presently unknown, although several lines of evidence (discussed in this chapter) suggest that free radical/oxidative damage plays a fundamental role in the pathogenesis of PD. Future studies should define more precisely the contribution provided by oxidative damage in the neurodegenerative changes characteristic of PD. The extent to which the intranigral iron infusion model reflects these neurodegenerative changes will be answered through further experimentation involving such questions as: Can modifications of this model provide a progressive degeneration of DA neurons? Does intranigral iron infusion involve increased reactive free radical formation and an alteration in mitochondrial Complex I activity in the SN? Can intranigral iron infusions induce a decrease in the nigral glutathione system—a system that normally protects against free radical formation? What are the neurochemical, neuropathological, and behavioral effects of iron infusions into the SN of primates? More detailed knowledge of iron's intracerebral effects and its mechanism(s) of action could further establish the intranigral iron infusion model as a viable model for idiopathic PD.

If oxidative reactions contribute to the neuronal damage that underlies PD, strategies designed to interfere with these reactions may have therapeutic value. In that context, an animal model of PD due to oxidative damage in the substantia nigra might be extraordinarily valuable in assessing antioxidant regimens. Future studies may indicate that intranigral iron infusions provide such a model.

ACKNOWLEDGMENTS. This work was supported by grants from the United Parkinson Foundation, the Southern Medical Association, and the University of South Florida College of Medicine.

7. References

Adams, J.D., Odunze, I.N., and Sevanian, A., 1990, Induction by 1-methyl-4-phenyl-1,2,3,6-tetrahydropyridine of lipid peroxidation *in vivo* in vitamin E deficient mice, *Biochem. Pharmacol.* 39:R5–R8.

Adams, J.D., and Odunze, I.N., 1991, Biochemical mechanisms of 1-methyl-4-phenyl-1,2,3,6-tetrahydropyridine toxicity, *Biochem. Pharmacol.* 41:1099–1105.

Banks, W.A., Kastin, A.J., Fasold, M.B., Barrera, C.M., and Augereau, G., 1988, Studies of the slow bidirectional transport of iron and transferrin across the blood brain barrier, *Brain Res. Bull.* 21:881–887.

Basset, P., Quesneau, Y., and Zwiller, J., 1986, Iron-induced L1210 cell growth: Evidence of a transferrin-independent iron transport, *Cancer Res.* 46:1644–1647.

Ben-Shachar, D., Pinhassi, B., and Youdim, M.B.H., 1991a, Prevention of neuroleptic-induced dopamine D2 receptor supersensitivity by chronic iron salt treatment, *Eur. J. Pharmacol.* 202:177–183.

Ben-Shachar, D., Eshel, G., Finberg, J., and Youdim, M.B.H., 1991b, The iron chelator desferrioxamine (Desferal) retards 6-hydroxydopamine-induced degeneration of nigrostriatal dopamine neurons, *J. Neurochem.* 56:1441–1444.

Braughler, J.M., Duncan, L.A., and Chase, R.L., 1986, The involvement of iron in lipid peroxidation, *J. Biol. Chem.* 261:10282–10289.

Breese, G.R., and Traylor, T.D., 1970, Effect of 6-hydroxydopamine on brain norepinephrine and dopamine: Evidence for selective degeneration of catecholamine neurons, *J. Pharmacol. Exp. Ther.* 174:413–420.

Brooks, W.J., Jarvis, M.F., and Wagner, G.C., 1989, Astrocytes as a primary locus for the conversion of MPTP to MPP+, *J. Neural Transm.* 76:85–90.

Burns, R.S., Chiueh, C.C., Markey, S.P., Ebert, M.H., Jacobowitz, D.M., and Kopin, I.J., 1983, A primate model of parkinsonism: Selective destruction of dopaminergic neurons in the pars compacta of the substantia nigra by N-methyl-4-phenyl-1,2,3,6-tetrahydropyridine, *Proc. Natl. Acad. Sci. USA* 80:4546–4550.

Cadet, J., Katz, M., Jackson-Lewis, V., and Fahn, S., 1989, Vitamin E attenuates the toxic effects of intrastriatal injection of 6-hydroxydopamine (6-OHDA) in rats: Behavioral and biochemical evidence, *Brain Res.* 476:10–15.

Carlsson, A., and Fornstedt, B., 1991, Possible mechanisms underlying the special vulnerability of dopaminergic neurons, *Acta Neurol. Scand.* 84 (Suppl. 136): 16–18.

Chomyn, A., Mariottini, P., Cleeter, M., and Ragan, C., 1985, Functional assignment of the unidentified reading frames of human mitochondrial DNA, *in* "Achievements and Perspectives in Mitochondrial Research" Vol. 4 (E. Quagliariello, E. Slater, F. Palmieri, C. Saccone, and A. Kroon, eds.), pp. 259–275, Elsevier Science Publishers, Amsterdam.

Ciuffi, M., Gentilini, G., Franchi-Micheli, S., and Zilletti, L., 1991, Lipid peroxidation induced "in vivo" by iron-carbohydrate complex in the rat brain cortex, *Neurochem. Res.* 16:43–49.

Cohen, G., and Mytilineou, C., 1985, Studies on the mechanism of action of 1-methyl-4-phenyl-1,2,3,6-tetrahydropyridine (MPTP), *Life Sci.* 36:237–242.

Connor, J.R., and Benkovic, S.A., 1992, Iron regulation in the brain: Histochemical, biochemical, and molecular considerations, *Ann. Neurol.* 32:S51–S61.

Connor, J.R., Menzies, S.L., St. Martin, S.M., and Mufson, E.J., 1990, Cellular distribution of transferrin, ferritin, and iron in normal and aged human brains, *J. Neurosci. Res.* 27: 595–611.

Corongiu, F.P., Dessi, M.A., Banni, S., Bernardi, F., Piccardi, M., Del Zampo, M., and Corsini, G., 1987, MPTP fails to induce lipid peroxidation *in vivo, Biochem. Pharmacol.* 36:2251–2253.

Corsini, G.U., Pintus, S., Chiueh, C.C., Weiss, J.F., and Kopin, I.J., 1985, 1-methyl-4-phenyl-1,2,3,6-tetrahydropyridine (MPTP) neurotoxicity in mice is enhanced by pretreatment with diethyldithiocarbamate, *Eur. J. Pharmacol.* 119:127–128.

Dexter, D.T., Wells, F.R., Agid, F., Agid, Y., Lees, A.J., Jenner, P., and Marsden, C.D., 1987, Increased nigral iron content in postmortem Parkinsonian brain, *Lancet* 2:1219–1220.

Dexter, D.T., Wells, F.R., Lees, A.J., Agid, F., Agid, Y., Jenner, P., and Marsden, C.D., 1989a, Increased nigral iron content and alterations in other metal ions occurring in brain in Parkinson's disease, *J. Neurochem.* 52:1830–1836.

Dexter, D.T., Carter, C.J., Wells, F.R., Agid, F., Agid, Y., Lees, A., Jenner, P., and Marsden, C.D., 1989b, Basal lipid peroxidation in substantia nigra is increased in Parkinson's disease, *J. Neurochem.* 52:381–389.

Dexter, D.T., Jenner, P., and Marsden, C.D., 1989c, Oxyferriscorbone elevates the total iron content of blood but not brain. *Mov. Disord.* 4:176–182.

Dexter, D.T., Carayon, A., Vidailhet, M., Ruberg, M., Agid, F., Agid, Y., Lees, A.J., Wells, F.R., Jenner, P., and Marsden C.D., 1990, Decreased ferritin levels in brain in Parkinson's disease, *J. Neurochem.* 55:16–20.

Dexter, D.T., Carayon, A., Javoy-Agid, F., Agid, Y., Wells, F.R., Daniel, S.E., Lees, A.J., Jenner, P., and Marsden, C.D., 1991, Alterations in the levels of iron, ferritin, and other trace metals in Parkinson's disease and other neurodegenerative diseases affecting the basal ganglia, *Brain* 114:1953–1975.

Di Chiara, G., Olianas, M., Del Fiacco, M., Spano, P., and Tagliamonte, A., 1977, Intranigral kainic acid is evidence that nigral non-dopaminergic neurons control posture, *Nature* 268:743–745.

Dwork, A.J., Schon, E.A., and Herbert, J., 1988, Nonidentical distribution of transferrin and ferric iron in human brain, *Neuroscience* 27:333–345.

Earle, K.M., 1968, Studies in Parkinson's disease including x-ray fluorescent spectroscopy of formalin fixed brain tissue, *J. Neuropathol. Exp. Neurol.* 27:1–14.

Ernster, L., and Nordenbran, K., 1982, *in* "Lipid Peroxides in Biology and Medicine," pp. 55–79, Academic Press, Orlando, FL.

Forno, L.S., 1982, Pathology of Parkinson's disease, *in* "Movement Disorders" (C.D. Marsden and S. Fahn, eds.), pp. 25–40, Butterworths, London.

Frank, D.M., Arora, P.K., Blumer, J.L., and Sayre, L.M., 1987, Model study on the bioreduction of paraquat, MPP^+, and analogs: Evidence against a "redox cycling" mechanism in MPTP neurotoxicity, *Biochem. Biophys. Res. Commun.* 147:1095–1104.

Gerber, M.R., and Connor, J.R., 1989, Do oligodendrocytes mediate iron regulation in the human brain?, *Ann. Neurol.* 26:95–98.

Good, P.F., Olanow, C.W., and Perl, D.P., 1992, Neuromelanin-containing neurons of the substantia nigra accumulate iron and aluminum in Parkinson's disease—A LAMMA study, *Brain Res.* 593:343–346.

Halliwell, B., 1989, Oxidants and the central nervous system: Some fundamental questions, *Acta Neurol. Scand.* 126:23–33.

Halliwell, B., 1991, Reactive oxygen species in living systems: Source, biochemistry, and role in human disease. *Am. J. Med.* 91 (Suppl. 3C):14S–22S.

Halliwell, B., and Gutteridge, J.M., 1984, Oxygen toxicity, oxygen radicals, transition metals and disease, *Biochem. J.* 219:1–14.

Halliwell, B., and Gutteridge, J.M., 1989, "Free Radicals in Biology and Medicine" (2nd ed.), Oxford University Press, New York.

Hallman, H., Olson, L., and Jonsson, G., 1984, Neurotoxicity of the meperidine analogue N-methyl-4-phenyl-1,2,3,6-tetrahydropyridine on brain catecholamine neurons in the mouse, *Eur. J. Pharmacol.* 97:133–136.

Hartley, A., Cooper, J.M., and Schapira, A.H.V., 1993, Iron induced oxidative stress and mitochondrial dysfunction—relevance to Parkinson's disease, *Brain Res.*, 627:349–353.

Haykal-Coates, N., O'Callaghan, J.P., Reinhard, J.F., and Jensen, K.F., 1991, Pargyline and butyrolactone enhance tyrosine hydroxylase immunostaining of nigrostriatal axons, *Brain Res.* 556:353–357.

Herrera-Marschitz, M., and Ungerstedt, U., 1984, Evidence that striatal efferents relate to different dopamine receptors, *Brain Res.* 323:269–278.

Herrera-Marschitz, M., and Ungerstedt, U., 1985, Evidence that apomorphine and pergolide induce rotation in rats by different actions on D1 and D2 receptors sites, *Eur. J. Pharmacol.* 98:165–176.

Hill, J.M., and Switzer III, R.C., 1984, The regional distribution and cellular localization of iron in the rat brain, *Neuroscience* 11:595–603.

Hirsch, E., Graybiel, A.M., and Agid, Y.A., 1988, Melanized dopaminergic neurons are differentially susceptible to degeneration in Parkinson's disease, *Nature* 334:345–348.

Hirsch, E., Brandel, J.-P., Galle, P., Javoy-Agid, F., and Agid, Y., 1991, Iron and aluminum increase in the substantia nigra of patients with Parkinson's disease: An x-ray microanalysis, *J Neurochem.* 56:446–451.

Iiie, S., and Tavassoli, M., 1987, Transferrin-mediated cellular iron uptake, *Am. J. Med. Sci.* 293:103–111.

Jarvis, M.F., and Wagner, G.C., 1985, Neurochemical and functional consequences following 1-methyl-4-phenyl-1,2,3,6,-tetrahydropyridine (MPTP), *Neuropharmacology* 24:581–583.

Jarvis, M.F., and Wagner, G.C., 1990, 1-methyl-4-phenyl-1,2,3,6,-tetrahydropyridine-induced neurotoxicity in the rat: Characterization and age-dependent effects, *Synapse* 5:104–112.

Javitch, J.A., and Snyder, S.H., 1984, Uptake of MPP^+ by dopamine neurons explains selectivity of parkinsonian inducing toxin, MPTP, *Eur. J. Pharmacol.* 106:455–456.

Jellinger, K., Paulus, W., Grundke-Iqbal, I., Riederer, P., and Youdim, M.B.H., 1990, Brain iron and ferritin in Parkinson's and Alzheimer's diseases, *J. Neural Transm.* 2:327–340.

Jellinger, K., Kienzl, E., Rumpelmair, G., Riederer, P., Stachelberger, H., Ben-Shachar, D., and Youdim, M.B.H., 1992, Iron-melanin complex in substantia nigra of Parkinsonian brains: An x-ray microanalysis, *J. Neurochem.* 59:1168–1171.

Jenner, P., 1991, Oxidative stress as a cause of Parkinson's disease, *Acta Neurol. Scand.* 84(Suppl. 136):6–15.

Jenner, P., 1992, What process causes nigral cell death in Parkinson's disease?, *Neurol. Clin.* 10:387–403.

Jenner, P., and Marsden, C.D., 1986, The action of 1-methyl-4-phenyl-1,2,3,6-tetrahydropyridine in animals as a model of Parkinson's disease, *J. Neural Transm., Suppl.* 20:11–39.

Kastner, A., Hirsch, E.C., Lejeune, O., Javoy-Agid, F., Rascol, O., and Agid, Y., 1992, Is the vulnerability of neurons in the substantia nigra of patients with Parkinson's disease related to their neuromelanin content?, *J. Neurochem.*, 59:1080–1089.

Kish, S.J., Morito, C., and Hornykiewicz, O., 1985, Glutathione peroxidase activity in Parkinson's disease brain, *Neurosci. Lett.* 58:343–364.

Kopin, I.J., and D.G. Schoenberg, 1988, MPTP in animal models of Parkinson's disease, *Mt. Sinai J. Med.* 55:43–49.

Langston, J.W., Ballard, J., Tetrud, J., and Erwin, I., 1983, Chronic parkinsonism in humans due to a product of meperidine-analog synthesis, *Science* 219:979–980.

Langston, J.W., Irwin, I., and Ricaurte, G.A., 1987, Neurotoxins, Parkinsonism and Parkinson's disease, *Pharmacol. Ther.* 32:19–49.

Levy, A., Kant, G., Meyerhoff, J., and Jarrard, L., 1984, Non-cholinergic neurotoxic effects of AF64A in the substantia nigra, *Brain Res.* 305:169–172.

Martinovits, G., Melamed, E., Cohen, O., Rosenthal, J., and Uzzan, A., 1986, Systemic administration of antioxidants does not protect mice against the dopaminergic neurotoxicity of 1-methyl-4-1,2,3,6-tetrahydropyridine (MPTP), *Neurosci. Lett.* 69:192–197.

McCrodden, J.M., Tipton, K.F., and Sullivan, J.P., 1990, The neurotoxicity of MPTP and the relevance to Parkinson's disease, *Pharmacol. Toxicol.* 67:8–13.

McGeer, P.L., Itagaki, S., Akiyama, H., and McGeer, E., 1988a, Rate of cell death in parkinsonism indicates active neuropathological process, *Ann. Neurol.* 24:574–576.

McGeer, P.L., Itagaki, S., Boyes, B.E., and McGeer, E., 1988b, Reactive microglia are positive for HLA-DR in substantia nigra of Parkinson's and Alzheimer's disease brains, *Neurology* 38:1285–1291.

Minotti, G., and Aust, S., 1987, The requirement for iron (III) in the initiation of lipid peroxidation by iron (II) and hydrogen peroxide, *J. Biol. Chem.* 262:1098–1104.

Mitchell, I.J., Cross, A.J., Sambrook, M.A., and Crossman, A.R., 1985, Sites of the neurotoxic action of 1-methyl-4-phenyl-1,2,3,6,-tetrahydropyridine in the macaque monkey include the ventral tegmental area and the locus coeruleus, *Neurosci. Lett.* 61:195–200.

Nagy, K., Floyd, F., Simon, P., and Zs.-Nagy, I., 1985, Studies on the effect of iron overload on rat cortex synaptosomal membranes, *Biochim. Biophys. Acta* 820:216–222.

Nedergaard, M., Goldman, S.A., Desai, S., and Puisinelli, W.A., 1991, Acid-induced death in neurons and glia, *J. Neurosci.* 11:2489–2497.

Nicklas, W.J., Vyas, I., and Heikkila, R.E., 1986, MPP^+ inhibits NADH-linked mitochondrial oxidation: Role in the mechanism of MPTP neurotoxicity, *in* "MPTP: A Neurotoxin Producing a Parkinsonian Syndrome" (S. Markey, N. Castagnoli, A. Trevor, and I. Kopin, eds.), pp. 591–596, Academic Press, New York.

Olanow, C.W., 1990, Oxidation reactions in Parkinson's disease, *Neurology* 40(Suppl. 3):32–37.

Olanow, C.W., and Calne, D., 1991, Does selegiline monotherapy in Parkinson's disease act by symptomatic or protective mechanisms?, *Neurology* 42:13–26.

Olanow, C.W., and Cohen, G., 1992, The pathogenesis of Parkinson's disease, *in* "The Scientific Basis for the Treatment of Parkinson's Disease" (C.W. Olanow and A. Lieberman, eds.), pp. 59–76, Parthenon, Park Ridge, NJ.

Parkinson's Study Group, 1989, The effect of deprenyl on the progression of disability in early Parkinson's disease, *N. Engl. J. Med.* 321:1364–1371.

Pasinetti, G.M., Osterburg, H.H., Kelly, A.B., Kohama, S., Morgan, D.G., Reinhard, J.F., Stellwagen, R.H., and Finch, C.E., 1992, Slow changes of tyrosine hydroxylase gene expression in dopaminergic brain neurons after neurotoxin lesioning: A model for neuron aging, *Mol. Brain Res.* 13:63–73.

Perry, T. L., and Yong, V.W., 1986, Idiopathic Parkinson's disease, progressive supranuclear palsy and glutathione metabolism in the substantia nigra of patients, *Neurosci. Lett.* 67:269–274.

Perry, T.L., Yong, V.M., Clavier, R.M., Jones, K., Wright, J.M., Foulks, J.G., and Wall, R.A., 1985, Partial protection from the dopaminergic neurotoxin N-methyl-4-phenyl-1,2,3,6-tetrahydropyridine by four different antioxidants in the mouse, *Neurosci. Lett.* 60:109–114.

Poirier, J., Donaldson, J., and Barbeau, A., 1985, The specific vulnerability of the substantia nigra to MPTP is related to the presence of transition metals, *Biochem. Biophys. Res. Commun.* 128:25–33.

Ricaurte, G.A., Langston, J.W., Delanney, L.E., Irwin, I., Peroutka, S., and Forno, L.S., 1986, Fate of nigrostriatal neurons in young mature mice given 1-methyl-4-phenyl-1,2,3,6-tetrahydropyridine: A neurochemical and morphological reassessment, *Brain Res.* 375:117–124.

Ricaurte, G.A., Delanney, L.E., Irwin, I., and Langston, J.W., 1987, Older dopaminergic neurons do not recover from the effects of MPTP, *Neuropharmacology* 26:97–99.

Richter, C., Park, J., and Ames, B., 1988, Normal oxidative damage to mitochondrial and nuclear DNA is extensive, *Proc. Natl. Acad. Sci. USA* 85:6465–6467.

Riederer, P., and Youdim, M.B.H., 1987, MPTP induced dopaminergic neurotoxicity—a useful model in the study of Parkinson's disease?, *Neurochem. Int.* 11:379–381.

Riederer, P., Sofic, E., Wolf-Dieter, R., Schmidt, B., Reynolds, G., Jellinger, K., and Youdim, M.B.H., 1989, Transition metals, ferritin, glutathione, and ascorbic acid in Parkinsonian brains, *J. Neurochem.* 52:515–520.

Roberts, R., Sandra, A., Siek, G.C., Lucas, J.J., and Fine, R.E., 1992, Studies of the mechanism of iron transport across the blood-brain barrier, *Ann. Neurol.* 32:S43–S50.

Robertson, H.A., 1992, Dopamine receptor interaction: Some implications for the treatment of Parkinson's disease, *Trends Neurosci.* 15:201–206.

Rossetti, Z., Sotgiu, A., and Sharp, D., Hadjiconstantinou, M., and Neff, H., 1988, 1-methyl-4-phenyl-1,2,3,6-tetrahydropyridine and free radicals *in vitro, Biochem. Pharmacol.* 37:4573–4574.

Sachs, C.H., and Jonsson, G., 1975, Mechanism of action of 6-hydroxydopamine, *Pharmacology* 24:1–8.

Schapira, A.H., Cooper, J.M., Dexter, D., Clark, J.B., Jenner, P., and Marsden, C.D., 1990, Mitochondrial Complex I deficiency in Parkinson's disease, *J. Neurochem.* 54:823–827.

Schwartz, W.J., Gunn, R.H., Sharp, F.R., and Evarts, E.V., 1976, Unilateral electrolytic lesions of the substantia nigra cause contralateral circling in rats, *Brain Res.* 105:358–361.

Sengstock, G.J., Olanow, C.W., Dunn, A.J., and Arendash, G.W., 1992a, Iron induces degeneration of nigrostriatal neurons, *Brain Res. Bull.* 28:645–649.

Sengstock, G.J., Zawia, N., Olanow, C.W., Dunn, A.J., and Arendash, G.W., 1992b, Short-term alterations of nigral malondialdehyde (MDA) levels and striatal dopaminergic markers induced by intranigral iron infusion, *Soc. Neurosci. Abstr.* 18:1444.

Sengstock, G.J., Olanow, C.W., Menzies, R.A., Dunn, A.J., and Arendash, G.W., 1993, Infusion of iron into the rat substantia nigra: Nigral pathology and dose-dependent loss of striatal dopaminergic markers, *J. Neurosci. Res.* 35:67–82.

Singer, T.P., Trevor, A.J., and Castagnoli, N., 1987, Biochemistry of the neurotoxic action of MPTP, *Trends Biochem. Sci.* 12:266–270.

Singer, T.P., and Ramsey, R.R., 1990, Mechanism of the neurotoxicity of MPTP: An update, *FEBS Lett.* 274:1–8.

Slater, T.F., 1984, Free-radical mechanisms in tissue injury, *Biochem. J.* 222:1–15.

Sofic, E., Riederer, P., Heinsen, H., Beckmann, H., Reynolds, G.P., Hebenstreit, G., and Youdim,

M.B.H., 1988, Increased iron (III) and total iron content in postmortem substantia nigra of parkinsonian brain, *J. Neural Transm.* 74:199–205.

Sofic, E., Paulus, W., Jellinger, K., Riederer, P., and Youdim, M.B.H., 1991, Selective increase of iron in substantia nigra zona compacta of parkinsonian brains, *J. Neurochem.* 56:978–982.

Sonsalla, P.K., and Heikkila, R.E., 1986, The influence of dose and dosing interval on MPTP-induced dopaminergic neurotoxicity in mice, *Eur. J. Pharmacol.* 129:339–345.

Stern, Y., 1990, MPTP-induced Parkinsonism, *Prog. Neurobiol.* 34:107–114.

Tanaka, M., Sotomatsu, A., Kanai, H., and Hirai, S., 1991, Dopa and dopamine cause cultured neuronal death in the presence of iron, *J. Neurol. Sci.* 101:198–203.

Tanner, C.M., and Langston, J.W., 1990, Do environmental toxins cause Parkinson's disease? A critical review, *Neurology* 40(Suppl. 3):17–30.

Tetrud, J.W., and Langston, J.W., 1989, The effect of deprenyl (selegiline) on the natural history of Parkinson's disease, *Science* 254:519–522.

Triggs, W.J., and Willmore, L.J., 1984, *In vivo* lipid peroxidation in rat brain following intracortical Fe^{++} injection, *J. Neurochem.* 42:976–980.

Wagner, G.C., Jarvis, M.F., and Carelli, R.M., 1985, Ascorbic acid reduces the dopamine depletion induced by MPTP, *Neuropharmacology* 24:1261–1262.

Wagner, G.C., Carelli, R.M., and Jarvis, M.F., 1986, Ascorbic acid reduces the dopamine depletion induced by methamphetamine and the 1-methyl-4-phenylpyridinium ion, *Neuropharmacology* 25:559–561.

Willmore, L.J., and Rubin, J.J., 1982, Formation of malonaldehyde and focal brain edema induced by subpial injection of $FeCl_2$ into rat isocortex, *Brain Res.* 246:113–119.

Willmore, L.J., Hiramatsu, M., Kochi, H., and Mori, A., 1983, Formation of superoxide radicals after $FeCl_3$ injection into rat isocortex, *Brain Res.* 277:393–396.

Yong, V.W., Perry, R.L., and Krisman, A.A., 1986, Depletion of glutathione in brainstem of mice caused by N-methyl-4phenyl-1,2,3,6-tetrahydropyridine is prevented by antioxidant pretreatment, *Neurosci. Lett.* 63:56–60.

Youdim, M.B.H., Ben-Shachar, D., and Riederer, P., 1989, Is Parkinson's disease a progressive siderosis of substantia nigra resulting in iron and melanin induced neurodegeneration?, *Acta Neurol. Scand.* 126:47–54.

Youdim, M.B.H., Ben-Shachar, D., Yehuda, S., and Riederer, P., 1990, The role of iron in the basal ganglion, *Adv. Neurol.* 53:155–162.

Youdim, M.B.H., Ben-Shachar, D., and Riederer, P., 1991, Iron in brain function and dysfunction with emphasis on Parkinson's disease, *Eur. Neurol.* 31 (Suppl. 1):34–40.

Zaleska, M.M., and Floyd, R.A., 1985, Regional lipid peroxidation in rat brain *in vitro:* Possible role of endogenous iron, *Neurochem. Res.* 10:397–410.

Zigmond, M.J., Abercrombie, E.D., Berger, T.W., Grace, A.A., and Stricker, E.M., 1990, Compensation after lesions of central dopaminergic neurons: Some clinical and basic implications, *TINS* 13:290–295.

8

Excitotoxic Lesions of the Neostriatum as an Animal Model of Huntington's Disease

DWAINE F. EMERICH, DAVID W. CAHILL, and PAUL R. SANBERG

1. Introduction

Huntington's disease (HD) is an autosomal dominant disorder characterized by a progressive dementia coupled with bizarre uncontrollable movements and abnormal postures. The disease is found in nearly all ethnic and racial groups with slight variations in prevalence rates. Overall, the prevalence rate of HD in the United States is approximately 50/1,000,000 (Reed and Chandler, 1958; Sanberg and Coyle, 1984). Although HD may occur during the juvenile years (Korenyi and Whittier, 1973) the manifestation of the disorder typically occurs in middle life, about 35–45 years of age. From the time of onset, an intractable course of mental deterioration and progressive motor abnormalities begins, with death usually occurring within 15 years.

Because of its dramatic symptomology, relentless course, and hereditary nature, HD has gained a substantial amount of attention. Despite this, little is known about the underlying etiology of the disorder. Although research is beginning to unravel the pathophysiology and molecular biology involved, a wide gap still exists between our understanding of the neural substrates of the disease and our ability to prevent or alleviate them. Our increasing knowledge of the neural pathology in HD has revealed a complex mosaic of related and interdependent neurochemical and histopathological alterations. In addition, genetic research has recently isolated the chromosome responsible for the expression of HD. The

---

DWAINE F. EMERICH • CytoTherapeutics, Inc., 2 Richmond Square, Providence, Rhode Island 02906 and **DAVID W. CAHILL and PAUL R. SANBERG** • Division of Neurological Surgery, Department of Surgery, University of South Florida College of Medicine, 12901 Bruce B. Downs Boulevard, MDC-16, Tampa, Florida 33612.

Toxin-Induced Models of Neurological Disorders, edited by Michael L. Woodruff and Arthur J. Nonneman. Plenum Press, New York, 1994.

elucidation of the genetic abnormality in HD together with a more complete understanding of the pathology associated with it provides new avenues of treatment for the disorder.

An important approach for understanding and treating neurodegenerative diseases such as HD is the development of appropriate animal models that closely mimic the behavioral and neurobiological sequelae of the disorder. Such a model would aid in the further elucidation of the biological and behavioral expression of HD, and would also suggest unique therapeutic strategies for its treatment. This chapter evaluates the current status of the development of animal models of HD and possible treatments for the disorder. First, the behavioral pathology of HD is discussed. Next, the organization of the basal ganglia is described with reference to its neurochemical and cellular pathology in HD. Animal models of HD are then described with emphasis on excitotoxin-mediated damage to the striatum. Because of the promise that excitotoxin models hold for understanding the etiology of HD, considerable attention is paid to the kainic and quinolinic acid models and what aspects of the disease they may be used to mimic. Finally, the suitability of pharmacological treatments and neural transplantation as possible therapies in excitotoxic models and HD is addressed with discussion of the advantages and disadvantages of each.

2. Behavioral Pathology of Huntington's Disease

The onset of the motor and mental features of HD is variable, and the symptoms typically manifest themselves over a protracted time-course. Although the onset of these symptoms typically occurs at approximately 35–40 years of age, it may range from childhood to the eighth decade. The variability of the clinical features of HD makes a family history crucial for correct diagnosis. A number of disorders present symptoms that may be mistakenly diagnosed as HD including cerebellar ataxia, Creutzfeldt-Jakob disease, Wilson's disease, tardive dyskinesia, and basal ganglia infarction.

Clinically, the physical features of HD consist of constant choreiform movements of the entire body, which are irregular and involuntary (Sanberg and Coyle, 1984; Hefter *et al.*, 1987). These movements may be unilateral for a time, but later recruit all the limbs (Pinel, 1976). The face often looks quite grotesque because of the constant writhing contortions of facial muscles. Dysarthia, dysphagia, and disturbances in ocular motility may also develop, and the individual's speech becomes quite unintelligible (Starr, 1967; Dix, 1970, Davis, 1976; Young *et al.*, 1986; Podoll *et al.*, 1988). Expression of the mental symptoms in HD is more variable than the motor abnormalities. The primary mental features of HD are generally considered similar to dementia (Burns *et al.*, 1990; Dewhurst *et al.*, 1979; Knopman and Nissen, 1991; Mann *et al.*, 1980; Sanberg and

Coyle, 1984; Cummings and Benson, 1988; Brandt *et al.*, 1988; Heindel *et al.*, 1988). The initial signs may consist of lack of grooming, eccentric traits, irritability, impairments in memory, emotional instability, and delusions of paranoid grandeur. Personality disorders, such as outbursts of rage or violent temper, may occur. Conversely, some patients may be slow and apathetic. Depression is the most common early psychological symptom and may be associated with changes in the ability to manage daily work and household-related activities. Patients also show a lack of initiative as well as decreased attention and communicativeness. These early symptoms of HD can mimic those of schizophrenia and bipolar affective illness, leading to many patients being misdiagnosed as having schizophrenia (Klawans *et al.*, 1972; Bowman and Lewis, 1980; Van Putten and Menkes, 1973). It is generally agreed that the mental symptoms occur prior to the motor symptoms, which probably also contributes to misdiagnosis (Bruyn, 1958; James *et al.*, 1969; Girotti *et al.*, 1988).

Often, the HD patient does not appear demented owing to a general preservation of intellect and language functions (Scholz and Berlemann, 1987; Brandt *et al.*, 1988; Heindel *et al.*, 1988; Saint-Cyr *et al.*, 1988). This is likely the result of the sparing of cortically mediated cognitive functions (Reynolds *et al.*, 1990). In other neurodegenerative disorders, such as Alzheimer's disease, the cognitive alterations occur against a backdrop of substantial cortical degeneration (Coyle *et al.*, 1983; Brandt *et al.*, 1988). Accordingly, the mnemonic impairments observed in HD are referred to as a subcortical dementia. McHugh and Folstein (1975) suggested that the dementia in HD consists of several features characteristic of subcortical pathology. These include a progressive loss of generalized cognitive powers, psychic apathy, and inertia, which progresses to an akinetic mute state. These symptoms were reported to appear in the absence of aphasia, alexia, cortical blindness, and Koraskoff-like amnesia.

The often subtle differences between the dementia in various neurodegenerative disorders illustrate the notion that the cortical and subcortical dementia distinction is a relative one. This distinction is based primarily on anatomical grounds, but much like the behavioral expression of different diseases the anatomical differences are not always clear-cut. Some Parkinson's patients have changes in cortical morphology that resemble those seen in Alzheimer's patients. Moreover, Alzheimer's patients have well-documented changes in subcortical nuclei including the nucleus basalis. Accordingly, the dementia associated with a given disease state is best considered to involve a relatively greater contribution of either cortical or subcortical pathology. Recent metabolic studies support this contention (Buell *et al.*, 1990; Jernigan *et al.*, 1991; Kuwert *et al.*, 1989, 1990; Leblhuber *et al.*, 1989).

More recent studies have supported the notion of a prominent subcortical dementia in HD. Toward this end, most studies have made direct comparisons of the cognitive changes in Alzheimer's and Huntington's patients (Hodges *et al.*,

1991; Salmon *et al.*, 1989; Pillon *et al.*, 1991; Randolph, 1991; Heindel *et al.*, 1990). Individuals afflicted with Alzheimer's disease have an inability to store or consolidate new material, show increased sensitivity to proactive interference, display poor recognition memory, and have rapid forgetting over time. In contrast, HD patients are impaired in their ability to initiate any systemic retrieval of already stored information, but exhibit superior retention of information and have normal recognition memory. The subcortical dementia in HD also appears to be different from that in other subcortical disorders including Parkinson's disease (PD). Massman and colleagues (1990) reported that, whereas many similarities existed in the dementia in Parkinson's and Huntington's patients, the Huntington's patients had poorer free recall, impaired learning over trials, abnormal serial position effects, and increased preservation. These results indicate that disorders characterized by a predominant subcortical pathology are still differentiated by subtle differences in clinical symptomatology.

Even in advanced cases, the pattern of dementia in HD is not diffuse and homogeneous, but is characterized by a relative sparing of many higher cortical functions. In addition, the cognitive deficits do not develop uniformly; rather, the memory disorders appear to proceed the more generalized intellectual deterioration. From the time of onset, the motor and mental symptoms of HD progress at an increasingly disabling rate. Initially, the individual may exhibit subtle shifts in personality or cognition that occur together with a mild chorea. As the neuronal degeneration of the disease progresses, the afflicted individual undergoes more intense and abrupt personality changes combined with substantial cognitive impairments. As the mental features of HD change over time, the motor effects change from a choreic dyskinesia to a more disabling dystonic and parkinsonian-like syndrome (Shoulson, 1986). Finally, the severity of the cognitive and mood disturbances do not correlate well with the motor impairments. Table I lists the psychological, psychiatric, and neuropsychological changes reported in HD.

Table I. Psychological, Psychiatric, and Neuropsychological Findings in Huntington's Disease

Psychiatric changes

- Patients are erratic, impulsive, and prone to emotional outbursts, rage, and violence. May also be apathetic.
- Lack of initiative, labile affect, decreased communicativeness.
- Emotional instability, severe fluctuations in mood, irritability, and anxiety.
- Eccentric traits, lack of grooming and delusions of grandeur and paranoia.
- Depression (more common in females) and inability to manage work and household-related tasks.
- Mental restlessness and feelings of discomfort.

(*continued*)

Table I. (*Continued*)

Psychological changes

- Loss of memory.
- Difficulty with organization, planning, and temporal ordering of information.
- Easily distracted and disturbed by excessive sensory input.
- Repetitive thought patterns.
- Decreased spontaneity in the absence of strong motivation.
- Mental apathy and inertial.
- Absence of deterioration of higher cortical functions (agnosia, aphasia, apraxia, disturbances of spatiotemporal and personal orientation).
- Loss of psychomotor inhibition.

Neuropsychological changes

- Impaired performance on neuropsychological test batteries
  California Verbal Fluency Test
  Wechsler Memory Scale
  Wechsler-Bellevue Scale
  Wechsler Adult Intelligence Scale
  Halstead Neuropsychological Test Battery
  Minnesota Multiphasic Personality Inventory
  Bender Visual-Motor Gestalt Test
  Shipley-Hartford Retreat Scale
  Dementia Rating Scale
  Boston Naming Test
- Impaired imagery recall
- Impaired trail making
- Increased anxiety
- Normal language performance, reading, writing, picture naming, and tactual performance also normal.
- Altered cognitive processes
  decreased procedural learning
  loss of immediate memory
  impaired semantic clustering
  impaired delayed recall
  loss of recognition memory
  increased susceptibility to priming effects
  poor improvement of learning over trials
  abnormal serial position effects
  higher rates of perseveration
  normal rates of intrusion errors
  normal vulnerability to proactive and retroactive interference
  normal retention of information
  decreased comprehension of prosody

3. Anatomy and Neurochemistry of the Basal Ganglia

The basal ganglia consists primarily of the caudate, putamen, and globus pallidus. The subthalamic nucleus, substantia nigra, nucleus accumbens, and amygdala, although not universally considered part of the basal ganglia, provide important anatomical links, and they likely play a vital role in its functional output under normal conditions and in disease states. The caudate and putamen (striatum) are fused at the anterior segment and contain cells that project to both the internal and external components of the globus pallidus as well as the pars reticulata of the substantia nigra. Both the globus pallidus, which lies medial to the putamen and lateral to the internal capsule, and the substantia nigra have diverse connections with the subthalamic nucleus. The subthalamic nucleus receives afferent connections from the external segment of the globus pallidus. In turn, the subthalamic nucleus projects to both segments of the globus pallidus and to the pars reticulata of the substantia nigra. Because the basal ganglia does not have direct afferent or efferent connections with the spinal cord, its influence on motor processes is indirect. Instead, the primary input to the basal ganglia arises in disperse neocortical regions, and the primary output is directed to the prefrontal and premotor area of the frontal cortex.

The striatum consists of at least six different cell types that are distinguishable on the basis of their size, dendritic spines and arborizations, and axonal trajectories. The four types of projection neurons are similar in that they all use GABA as their neurotransmitter, but they differ in their peptide content as well as the region to which they project. The projection neurons, in contrast to the smaller, less prevalent interneurons, make up approximately 90% of all the striatal cells. The larger aspiny interneurons contain acetylcholine (ACh), whereas the small aspiny neurons are identifiable by the presence of somatostatin and neuropeptide Y.

The striatum is inhomogeneous and consists of islands of cells or striosomes, which stain weakly for acetylcholinesterase and are interspersed by a stronger staining matrix (Graybiel, 1983; Graybiel *et al.*, 1981, 1989; Joyce *et al.*, 1986; Rhodes *et al.*, 1987; Gerfen *et al.*, 1987). Striosome and matrix neurons are also differentiated on the basis of their innervation with the striosomal neurons receiving afferent input from the medial substantia nigra and medial frontal cortex and the matrix neurons receiving input from the substantia nigra, ventral tegmental area, and the motor, sensory, and association cortices. In addition to their differential staining for acetylcholinesterase, the striosome and matrix neurons contain different neurotransmitters. Both cell types utilize GABA, but the matrix neurons also contain somatostatin and neuropeptide Y, whereas the striosome neurons contain dynorphin and neurotensin (Graybiel, 1983).

All of the afferent inputs to the basal ganglia terminate in the striatum. The largest of these connections, the corticostriate projection, uses glutamate as its major neurotransmitter and innervates the majority of the striatum, exerting an excitatory influence over the numerous small spiny interneurons. Other inputs to the striatum include: (1) a dense dopaminergic projection from the substantia nigra (Fibiger *et al.*, 1972; McGeer *et al.*, 1987); (2) the thalamostriatal tract, which has been suggested to use either ACh, glutamate, or aspartate as its major neurotransmitter (McGeer *et al.*, 1987): (3) a cholecystokinin (CCK) projection from the pryiform cortex and amygdala (Meyer *et al.*, 1982): (4) small projections from both the raphe nucleus (serotonin) (Chan-Palay, 1977; Steinbusch, 1984) and the locus ceruleus (norepinephrine) (Moore and Card, 1984); and (5) the GABAergic pallidostriatal projection (McGeer *et al.*, 1987).

The efferent connections of the striatum are relatively simple compared to its afferent input. These are enkephalin projections from the striatum to the globus pallidus (Brann and Emson, 1980; Del Fiacco *et al.*, 1982; Yang *et al.*, 1983), as well as a substance P projection to the globus pallidus and substantia nigra (Jessell *et al.*, 1978; Staines *et al.*, 1980). Striatal efferents also terminate in the substantia nigra, using GABA, dynorphin, and substance P as their neurotransmitters (Vincent *et al.*, 1982; Araki *et al.*, 1985). Striatal inputs to the globus pallidus have also been reported to use GABA and dynorphin (McGeer *et al.*, 1987).

The globus pallidus (pars interna) gives rise to the major efferents from the basal ganglia and projects to the thalamus via the ansa lenticularis and lenticular faciculus. The thalamic nuclei that receive these inputs also project back to the prefrontal cortex. Therefore, the basal ganglia receives an enormous cortical and thalamic input and, in turn, projects back to these same regions. Because there are no nuclei in the basal ganglia that exert a direct motor influence on the spinal cord, the basal ganglia modulates this activity through the cortex and thalamus. In this way, the various structures of the basal ganglia act as integrators and modulators of information to and from the cortex as well as from the cerebellum.

4. Pathology of Huntington's Disease

4.1. Anatomical Pathology

Classically, HD is associated with a gross generalized atrophy of the cortex and basal ganglia affecting both the gray and white matter (Dulap, 1927). Histologically, this is accompanied by an extensive gliotic reaction and loss of small neurons in both the striatum and in layers 3, 5, and 6 of the frontal and parietal cortices. The severe damage to the neostriatum results in a compensatory, secondary hydrocephalus with a gross dilation of the ventricular system (Dulap, 1927; Barr *et al.*, 1978; Bruyn *et al.*, 1979; Lange, 1981; Vonsattel *et al.*, 1985).

The severe atrophy of the striatum is uniformly exhibited among Huntington's patients with approximately a 60% to 90% decrease in mass occurring in cases of juvenile onset. These decreases in striatal volume are related to a severe loss of the medium-sized spiny projection neurons, the major output neurons of the striatum (Reiner *et al.*, 1988). Neurochemically, the loss of these neurons is associated with decreases in GABA, substance P, dynorphin, and enkephalin (Bird, 1980; Buck *et al.*, 1981). On the other hand, local circuit aspiny neurons reactive for NADPH-diaphorase and somatostatin are relatively spared (Beal, 1988b, 1989). The large aspiny neurons, which are likely cholinergic, although spared in the early stages of the disorder, may exhibit degenerative changes as the disease progresses (Ferrante *et al.*, 1987). Ferrante and colleagues (1991) reported that striatal spiny neurons exhibited both proliferative and degenerative changes in the HD brain. Proliferative changes included increases in the size and numbers of dendritic spines, whereas degenerative changes included dendritic swelling and spine loss. Interestingly, the proliferative changes were found in the more moderate cases of HD, while the degenerative changes were observed primarily in the most severe cases. Although the nature of the plasticity of these neurons is not understood it was suggested that early, proliferative changes may play a prominent role in the loss of spiny neurons in HD.

Magnetic resonance imaging and photon emission tomography make it possible today to evaluate pathological changes in HD patients during life. Cerebral blood flow and glucose utilization are decreased in both cortical and basal ganglia structures. Kuwert and colleagues (1989, 1990) reported that cortical and subcortical glucose utilization was markedly decreased in HD patients with significant correlations occurring between both the duration and severity of the chorea and the decreases in cortical glucose consumption.

The globus pallidus also exhibits a marked atrophy (typically 50%), which is a result of loss of strio-pallidal fibers as well as a substantial decrease in the number of pallidal neuronal perikarya (Lange *et al.*, 1976). Given that the globus pallidus is a major output from the caudate and putamen, the atrophy of this nucleus may reflect in part a transneuronal degeneration subsequent to striatal degeneration.

The sulcal widening and the loss of neuronal weight in the cerebral cortex indicate that the cortex also undergoes a substantial deterioration (Forno and Jose, 1973). However, unlike the relatively ubiquitous cell loss observed in the striatum, the degenerative process appears to be more restricted in the cortex with cell loss occurring in the third, fifth, and sixth cortical layers. Interestingly, studies using retrograde-tracing techniques indicate that the small pyramidal cells of these deep cortical layers are the source of cortical innervation to the striatum (Oka, 1980; Cudkowicz and Kowall, 1990; Hedreen *et al.*, 1991; Sortel *et al.*, 1991). Together with the observation by Cross *et al.* (1986) that high-affinity glutamate uptake is reduced in the striatum of HD patients, it appears that the disease is associated with severe striatal atrophy concomitant with degeneration of both its major inputs and outputs.

Table II. Regional Pathology in Huntington's Disease Brain

Region	Pathology
Gross brain	Weight and size reduced
Ventricular system	Dilation
Caudate/putamen	Reduction in volume, neuronal loss, gliosis decrease in glucose/oxygen utilization
Lentiform nucleus	Atrophy
Globus pallidus and claustrum	Reduction in volume, neuronal loss, gliosis
Nucleus accumbens	Slight neuronal loss and gliosis
Thalamic nuclei	
Centromedian	Neuronal loss and gliosis
Dorsomedial	Slight gliosis
Ventral anterior	Neuronal loss
Ventral lateral	Neuronal loss
Hypothalamic nuclei	
Lateral	Neuronal loss
Ventromedial	Cell shrinkage and neuronal loss
Paraventricular	Neuronal loss
Supraoptic	Neuronal loss
Cortex	Atrophy, neuronal loss, gliosis, decrease in glucose utilization
Hippocampus	Atrophy, gliosis, neurofibrillary tangles
Cerebellum	
Dentate nucleus	Neuronal loss and gliosis
Purkinje layer	Neuronal loss and gliosis
Inferior olive	Atrophy, neuronal loss, gliosis
Subthalamic nucleus	Neuronal loss and gliosis
Substantia nigra pars reticulata	Atrophy, neuronal loss, gliosis
Brainstem nuclei	
Superior olive	Neuronal loss
Occulomotor	Gliosis
Midbrain tegmentum	Gliosis
Paramedian reticular	Neuronal loss
Dorsal vagus	Gliosis
Aquaduct area	Gliosis
Red nucleus	Neuronal loss
Spinal cord	Gliosis

Although HD has been generally considered to be a neurodegenerative disorder characterized by the restricted loss of certain populations of neurons, it is becoming clear that the neural degeneration is actually quite widespread (see Table II). The pathological changes in neural regions removed from the basal ganglia are, however, typically not as severe. For example, the pars reticulata of the substantia nigra (Gebbink, 1968; Oyanagi and Ikuta, 1987; Oyanagi *et al.*, 1989), thalamic nuclei (Forno and Jose, 1973; Dom *et al.*, 1976), subthalamic nucleus (Lange *et al.*, 1976), cerebellum (Jervis, 1963; Bruyn, 1968; Castaigne *et al.*, 1976; Rodda, 1981), hypothalamus (Bruyn, 1968; Kremer *et al.*, 1990), hippocampus (Forno and Jose, 1973; Mattson *et al.*, 1974; McIntosh *et al.*, 1978; Averback, 1980), and a variety of brainstem regions, including the superior olive and red nucleus (Roizin *et al.*, 1976), have all been shown to exhibit reactive gliosis and evidence of neural degeneration. On the other hand, the nucleus accumbens and pars compacta of the substantia nigra appear relatively normal (Bots and Bruyn, 1981).

4.2. Neurochemical Pathology

The discovery of dopamine in the striatum and the depletion of this transmitter in PD, coupled with the suggestion that the symptomatologies of PD and HD are at opposite ends of a behavioral continuum, suggested that increased levels of dopamine may underlie the motor abnormalities in HD. Unfortunately, attempts to determine whether dopaminergic parameters are altered in HD have met with difficulty. Initial studies suggested either little change or a slight elevation of dopamine levels in striatum, substantia nigra, and nucleus accumbens (Bernheimer *et al.*, 1973; Chase, 1973; Bird and Iversen, 1974; McGeer and McGeer, 1976a; Mann *et al.* 1980; Bird, 1980; Spokes, 1980). It was suggested that this was the result of the generalized atrophy of the striatum, which resulted in an apparent preservation of dopaminergic neurons. However, other factors, such as the loss of GABA feedback fibers from the striatum to substantia nigra, may contribute to the apparent preservation of dopamine. Interestingly, Kish *et al.* (1987) recently examined dopamine levels in the regionally subdivided striatum of Huntington's patients. Both dopamine and Homo Vanillic acid (HVA) were found to be significantly reduced in the caudal caudate. It is therefore conceivable that subtle alterations in striatal dopaminergic parameters are masked by the tendency to examine neurochemical changes in the striatum as a whole. Taken together, it appears that striatal dopaminergic indices may not be dramatically altered, but compared to the marked loss of other neurochemical parameters, the net effect may be a relative overactivity of striatal dopamine.

Because dopaminergic antagonists have been used clinically to ameliorate chorea, it has been suggested that dopamine receptors are supersensitive. However, radioligand studies indicate that the number of striatal dopaminergic receptors are actually reduced in HD (Reisine *et al.*, 1977; Filloux *et al.*, 1990). The reduction in the number of dopamine binding sites could reflect either a compensatory action in response to elevated levels of dopamine in the striatum or the loss of receptors located on dying striatal neurons. Alternatively, a reduced inhibition of dopaminergic transmission as a result of degenerating GABA systems may partially account for the altered pharmacological sensitivity. Melamed *et al.* (1982), however reported that the nigrostriatal pathway adequately adapts to the loss of striatal neurons, which normally inhibit dopamine activity both in HD and kainate-lesioned rats. It therefore remains unclear why dopaminergic compounds exert such powerful effects.

The GABAergic neurons of the basal ganglia undergo a severe deterioration as indexed by measures of transmitter levels and enzymatic activity. Perry and colleagues (1973) reported that levels of GABA are reduced in the substantia nigra, caudate, putamen, and globus pallidus in HD. Ever since these studies were conducted, decreases in glutamate decarboxylase have also been found in the choreic striatum, further suggesting decreases in striatal GABA concentrations. Decreases also occur in glutamate decarboxylase in the substantia nigra and globus pallidus, likely due to degeneration of striatonigral and striatopallidal neurons (McGeer *et al.*, 1973; Bird and Iversen, 1974). Paralleling the decreases in striatal GABA and glutamate decarboxylase are decreases in GABAergic receptors.

Both the cerebral spinal fluid (CSF) and the striatal content of ACh, as measured by the activity of the synthetic enzyme choline acetyltransferase (ChAT), are also greatly reduced in HD (McGeer *et al.*, 1973; Bird and Iversen, 1974; Stahl and Swanson, 1974; Aquidonius *et al.*, 1975; McGeer and McGeer, 1976a; Spokes, 1980; Manyam *et al.*, 1990), as are the numbers of muscarinic receptors in the striatum (Hiley and Bird, 1974; Enna *et al.*, 1976a,b). It is unclear, however, whether the decreases in ChAT represent the actual degeneration and death of striatal cholinergic neurons. Ferrante *et al.* (1987, 1991) reported that, although ChAT activity was decreased in striatum of Huntington's patients, there was a relative preservation of acetylcholine esterase (AChE)-reactive cells. Because ChAT and AChE co-localize within cholinergic neurons, it was suggested that AChE-containing cell bodies are preserved, but ChAT-rich cholinergic terminals degenerate in the disease. The decreases in cholinergic parameters may also be the consequence of alterations in nearby dopaminergic neurons. As the action of dopamine on striatal cholinergic neurons is likely inhibitory, it may be that a general overactivity of dopaminergic neurons decreases cholinergic indices. Sanberg and Creese (1981) reported that a single injection of the dopaminergic antagonist bromocriptine produced a persistent decrease in striatal ChAT activity. If a similar situation occurred in the brain of Huntington patients (i.e., dopamine overactivity and decreased ChAT activity) the decreases in cholinergic indices may not have reflected the actual death of those neurons, but rather a dopamine-mediated decrease in activity. Significant decreases in muscarinic receptors have also been reported in the HD brain (Wastek and Yamamura, 1978; Wastek *et al.*, 1976).

Various other neurochemical alterations are prevalent in HD (see Bird, 1980, for a review). The undecapeptide substance P was one of the first peptides examined in HD. The basal ganglia contains one of the highest levels of substance P of any brain region, where it appears to exert an excitatory influence over striatal afferents projecting to the globus pallidus and substantia nigra (Mroz *et al.*, 1977). Substance P is decreased by as much as 90% in the pallidum and substantia nigra (Kanazawa *et al.*, 1979; Buck *et al.*, 1981). On the other hand, striatal levels of substance P have been reported to be either normal or slightly decreased (Gayle *et al.*, 1978; Beal *et al.*, 1988b). Likewise, substance P is unaffected in the cortex, thalamus, and hypothalamus.

Cholecystokinin (CCK) levels have also been evaluated in the HD brain. Cholecystokinin was once thought to be predominantly a gastrointestinal peptide, but was subsequently found to be present in high concentrations in the central nervous system, especially the cortex and striatum (Innis *et al.*, 1979; Innis and Snyder, 1980). An initial study by Emson *et al.* (1980) revealed that CCK immunoreactivity was decreased by approximately 50% in the globus pallidus and substantia nigra. A subsequent study by Hays *et al.* (1981) demonstrated that the number of CCK receptors was decreased in the basal ganglia and cerebral cortex of Huntington's patients.

Enkephalin is an endogenous neurotransmitter found in particularly high concentrations in the globus pallidus. The cell bodies of origin for these enkephalinergic processes are localized within the striatum. Arregui *et al.* (1979) conducted a detailed analysis of enkephalin levels in the human brain and observed that the lateral pallidum and nucleus accumbens contained the highest levels of any region examined. Although the levels in the medial pallidum and substantia nigra were approximately 50% lower, they still were higher than other brain regions examined. It was further demonstrated that enkephalin levels were decreased by as much as 50% in the medial and lateral pallidum as well as the substantia nigra. Enkephalin levels were not altered in either the caudate or putamen.

Angiotensin II is an octapeptide that has been reported to be decreased in HD (Arregui *et al.*, 1977, 1978). The activity of angiotensin II-converting enzyme is decreased by as much as 60% to 90% in the caudate, putamen, globus pallidus, and substantia nigra of HD brains. In contrast, several neurotransmitters and neuropeptides, including norepinephrine, somatostatin, neurotensin, neuropeptide Y, and thyrotropin-releasing hormone, are present at normal or slightly elevated levels in HD.

Together, these data indicate that every neuronal system known to originate in the striatum exhibits decrements in the concentration of its presynaptic neuronal markers in HD. Furthermore, when the degree of striatal atrophy is taken into account, the total deficits of the markers for these neuronal systems support a profound loss consistent with the known cellular pathology. The specificity of these synaptic neurochemical changes, however, has been confirmed by the demonstration that the spared dopaminergic afferent system does not show comparable decrements in either the substantia nigra or the striatum. Levels of serotonin and somatostatin appear normal (but see Kish *et al.*, 1987); however, recent reports indicate a reduced density of striatal somatostatin, serotonin, and neurotensin receptors (Waeber and Palacios, 1989; Palacios *et al.*, 1990; Gonzalez-Heydrich and Peroutka, 1991; Palacios *et al.*, 1991). These findings clearly demonstrate that the neuronal vulnerability of HD is not specified by any one striatal neurotransmitter but affect virtually all of the identified intrinsic and efferent systems of the corpus striatum.

5. Excitotoxin Models of Huntington's Disease

An important approach for understanding neurological diseases is to develop appropriate animal models. Such models help to elucidate the biological and behavioral manifestations of clinical disorders, and they suggest novel therapeutic strategies for their prevention and/or treatment. Although animal models may not reproduce the complex etiologies, pathophysiology, or behavioral abnormalities associated with diseases of the nervous system, they do provide a practical way to explore specific questions concerning structure and function.

Various avenues have been explored to develop an animal model of HD (see Sanberg and Coyle, 1984, for a review). Although models focusing on hyperkinesias in animals resulting from genetic conditions and various types of brain lesions have been used, the majority of models have centered around motor abnormalities produced by short-term pharmacological stimulation of some of the neurochemical changes found in HD. Although useful, there are several inherent problems with pharmacological models of HD or chorea-like syndromes. Many of the compounds used in these studies have limited CNS specificity, and they affect more than one transmitter or receptor type. These compounds may also exhibit a high degree of behavioral toxicity or have a short duration of action and therefore cannot reproduce the chronicity of the pathology in HD. Finally, the dyskinesia produced in these models often bears little homology to the motor impairments observed in HD. This is not to say that the behavioral alterations observed following these treatments are of no relationship to those in HD. In fact, it is unlikely that any rodent model of HD would be able to reproduce with precision the chorea found in HD. Considering that rats are quadrupeds and humans are bipeds, it is not surprising that chorea found in HD as such does not develop in these models. Although pharmacological manipulations of striatal function do not reproduce the chorea of HD, the resulting locomotor abnormalities may be generally considered to resemble those in the disease.

Given the above considerations, it is difficult to establish causal or even suggestive relationships between drug-induced neural changes and specific behavioral alterations. Another approach to mimic the symptomatology of HD is to mechanically or electrolytically lesion various brain areas, such as the striatum. However, this is an unsatisfactory approach in that lesion techniques almost invariably damage supportive structures as well as fibers that pass through and terminate in the damaged area. A better strategy for investigating the relationship between striatal damage and locomotor abnormalities might involve the use of selective cytotoxic compounds. Selective toxic compounds have been widely used in neurobiology to examine the functional properties of the nervous system. Toxins have been used to examine the function and molecular biology of ion channels, axoplasmic transport processes, neurotransmitter systems, and the principles of synaptic transmission (McGeer *et al.*, 1987). In a related context, neurotoxins have been successfully used to examine the covariation between altered neurotransmitter dynamics and behavior, and to develop animal models of neurological disorders.

Glutamate is one of the major excitatory neurotransmitters found in the CNS. It can act, however, as a potent neurotoxin. Because the corticostriate projections utilize glutamate (Divac *et al.*, 1977; Fonnum *et al.*, 1981), a number of attempts have been made to develop animal models of HD based on the relatively specific cytotoxic effects of excitotoxic compounds (Olney and de Gubareff, 1978). These compounds include structural analogs of glutamate, such as kainic acid (KA), ibotenic acid (IA), and the endogenous tryptophan metabolite quinolinic acid (QA). When injected into the brains of rats, in extremely small doses, these compounds produce a marked and locally restricted toxic effect while sparing axons of passage and afferent nerve terminals. The behavioral, neurochemical, and anatomical consequences of excitotoxicity resemble those observed in HD and have led to the speculation that an aberrant overproduction or breakdown of endogenous excitotoxic compounds is an etiological factor in HD.

5.1. The Kainic Acid Model

5.1.1. Neuropathology

Intrastriatal injections of nanomolar quantities of KA result in the degeneration of local, intrinsic neurons while sparing axons of passage and terminals of extrinsic origin. The consequent neurochemical and pathological alterations bear a considerable analogy to that described in the striatum of HD patients upon postmortem examination (Coyle and Schwarcz, 1976; McGeer and McGeer, 1976b; Zaczek *et al.*, 1978; Fields *et al.*, 1978; Coyle, 1979; McGeer *et al.*, 1989; McGeer and McGeer, 1982). The neurochemical alterations induced by KA injections include decreases in GABA, ACh, and angiotensin levels, as well as decreases in receptors for GABA, serotonin, and dopamine. In contrast, striatal levels of dopamine and serotonin are unaffected. These biochemical changes are accompanied by reductions in the number of intrinsic striatal neurons, ventricular dilation, and decreases in striatal volume. Secondary to these alterations is a transneuronal degeneration of the globus pallidus and the pars reticulata of the substantia nigra.

Table III. Biochemical and Morphological Changes in the Striatum Reported in Huntington's Disease and Rats following Intrastriatal Injections of Kainic or Quinolinic Acid

	Huntington's disease	Kainic acid	Quinolinic acid
Neurochemical indices			
GABA	Decreased	Decreased	Decreased
Dopamine	Normal	Normal	Normal
Serotonin	Normal	Normal	Normal
Acetylcholine	Normal/decreased	Decreased	Normal/decreased
Somatostatin	Normal/increased	Decreased	Normal/increased
Neuropeptide Y	Normal/increased	Decreased	Normal/increased
Substance P	Decreased	Decreased	Decreased
Angiotensin	Decreased	Decreased	?
Enkephalin	Decreased	Decreased	Increased
Receptor binding			
GABA	Decreased	Decreased	Decreased
Muscarinic	Decreased	Decreased	Decreased
Serotonin	Decreased	Decreased	Decreased
Dopamine	Decreased	Decreased	Decreased
Somatostatin	Decreased	?	?
Excitatory amino acids	Decreased	?	Decreased

Importantly, KA does not selectively spare neurons that are reactive for somatostatin or neuropeptide Y (Table III). Because this population of neurons seems to be spared in HD, the KA model does not appear to reproduce faithfully the entire constellation of histological changes observed in the disease. The specificity of KA has also been criticized because it can produce hippocampal damage (Pisa *et al.*, 1980, 1981; Sanberg *et al.*, 1979, 1981b). However, this occurs as a result of the epileptic sensitivity of hippocampal neurons and may be prevented by prior treatment with the anticonvulsant diazepam (Sanberg, 1980; Ben-Ari *et al.*, 1980). Other excitotoxins, such as IA, produce a profile of striatal damage that is similar to that produced by KA without damage to the hippocampus or neighboring structures. Because the behavioral consequences of IA are similar to those following KA, it is likely that the effects of KA are the result of striatal and not hippocampal damage. Given these considerations, it appears that intrastriatal injections of KA produce a profile of histological and neurochemical alterations that closely mimic some, but not all, of those observed in patients afflicted with HD.

5.1.2. Behavioral Pathology

The behavioral hallmarks of HD are abnormal, uncontrollable, and constant choreiform movements. Unfortunately, KA-induced chorea is not as recognizable as that observed in HD. However, the KA-lesioned rat does exhibit an abnormal profile of locomotor activity, including increases in swing time and a decrease in stance time compared to control rats. Thus, the lesioned rats put their paws on the ground for shorter periods of time and swing them longer than do normal rats. This has been suggested to be analogous to the locomotor pattern observed in Huntington's patients (Hruska and Silbergeld, 1979). Another similarity to HD that exists in the motor activity of KA-lesioned rats is that, compared to controls, their locomotor activity is markedly potentiated during arousal (Mason and Fibiger, 1979). Huntington's patients also show greater choreiform movements during their awake period and when aroused (Sanberg and Johnston, 1981; Sanberg and Coyle, 1984). In addition, rats with striatal KA lesions tested in photoactometers have shown an exaggerated motor response to amphetamine, whereas the response to apomorphine did not differ significantly from controls. These animals also exhibit a greater locomotor response to the cholinergic antagonist scopolamine and a decreased cateleptic response to the dopaminergic antagonist haloperidol, as compared to controls (Sanberg, 1980; Sanberg *et al.*, 1981b). Clinical studies have suggested similar pharmacological responses to comparable drug challenges in HD patients (see Sanberg and Johnston, 1981, for a review).

Various psychological and psychiatric symptoms occur against the neurobiological backdrop of HD. As described previously, the behavioral symptomatology of HD occurs primarily as a result of subcortical, not cortical, damage. Similarly, studies with KA-lesioned rats indicate that marked impairments in learning, memory, and affect occur without concomitant cortical damage. Specifically, intrastriatal injections of KA have resulted in impaired acquisition and retention of a passive avoidance and spatial alternation task, impaired retention of delayed alternation and short-term memory of reward alternation, resistance to extinction operant learning, and failure to initiate activity spontaneously in a spontaneous alternation task (Sanberg *et al.*, 1978, 1979; Pisa *et al.*, 1980, 1981; Dunnett and Iversen, 1981). These alternations occur in conjunction with decreases in body weight and regulatory deficits that are symptomatic in HD patients (Sanberg *et al.*, 1981a; Sanberg and Fibiger, 1979).

5.2. The Quinolinic Acid Model

5.2.1. Neuropathology

Quinolinic acid (2,3-pyridine dicarboxylic acid), a metabolite of tryptophan, has attracted a great deal of attention because of its powerful excitotoxic properties and wide distribution in both rat and human brain (Schwarcz and Kohler, 1983; Foster *et al.*, 1983; Schwarcz *et al.*, 1983, 1984; Schwarcz and Shoulson, 1987; Du *et al.*, 1991). High concentrations of its catabolic enzyme, quinolinic acid phosphoribosyltransferase (QPRT), and immediate anabolic enzyme, 3-hydroxyanthranilic acid (3HAO), have been detected within the caudate, suggesting that it normally serves a role in striatal functioning (Foster *et al.*, 1985; Okuno *et al.*, 1987; Reynolds *et al.*, 1988). The striatum is also among the structures most vulnerable to the excitotoxic effects of quinolinic acid (QA) with the cerebellum, amygdala, substantia nigra, septum, and hypothalamus being significantly less susceptible (Schwarcz and Kohler, 1983). Although these structures are affected in HD, they are less so than is the striatum, providing further support for a role of QA or some endogenous excitotoxin in the selective neuropathology of HD. Neonatal, but not mature, animals also appear to be resistant to the toxic effects of QA, possibly corresponding roughly to the typical onset of HD in middle age. However, there do not appear to be any differences in the neurotoxicity of QA in the lifespan of adult animals. Substance P and GABA concentrations show greater depletions in 1-month-old than 4- and 20-month-old rats, arguing against an increasing susceptibility to excitotoxicity in aged animals (Finn *et al.*, 1991).

Quinolinic acid has been reported to exert a more selective degenerative effect in the striatum than KA or IA, which more closely resemble the pathology of HD (Ellison *et al.*, 1987; Beal *et al.*, 1986, 1988a, 1989). Like KA, QA injections cause depletions of GABAergic neurons while relatively sparing cholinergic neurons and axons of extrinsic origin. Beal and colleagues (1986, 1988a, 1989) have reported that unlike KA or IA, intrastriatal injections of QA spare somatostatin- and neuropeptide-Y-containing neurons. Because this pattern of cell loss closely mimics that observed in HD, it has been suggested that this model most nearly reproduces the neuropathology observed in the disease (Table III). However, controversy exists over whether these cell populations are actually spared following QA. Davies and Roberts (1987, 1988) injected QA into the striatum of rats and found no evidence for a sparing of somatostatin-containing neurons. Likewise, Boegman *et al.* (1987) reported that QA did not spare neuropeptide Y-immunoreactive neurons following intrastriatal QA. Finally, Sanberg and colleagues (1990) were unable to find any evidence of a sparing of NADPH-diaphorase-containing neurons following QA. In an attempt to reconcile these discrepancies, Beal and colleagues (1989) conducted a detailed analysis of QA's toxicity and compared its profile of neurotoxicity to that produced by other toxins, including KA, quisqualic acid, n-methyl-d-asparate (NMDA), and homocysteic acid. They reported that doses of QA, NMDA, and homocysteic acid, which decreased striatal GABA levels by 50% or more, produced a sparing of somatostatin-neuropeptide Y levels. On the other hand, KA, quisqualic acid, and AMPA, which are not NMDA agonists, produced a more generalized toxicity and did not spare somatostatin-neuropeptide Y concentrations. Although neuronal profiles were decreased within the primary lesion region, somatostatin and neuropeptide-Y neurons were relatively spared within the transition zone.

Experiments using cultured striatal neurons support the selective nature of QA. Koh and Choi (1988) incubated cultured striatal neurons with various doses of either NMDA or QA and evaluated the possibility that certain populations of striatal neurons would be more resistant than others. Although high doses of NMDA or QA produced a general neurotoxic effect, low to moderate doses exerted a more selective pattern of toxicity. At these doses, only striatal neurons containing NADPH-d or ACh were resistant to NMDA and QA. Similar effects have been reported in cortical as well as striatal neurons (Koh *et al.*, 1986; Whetsell and Schwarcz, 1989).

Roberts and DiFiglia (1989) recently reported that progressive pathological changes follow QA that may bear a homology to the ongoing deterioration found in HD. Rats received a single injection of QA and were sacrificed 2, 7, or 30 weeks later. Histological analysis revealed that there was a progressive shrinkage of the striatum over time with cell loss maximal in the region of injection and declining within the proximal transition zone. Although large, presumably cholinergic, neurons were relatively spared, there was a dramatic reduction in the number of these neurons between 2–30 weeks postlesion. Despite this progressive neuronal decline, those neurons that did survive retained their normal ultrastructural appearance and maintained some synaptic inputs. These time-related changes suggested a prolonged neuronal death following excitotoxic damage and that the pattern of progressive neuronal loss may be similar to that observed in HD. The long-term selective neuropathology produced by QA has recently been observed one year following injection (Beal *et al.*, 1991).

In an attempt to mimic the progressive neurodegeneration in HD, low doses of QA have been chronically infused into the striatum. Susel and co-workers (1991) administered various doses of QA into rat striatum via osmotic minipumps for 14 consecutive days. Continuous infusion of QA produced dose-dependent decreases in striatal ChAT activity, Glutamic acid decarboxylase (GAD) activity, and somatostatin content. NADPH-d histochemistry further revealed a sparing of NADPH-d neurons at the lowest doses of QA used. In contrast, intraventricular QA had no effect on striatal markers including ChAT and GAD activity (Yamada *et al.*, 1990).

If an abnormal production of QA or similar excitotoxic agent underlies the pathological consequences of HD, then it would be important to demonstrate altered levels of these compounds in the HD brain. Foster *et al.* (1985) reported a trend toward elevated QPRT in the striatum and substantia nigra of HD brains. Moreover, Schwarcz *et al.* (1989) reported similar increases in rat striatum following QA lesioning, suggesting a possible altered metabolism in response to abnormally high levels of QA in the CNS. In the studies by Schwarcz *et al.*

(1989), IA injections and ablations of the cortico-striatal pathway produced significant elevations of striatal QPRT and 3HAO, whereas the local administration of 6-OHDA did not affect QA metabolism. These data suggested that the source of endogenous QA is nonneuronal, and the authors speculated that glial cells may be the striatal source of QA. Several lines of evidence support this notion. For instance, 3-HAO has been detected in astroglial cells (Okuno *et al.*, 1987). In addition, the increases in QPRT and 3-HAO reported by Schwarcz and colleagues (1989) paralleled the onset of reactive gliosis following striatal damage. Finally, HD is associated with a marked gliotic reaction that is not dissimilar to that observed in the excitotoxin-lesioned striatum, suggesting that the initial insult in HD may trigger a reactive gliosis that perpetuates and maintains the ongoing neural degeneration observed in the disease. Although these findings suggest a role for QA as a possible etiologic agent in HD, more recent evidence suggested that QA is not elevated in either the CSF or postmortem striatum of HD patients. Interestingly, however, a reduction in kynurenic acid was observed in the striatum of HD patients (Beal *et al.*, 1990). Because kynurenic acid is found endogenously and prevents excitotoxic damage *in vivo*, a deficiency of this compound may contribute to the pathology of HD by producing a relative increase in QA levels.

5.2.2. *Behavioral Pathology*

Although histological and neurochemical experiments support the use of QA as a model for HD, there have been few behavioral experiments to examine the suitability of the model. Sanberg and colleagues (1989a) examined the locomotor effects of intrastriatal injections of QA. Although injections of 75 nmole of QA did not alter locomotor activity, 150 and 225 nmole produced significant increases in activity at both 2 and 4 weeks postsurgery. These same animals exhibited a transient weight loss following both 150 and 225 nmole of QA. The alterations in locomotor activity and weight were qualitatively similar to those following 3 nmole of KA, but were smaller in overall magnitude.

Calderon *et al.* (1988) reported that QA lesions abolished dopamine-mediated catalepsy. Bilateral QA (150 and 225 nmole) decreased the cataleptic response to the D_1 receptor antagonist SCH23390 and the D_2 receptor antagonist haloperidol at 14–16 weeks following surgery. These data are similar to those observed following KA and indicate that the D_1 and D_2 dopaminergic receptors, which mediate the catalepsy response, are sensitive to the neurotoxicity of QA.

There appears to be a sex-dependent regulation of the behavioral effects produced by QA administration. In one study, male, female, or ovariectomized female rats received bilateral intrastriatal injections of QA, and the extent of body weight loss was examined for 30 days (Zubrycki *et al.*, 1990). Although all rats injected with QA exhibited an equivalent initial loss of body weight, the female rats rapidly regained weight and did not differ from controls after the first postsurgical day. In contrast, the male and the female ovariectomized rats that received QA exhibited a significant loss of body weight that did not recover until 19 days following surgery. Similarly, QA has been reported to produce locomotor hyperactivity in female but not male or ovariectomized female rats (Emerich *et al.*, 1991). These data suggested that sex and hormonal variables play an important role in the regulatory and locomotor changes following QA.

Although these data are similar to those obtained following KA and resemble the behavioral pathology observed in HD, they do not adequately permit the evaluation of the behavioral specificity of QA. The existence of no qualitative differences between KA and QA in these experiments may (1) support the lack of neural specificity of QA observed by some investigators (Davies and Roberts, 1987, 1988; Boegman *et al.*, 1987) or (2) indicate that the behavioral analysis employed to date is insufficient, and that additional, more precise testing paradigms are required to validate the utility of the QA model. Regardless, the validity of QA as a model of HD must be based in part on the behavioral consequences of the compound and the resulting homology to HD itself.

6. The Role of Excitotoxicity in Huntington's Disease: Glutamate, NMDA Receptors, and Huntington's Disease

Similarities between the KA and QA models and HD itself suggest the existence of an endogenous molecule with excitotoxic properties in the disease. The strongest candidate at present is glutamate or an analog. Glutamate is a potent agonist for all excitatory amino acid receptors and is present in virtually all brain regions. In support of the excitotoxic hypothesis, the striatum receives a substantial innervation of glutamatergic fibers and contains one of the highest densities of excitatory amino acid receptors of any major brain region (Cotman *et al.*, 1987; Cotman and Monaghan, 1988). Postmortem studies have also demonstrated significant reductions in the numbers of NMDA, KA, and quisqualate receptors in Huntington's patients. Notably, the binding sites for KA in the striatum have been reported to decrease by 50% to 60%, whereas the number of NMDA receptors in the putamen may be decreased by as much as 90% (Henke, 1979; Beaumont *et al.*, 1979; London, 1981; Albin *et al.*, 1990; Dure *et al.*, 1991).

To determine whether any glutamatergic abnormalities occur in HD, Carter (1982) measured the activity of glutamine synthetase in HD patients and age-matched controls. He demonstrated a 20% to 30% decrement in the activity of the enzyme in the frontal cortex, putamen, temporal cortex, and cerebellum. Because this enzyme plays an important part in the process of glial uptake and conversion of glutamate to glutamine, it was argued that these decreases may relate to an excitotoxic-induced neuronal degeneration in HD.

Because ornithine aminotransferase may be a synthesizing enzyme for glutamate from ornithine, and this enzyme may be partially concentrated in glutamatergic neurons, Wong *et al.* (1982) investigated ornithine aminotransferase activity in postmortem cortical and striatal tissues from HD patients. They found that the specific activity was reduced by 34% to 49% in the frontal cortex, parietal cortex and striatum, possibly reflecting deterioration of the corticostriatal glutamatergic neurons in HD. In this regard, it is noteworthy that Kim *et al.* (1980) showed a decrease in the amount of glutamate in the CSF of long-term HD patients.

Gray *et al.* (1980) demonstrated that cultured HD fibroblasts degenerated at concentrations of glutamate that do not significantly affect fibroblasts from age-matched controls. This finding, together with those of Wong *et al.* (1982), is important, and it supports an increased sensitivity to the toxic effects of glutamate in HD patients.

More recently, Cross *et al.* (1986) reported marked reductions in high-affinity glutamate uptake sites in the HD striatum. It is conceivable that an impaired uptake system would locally increase glutamate until it reached toxic levels. Interestingly, McBean and Roberts (1985) demonstrated that disruption of glutamate uptake following damage to the corticostriatal tract rendered striatal cells more vulnerable to a subsequent injection of glutamate. Although there is comparatively little direct evidence supporting a role for glutamate in the pathology of HD, this theory remains one of the most attractive etiological considerations in the disorder. This owes, in part, to the discovery and elucidation of specific populations of receptors for glutamate and related excitotoxic compounds.

A great deal of attention has been focused on the N-methyl-D-aspartate (NMDA) receptor as a mediator of neuroplastic processes, as well as the cell loss observed in several neurodegenerative disorders. The NMDA receptors are one of at least three receptor subtypes that mediate the responsiveness to excitatory amino acids, such as glutamate and aspartate, and are found throughout the brain, with the highest concentration occurring in limbic regions and the basal ganglia (Cotman *et al.*, 1987; Cotman and Iversen, 1987). QA has a high affinity for the NMDA receptor and, consequently, produces its toxic effects via an interaction with these receptors.

The similarity of QA lesions to the neuropathology of HD together with the characterization of an endogenous receptor for QA strongly suggests either overproduction of the endogenous neurotoxic molecule or an abnormality in the postsynaptic membrane that renders neurons vulnerable (Koh and Choi, 1988; Koh *et al.*, 1986). Although direct evidence for QA or another endogenous NMDA agonist contributing to the pathology of HD is scarce, massive (up to 93%) reductions in NMDA receptors have been found in the putamen of relatively young HD patients (Young *et al.*, 1988). Further analyses showed reductions in NMDA, MK-801, glycine, kainate, and AMPA receptors, which suggest that HD is not characterized by the loss of any single excitatory amino acid receptor.

7. Therapeutic Strategies with Excitotoxin Models

7.1. Pharmacological Treatments

The possible involvement of NMDA receptors in the neuropathology of HD has provided new avenues of therapeutic intervention. It may be possible to negate the toxic effects of overactivated NMDA receptors by at least two general methods: (1) by preventing the presynaptic release of glutamate or (2) by administering select NMDA receptor antagonists (see Table IV). These possible treatments are outlined below.

The excitotoxicity of compounds such as QA may be partially dependent on the integrity of the glutamatergic input to the striatum. Lesioning the corticostriatal pathway prevents the KA- and QA-induced alterations in striatal morphology, implying that removal of the endogenous source of glutamate underlies the neural protection afforded by prior decortication (Coyle *et al.*, 1978). It may be possible to modify striatal glutamate levels pharmacologically and prevent excitotoxicity. Adenosine or adenosine analogs decrease presynaptic glutamate re-

Table IV. Excitatory Amino Acid Receptor Classes

	NMDA	Quinolinic acid	Kainic acid	L-AP4
Agonists	NMDA Aspartate Ibotenate Glutamate Glycine	Quinolinic acid AMPA L-Glutamate	Kainic acid Domoate L-Glutamate	
Antagonists	MK-801 D-AP5 D-AP7 CPP Glu-Amp PCP SKF10047	GDEE GAMS pCBPzDA	None available	L-AP4

Note: L-SOP, L-serine-O-phosphate; Glu-AMP, gamma—D-glutamylaminomethylphosphonate; GDEE, diethyl glutamate; MK-801, (+)-5-methyl-10,11-dihydro-5H-dibenzo[a,d]cyclohepten-5,10-imine maleate; NMDA, N-methyl-D-aspartate; GAMS, alpha-D-glutamylaminomethylsulphonate; pCBPzDA, 1-(p-chlorobenzoyl)-piperazine-2,3-dicarboxylate; L-AP4, L-2-amino-4-phosphonobutyrate; AMPA, alpha-amino-3-hydroxy-5-methyl-4-isoxazolepropionate; SKF10047, n-allylnormetazocine; CPP, 3-3(2-carboxypiperazine-4-yl)propyl-1-phosphonate; D-AP5, D-2-amino-5-phosphonovalerate; D-AP7, D-2-amino-7-phosphonoheptanoate; PCP, phencylclidine.

lease. Arvin *et al.* (1988) reported that prior treatment with 2-chloroadenosine prevents KA-induced toxicity in rat striatum.

Blockade of the NMDA receptors is another promising therapeutic approach for excitotoxicity. The NMDA receptor may be completely blocked by the administration of 2-amino-7-phosphonohepanate (AP7) or 2-amino-5-phosphovalerate (AP5) (Schwarcz *et al.*, 1984). The neuropathological and seizure-induced effects of QA following intrahippocampal injections are prevented by prior administration of AP7 (Schwarcz *et al.*, 1984). In addition, both AP5 and AP7 prevent QA-induced striatal damage, whereas other compounds, such as allopurinol, ketamine, nimodipine, baclofen, are without effect (Beal *et al.*, 1989). Unfortunately, the clinical value of those compounds may be limited in that they are highly charged and cross the blood–brain barrier poorly.

MK-801 is an anticonvulsant agent that is a potent NMDA receptor antagonist (Wong *et al.*, 1986; Foster and Wong, 1987; Kemp *et al.*, 1987; Cotman and Iversen, 1987; Kloog *et al.*, 1988; Huettner and Bean, 1988). Following systemic or oral administration, MK-801 is readily absorbed and easily crosses the blood–brain barrier. Recent evidence indicates that MK-801, in a dose-dependent manner, effectively prevented QA-induced striatal and hippocampal cell loss (Gill *et al.*, 1987; Foster *et al.*, 1987; Beal *et al.*, 1988a, 1989; Kochhar *et al.*, 1988). Moreover, MK-801 prevented the neuronal damage and decreases in ChAT following injection of NMDA into either of these structures. MK-801 also prevented the increased locomotor activity, decreased catalepsy response to haloperidol, and alterations in feeding behavior following intrastriatal QA (Giordano *et al.*, 1990a,b). However, because this compound acts at the phencyclidine site, it may produce undesirable psychotomimetic effects that severely question its clinical efficacy.

Norman *et al.* (1990) recently reported that chronic treatment with MK-801 increased the QA-induced loss of D_1 dopamine receptors in the striatum. If the blockade of NMDA receptors with MK-801 were a useful treatment for HD, then long-term use of the compound would be required. However, if chronic treatment with MK-801 results in supersensitivity of NMDA receptors, the neurotoxic actions of endogenous ligands for these receptors might be enhanced.

Glycine has also been reported to potentiate the response of NMDA receptors in cultured mouse cortical neurons and in oocytes. Johnson and Ascher (1987) reported that glycine increased NMDA-mediated responses through a strychnine-insensitive glycine recognition site that is part of the NMDA receptor/ion channel complex. Klechner and Dingledine (1988) further demonstrated that glycine is required for the activation of NMDA receptors, suggesting that the use of glycine antagonists may represent a useful strategy for blocking the neuropathology subsequent to NMDA activation. The endogenous tryptophan metabolite kynurenic acid also blocks the activation of NMDA receptors by acting as a selective antagonist at the glycine modulatory site (Foster *et al.*,

1984). Kemp and colleagues (1988) reported that 7-chlorokynurenic acid antagonizes NMDA-induced depolarizations in rat cortical slices, and this antagonism was reversed by glycine administration. Furthermore, kynurenic acid blocks both the neurotoxic and seizure-inducing effects of intrahippocampal injections of QA. These findings are particularly intriguing because they suggest that kynurenic acid modulates the delicate balance between glutamate transmission and excitotoxicity.

7.2. Neural Transplantation

Transplantation of fetal neural tissue into the adult brain was initially seen as a way to study the development and regenerative capacity of the nervous system. However, it was discovered that neural grafting could attenuate the functional deficits induced by a variety of central lesions. Based on these encouraging observations, it was proposed that neural grafting might be an appropriate treatment for various neurodegenerative disorders. The logic of replacing lost populations of neurons with fetal neurons was inherently appealing, and animal studies clearly indicated that grafted fetal tissue could integrate with the host brain and promote functional recovery following acute injury or in animal models of Alzheimer's, Parkinson's, or Huntington's disease (Gage and Bjorklund, 1985; Sanberg *et al.*, 1986, 1989b; Norman *et al.*, 1988a,b; Sladek *et al.*, 1988; Bjorklund *et al.*, 1988).

Various studies have indicated that transplanted striatal tissue readily survives and grows within the excitotoxin-lesioned striatum (Clarke *et al.*, 1988a,b; Deckel *et al.*, 1983, 1986a,b, 1988a,b; Deckel and Robinson, 1987; Emson *et al.*, 1985; McGeer *et al.*, 1984; Isacson *et al.*, 1984, 1985, 1986, 1987a,b; McAllister *et al.*, 1984, 1985; McAllister, 1987; Sanberg *et al.*, 1987a–c; Norman *et al.*, 1988a,b; DiFiglia *et al.*, 1988; Wictorin *et al.*, 1989a–c, 1990; Helm *et al.*, 1990; Zhou and Buchwald, 1989). Within approximately 6–10 weeks, striatal grafts appear as well-delineated, mature-looking tissue (McAllister *et al.*, 1984, 1985; DiFiglia *et al.*, 1988). Morphological analysis also indicated that the grafted tissue displayed a patchy distribution of AChE, met-enkephalin, and substance P–positive regions, which resemble that normally observed in the intact striatum (Isacson *et al.*, 1987a,b; Graybiel *et al.*, 1989). The host-transplant anatomical similarity is paralleled by a significant recovery of both ACh and GABA within the previously depleted striatum (Schmidt *et al.*, 1981; Isacson *et al.*, 1986; Segovia *et al.*, 1989). GABA release is also partially restored in the globus pallidus and substantia nigra by striatal grafts (Sirinathsinghji *et al.*, 1988). Moreover, these grafts may be innervated by efferents originating in the host neuropil, generating substantial interest for the use of fetal striatal transplants as a possible treatment for the behavioral sequelae associated with HD (McGeer *et al.*, 1984; Pritzel *et al.*, 1986; Clarke *et al.*, 1988b;

Table V. Neurochemical, Anatomical, and Behavioral Effects of Transplants into Excitoxically Lesioned Striatum

Neurochemical/anatomical effects
Tissue survives, grows, and forms appropriate afferent and efferent synaptic connections with host tissue.
Presence of normal intrinsic striatal neurons and receptors in graft variable.
Restoration of neurochemical markers and metabolic activity in lesioned striatum by grafts.
Prevention of subsequent excitoxocity.
Behavioral effects
Reduction of hyperactivity.
Reversal of apomorphine/amphetamine-induced rotation and hyperactivity.
Recovery of haloperidol- and SCH23390-induced catalepsy.
Reversal of delayed alternation learning.
Improvement in skilled paw reaching.
Induced hyperactivity and cognitive impairments in nonlesioned animals.
Recovery of apomorphine-induced dyskinesia and hyperactivity in nonhuman primates.

Wictorin *et al.*, 1988a,b; Wictorin and Bjorklund, 1989a–c, 1990). Table V summarizes the anatomical, biochemical, and behavioral consequences of striatal transplants into excitotoxic-lesioned striatum.

Reversal of the behavioral abnormalities resulting from excitotoxic lesions has been demonstrated in several studies. In particular, investigators have focused on the normalization of the increases in locomotor behavior observed in lesioned rats. Unfortunately, the extent of behavior varies across studies and likely reflects inconsistencies in testing paradigms. Deckel and colleagues (1983, 1986a,b) examined changes in spontaneous locomotion during the daytime, whereas Sanberg *et al.* (1986, 1987a, 1988) and Isacson *et al.* (1984, 1986, 1987a,b) examined changes in nocturnal activity. Although all of these studies reported that striatal grafts reversed the increases in activity following excitotoxic lesions, the improvements observed during daytime testing were often smaller in magnitude and sometimes failed to reach statistical significance.

The dependent measures used to assess locomotion activity may differentially influence the interpretation of transplant-induced recovery. Normalization of locomotor behavior is usually reflected in the number of photobeams interrupted. However, locomotion may be reflected in a multitude of behavioral indices, such as the velocity of movement and the total distance traveled. With this in mind, the Digiscan Animal Activity Monitoring System has been used to provide a detailed description of the topography of the animal's locomotor profile (see Sanberg *et al.*, 1985, 1986, for details). This system provides a detailed account of a variety of locomotor indices, including the velocity of movement, distance traveled, number of movements, stereotypy, and so forth. The importance of using a multivariate system is reflected in the report that KA and QA differentially affect these locomotor variables and that there is a differential recovery of these alterations following fetal striatal transplants (Giordano *et al.*, 1988). Together, these studies clearly indicated that the locomotor abnormalities in excitotoxic-lesioned animals were at least partially recovered, but also that additional investigations are required to determine the extent and nature of this recovery. It is apparent that, when evaluating behavioral recovery, the behavior under investigation must also be clearly defined.

Isacson *et al.* (1989, 1990) and Hantraye *et al.* (1990) have recently described the dyskinesia and hyperactivity that resulted from intrastriatal injections of ibotenic acid into the primate striatum. As in the rodent studies to date, grafted fetal tissue survived, grew, and received afferent innervation from the host. Moreover, a significant reduction was reported in apomorphine-induced dyskinesia, choreiform movements, and locomotor hyperactivity. Although these studies are preliminary, they offer a potentially valuable primate model for studying the dyskinesia in HD and the use of potential therapeutic treatments for such symptoms.

Following excitotoxic damage to the striatum, animals exhibit not only locomotor abnormalities but also impairments in learning and memory (Sanberg *et al.*, 1978; Sanberg and Johnston, 1981). It is therefore important to determine whether the behavioral recovery observed in locomotor testing extends to cognitively oriented tasks. Fetal striatal transplants partially reversed the impairments in delay-rewarded alternation in a T-maze following striatal damage, and therefore appeared to improve learning and memory abilities (Deckel *et al.*, 1986a,b; Isacson, 1987a,b). Interestingly, rats exhibiting a partial reversal of locomotor hyperactivity and T-maze performance may show no recovery of sensorimotor changes, including vibrissae orientation, responsiveness to olfactory stimuli, blunt and sharp touch, muscle tone, or limb strength (Deckel *et al.*, 1986b). Apparently, the extent of behavioral recovery following fetal striatal transplants is task-dependent.

The excitotoxin-lesioned striatum provides a useful model for evaluating the pharmacological properties of striatal transplants. The utility of this model is founded on the observation that dopamine exerts a tonic influence on striatal neurons, and drugs that alter dopaminergic transmission also modulate locomotor activity in rats. Following bilateral lesions of the striatum, rats demonstrated an exaggerated hyperactivity in response to amphetamine (Mason *et al.*, 1978; Sanberg *et al.*, 1986) that was reversed following striatal transplants (Sanberg *et al.*, 1986). Similar results were found following unilateral striatal lesions. In this model, striatal transplants produce a time-dependent decrease in rotational behavior following both amphetamine and apomorphine administration (Schwarcz *et al.*, 1979; Norman *et al.*, 1987, 1988a,b; Dunnett *et al.*, 1988). It is important to point out, however, that although the basic findings in these studies are similar, there are methodological differences that may produce somewhat different results. The studies by Norman *et al.* (1987, 1988a,b) assessed rotational behavior in an open field, which was in contrast to the rotometer used by Dunnett *et al.* (1988). Use of these two different procedures provides different information (i.e., the open field may be used to examine both the number and topography of rotations, whereas the rotometer restricts movement and assesses pivotal-type rotations). Interestingly, the animals tested in the open-field exhibited changes from pivotal-type rotations to a wide circle ambulation as early as 3–5 weeks posttransplantation. On the other hand, changes in rotometer testing were not found for 8 months posttransplantation (Dunnett *et al.*, 1988). Because changes in the topography of rotation are not detectable in the rotometer, it is possible that topographical changes in locomotion occur earlier than actual reductions in rotation following striatal transplants.

Until recently, no studies had evaluated the intracellular biochemical characteristics of transplanted striatal neurons. Dragunow and colleagues (1990) examined the induction of Fos and Fos-related antigens (FRA) in fetal tissue grafts following systemic haloperidol administration. The induction of Fos and FRA had a time-course that was similar in both the normal striatum and in the transplanted tissue. Moreover, the numbers of Fos and FRA-positive neurons were the same in the normal striatum and the striatal transplant. These findings led to the suggestion that transplanted fetal striatal neurons are capable of transducing extracellular signals and generating third messengers in a manner similar to the normal striatum.

Because transplanted tissue reverses the enhanced sensitivity to dopaminergic agonists following excitotoxin lesions of the striatum, it might be expected that the transplanted tissue would express dopamime receptors. However, the studies conducted to date have produced dissimilar results. Isacson and colleagues (1987a,b) and Deckel *et al.* (1986b, Deckel and Robinson, 1987) have demonstrated that transplanted tissue contains patches of D_2 dopamine receptors resembling those observed in the intact striatum. Deckel *et al.* (1986b; Deckel and Robinson 1987) also reported a correlation between the lack of behavioral recovery and sparse presence of dopamine receptors revealed by [^{3}H] spiperone autoradiography in transplanted tissue. Conversely, Norman *et al.* (1988a,b, 1989a,b) reported a transplant-related recovery of apomorphine-induced rotational behavior in transplant animals without associated D_1 or D_2 receptors. These studies also revealed a deficit in [^{3}H] forskolin binding to the stimulatory guanine nucleotide regulatory subunit/adenylate cyclase complex within the transplanted tissue. These results led Norman *et al.* (1988a,b, 1989a,b) to conclude that tissue that is neurochemically dissimilar to the host is capable of reversing the functional deficits associated with striatal damage.

The above discussion illustrates the difficulty in correlating behavioral and neurobiological recovery processes. A number of possible mechanisms for functional recovery following fetal striatal transplants might be postulated. One of the most frequently cited mechanisms is an anatomical connectivity between the host and the transplant. Although intuitively appealing, attempts to demonstrate these anatomical connections have been controversial. Initial studies revealed that transplanted tissue grows and differentiates into mature-looking tissue, but it does not make afferent or efferent connections with the surrounding tissue (Schmidt *et al.*, 1981; McAllister *et al.*, 1985; McAllister, 1987; Walker *et al.*, 1987; Zemanick *et al.*, 1987). Horseradish peroxidase (HRP) injected into the host striatum or surrounding tissue at 5 weeks posttransplantation is not transported into either the transplant or the host tissue, respectively (Walker and McAllister, 1987). On the other hand, tyrosine hydroxylase (TH)-positive neurons have been demonstrated to innervate grafted striatal tissue at 8–10 months posttransplantation (Clarke *et al.*, 1988b). The use of antibodies against the dopamine receptor-related phosphoprotein DARPP-32 has confirmed the existence of host-efferent projections from the host brain to grafted fetal tissue (Labandeira-Garcia *et al.*, 1991; Roberts and DiFiglia, 1990). Fluorescent beads are also retrogradely transported from the transplant to the surrounding neuropil. Similarly, connections between the host cortex or thalamus and striatal transplants have been identified following injections of the retrograde tracer phaseolus vulgaris leucoagglutinin (Wictorin *et al.*, 1988b). These studies therefore indicate that anatomical connections may indeed form between the host and the transplanted tissue. However, it remains difficult to reconcile these data with the available behavioral data. For example, behavioral recovery has been reported to occur at time-points much earlier than those at which reorganization occurs (Norman *et al.*, 1988a,b). Even though TH-positive neurons have been reported to penetrate the grafted tissue as early as 6–7 weeks posttransplantation (Zhou *et al.*, 1986), behavioral recovery may still occur substantially earlier.

Electrophysiological studies may aid in determining the time-course of anatomical integration and its relation to functional recovery. Several convergent lines of data indicate physiological connections between host and transplanted striatal neurons. These include (1) recordings of inhibitory and excitatory postsynaptic potentials in grafts following stimulation of the host striatum (Walsh *et al.*, 1988); (2) stimulation of the corpus callosum, producing electrophysiological responses in transplanted tissue in a slice preparation (Rutherford *et al.*, 1987); (3) frontal cortex stimulation, producing similar responses in both the normal striatum and in striatal transplants following both KA and IA (Wilson *et al.*, 1987); and (4) evoked synaptic potentials in striatal grafts following cortical and thalamic stimulation (Xu *et al.*, 1991a,b).

The majority of data indicates that anatomical integration between host and transplanted tissue occurs. However, this is not a universal phenomenon, and not all studies find evidence of integration at the time when behavioral recovery occurs. These findings have led to the suggestion that a diffusible neurotrophic factor released by or within the vicinity of the transplant partially underlies functional recovery. Supporting this contention is the observation that adrenal medulla grafts promote behavioral recovery in animals models of PD in a manner not dependent on the reconstitution of anatomic circuitry or the release of catecholamines (Stromberg *et al.*, 1985; Freed *et al.*, 1986; Becker and Freed, 1988; Norman *et al.*, 1989a,b). Interestingly, these grafts may be equally effective at promoting recovery whether placed into the lesioned striatum or the adjacent lateral ventricle. Similar degrees of recovery have been accounted for by postulating that the adrenal tissue releases a trophic factor (such as nerve growth factor) and that placement into the lateral ventricle allows the factor to diffuse readily into the nearby striatum. Studies by Giordano *et al.* (1988) and Sanberg *et al.* (1987a, 1989b) demonstrated that striatal tissue placed into the lesioned striatum, but not the lateral ventricles, was required for behavioral recovery. It was suggested that the ventricular placements provided a limited behavioral recovery, because the levels of neurotrophic activity may have been insufficient to promote higher levels of recovery. The transplants themselves may not be the source of trophic activity. Rather, reactive macrophages or glial cells may be providing the necessary trophic support. Glial cells have been shown to promote neurite outgrowth, restore ionic balance, and decrease levels of excitotoxins, such as glutamate (Muller and Seifert, 1982; Bridges *et al.*, 1985; Kromer and Cornbrooks, 1984; Neito-Sampedro and Cotman, 1985). Kesslak *et al.* (1986) recently reported that the transplantation of cultured astrocytes is as effective as cortical tissue in promoting behavioral recovery following cortical aspirations. However, a recent report by Giordano and colleagues (1991) suggested that fetal striatum, but not cultured fetal astrocytes, was capable of producing recovery of apomorphine-induced rotational behavior.

Although the mechanism by which transplanted tissue exerts its beneficial effects is largely unknown, it remains a viable approach for promoting behavioral and neurobiological recovery. As encouraging as initial studies have been, evidence suggests that transplanted tissue may exert a deleterious as well as beneficial effect. Hagenmeyer-Houser and Sanberg (1987) injected striatal tissue into intact host striatum and observed a profound hyperactivity. Deckel and Robinson (1987) reported that, although striatal transplants ameliorated the behavioral deficits resulting from excitotoxic striatal lesions, transplants into normal striatum produced a "lesion-like effect" and disrupted locomotor activity. Moreover, Lu *et al.* (1990, 1991) have recently confirmed these results and further suggested transplanted tissue from the striatum disrupts locomotor behavior in a manner dependent on the initial volume of tissue injected. These studies also indicated that transplanted tissue from the striatum as well as other neural regions is capable of producing locomotor abnormalities following injection into intact striatum. Whether these transplant-induced abnormalities are the result of the physical or neurochemical characteristics of the transplants is unknown. Regardless of the underlying mechanism of these unwanted consequences, the possible contribution of the region from which the transplant is derived as well as the amount of tissue used should be given serious consideration. This is especially true given the current debates centering around the continuation of tissue transplantation in human neurodegenerative disorders. The detrimental effects of transplants may be especially difficult to detect in individuals with an already severe disruption of function. It is conceivable that the negative effects of transplantation would not manifest themselves directly, but might slow the progression of or limit the extent of behavioral recovery in laboratory animals or humans. It seems then further research is necessary to decipher both the negative and positive effects of neural transplantation. This research should also seriously consider the refinement of existing behavioral paradigms so as to evaluate the subtle topographic changes in behavior resulting from transplants of fetal tissue.

8. Conclusions

Ever since the pioneering studies of Bird and Iversen (1974) and Perry *et al.*(1973) demonstrating a selective alteration in GABAergic parameters in HD, there has been a remarkable escalation in our understanding of the pathophysiology of the disease. Our concepts of the neurochemical and morphological sequelae of HD have grown from these early observations to the present-day complicated mosaic of neurobiological changes in the striatum and related structures. Although the cooperation of clinicians, basic neuroscientists, and families of HD patients has contributed to this progress, an equally important contribution has been made from the development of animal models of HD. Investigators have taken advantage of new excitotoxic models that appear to mimic the neurobiological and behavioral characteristics of HD with remarkable homology. This research has substantially bolstered the hypothesis that a fundamental deficit in HD is a dysfunction of glutamatergic transmission, which results in the slow, progressive neural degeneration characteristic of HD.

The development of these models holds great promise both for understanding the etiology of HD and for the development of therapeutic strategies (i.e., pharmacological alterations of NMDA receptors and/or glutamate transmission, or neural transplantation). However, because HD is a genetically transmitted disorder, these animal models will not provide information concerning the genetic defect in HD. Uncovering of the genetic abnormality in HD will occur through molecular genetic research (Gilliam *et al.*, 1988; Hayden *et al.*, 1988). In this way, we are most likely to reach the ultimate goal in the treatment of HD, which is to prevent its expression. Recombinant DNA techniques have identified a restriction-fragment-length polymorphism, which is mapped to chromosome 4 and is linked to the HD gene (Gusella *et al.*, 1983). This fragment identifies a neighboring piece of DNA with a 5% recombination rate with the gene deficit, and may soon help determine the exact identity of the normal counterpart of the HD gene and with it the ensuing knowledge of its pathology in the disease. Based on these findings, tests have been developed that can identify a presymptomatic individual. This knowledge will suggest therapies specifically designed to compensate for the changes induced by the aberrant HD gene.

For example, if the genetic abnormality in HD is the result of a loss of function, then treatment could revolve around the replacement of a missing constituent. Recent advances in the development of "designer genes," which produce specific gene products, such as growth factors or enzymes, hold great promise as replacement strategies. These neurons could be implanted into the appropriate neural site, perhaps reverse or slow the progression of the disease state, and afford the individual a higher level of functioning. Alternatively, if the HD allele produces an increased or inappropriate expression of a specific gene product, then an appropriate therapeutic strategy would be to reduce the expression of the gene for Huntington's disease.

9. References

Albin, R.L., Young, A.B., Penney, J.B., Handelin, B., Balfour, R., Anderson, K.D., Markel, D.S., Tourtellotte, W.W., and Reiner, A., 1990, Abnormalities of striatal projection neurons and N-methyl-D-aspartate receptors in presymptomatic Huntington's disease, *N. Engl. J. Med.* 322 (18):1293–1298.

Aquidonius, S.M., Eckernas, S.A., and Sundwall, A., 1975, Regional distribution of choline acetyltransferase in the human brain: Changes in Huntington's chorea, *J. Neurol. Neurosurg. Psychiatry* 38:669–677.

Araki, M., McGeer, P.L., and McGeer, E.G., 1985, Striatonigral and pallidonigral pathways studied by a combination of retrograde horseradish peroxidase tracing and pharmacohistochemical method for gamma aminobutyric acid transaminase, *Brain Res.* 331:17–24.

Arregui, A., Bennett, J.P., Jr., Bird, E.D., Yamamura, H.I., Iversen, L.I., and Snyder, S.H., 1977, Huntington's chorea: Selective depletion of activity of angiotensin converting enzyme in the corpus striatum, *Ann. Neurol.* 2: 294–298.

Arregui, A., Emson, P.C., and Spokes, E.G., 1978, Angiotensin-converting enzyme in substantia nigra: Reduction of activity in Huntington's disease and after intrastriatal kainic acid in rats, *Eur. J. Pharmacol.* 52:121–124.

Arregui, A., Iversen, L.L., Spokes, E.G.S., and Emson, P.C., 1979, Alterations in post-mortem brain angiotensin-converting enzyme activity and some neuropeptides in Huntington's disease, *Adv. Neurol.* 23:517–525.

Arvin, B., Neville, L.F., and Roberts, P.J., 1988, 2-Chlorodenosine prevents kainic acid-induced toxicity in rat striatum, *Neurosci. Lett.* 93:336–340.

Averback, P., 1980, Histopathology of acute cell loss in Huntington's chorea brain, *J. Pathol.* 132:55–61.

Barr, A. N., Heinze, W.J., Dobben, G.D., Valvassori, G., and Sugar, O., 1978, Bicaudate index in computerized tomography of Huntington's disease and cerebral atrophy, *Neurology* 28:1196–1200.

Beal, M.F., Kowall, N.W., Ellison, D.W., Mazurek, M.F., Swartz, K.J., and Martin, J.B., 1986, Replication of the neurochemical characteristics of Huntington's disease by quinolinic acid, *Nature* 321:168–171.

Beal, M.F., Kowall, N.W., Swartz, K.J., Ferranti, R.J., and Martin, J.B., 1988a, Systemic approaches to modifying quinolinic acid striatal lesions in rats, *J. Neurosci.* 8:3901–3908.

Beal, M.F., Ellison, D.W., Mazurek, M.F., Swartz, K.J., Malloy, J.R., Bird, E., and Martin, J.B., 1988b, A detailed examination of substance P in pathologically graded cases of Huntington's disease, *J. Neurol.* Sci. 84:51–61.

Beal, M.F., Kowall, N.W., Swartz, K.J., Ferranti, R.J., and Martin, J.B., 1989, Differential sparing of somatostatin-neuropeptide Y and cholinergic neurons following striatal excitotoxin lesions, *Synapse* 3:38–47.

Beal, M.F., Matson, W.R., Swartz, K.J., Gamache, P.H., and Bird, E.D., 1990, Kynurenine pathway measurements in Huntington's disease striatum: Evidence for reduced formation of kynurenic acid, *J. Neurochem.* 55(4):1327–1339.

Beal, M.F., Ferrante, R.J., Swartz, K.J., and Kowall, N.W., 1991, Chronic quinolinic acid lesions in rats closely resemble Huntington's disease, *J. Neurosci.* 11(6):1649–1659.

Beaumont, K., Maurin, Y., Reisine, T.D., Fields, J.Z., Spokes, E., Bird, E.D., and Yamamura, H.I., 1979, Huntington's disease and its animal model alterations in kainic acid binding, *Life Sci.* 24:809–816.

Becker, J.B., and Freed, W.J., 1988, Adrenal medulla grafts enhance functional recovery of the striatal dopamine system following substantia nigra lesions, *Brain Res.* 462:401–406.

Ben-Ari, Y., Tremblay, E., Ottersen, O.P., and Meldrum, B.S., 1980, The role of epileptic activity in hippocampal and "remote" cerebral lesions induced by kainic acid, *Brain Res.* 191:79–97.

Bernheimer, H., Birkmayer, W., Hornykiewicz, O., Jellinger, K., and Seitelberger, F., 1973, Brain dopamine an syndromes of Parkinson and Huntington-clinical morphological and neurochemical correlation, *J. Neurol. Sci.* 20:415–455.

Bird, E.D., 1980, Chemical pathology of Huntington's disease, *Ann. Rev. Pharmacol. Toxicol.* 20:533–551.

Bird, E.D., and Iversen, L.L., 1974, Huntington's chorea: Postmortem measurement of glutamic acid decarboxylase, choline acetyltransferase, and dopamine in basal ganglia, *Brain* 97:457–461.

Bjorklund, A., Lindvall, O., Isacson, O., Brundin, P., Wictorin, K., Strecker, R.E., Clarke, D.J., and Dunnett, S.B., 1988, Mechanisms of action of intracerebral neural transplants: Studies on nigral and striatal grafts to the lesioned striatum, *Trends Neurosci.* 10:509–516.

Boegman, R.J., Smith, Y., and Parent, A., 1987, Quinolinic acid does not spare striatal neuropeptide Y-immunoreactive neurons, *Brain Res.* 415:178–182.

Bots, G., Th, A.M., and Bruyn, G.W., 1981, Neuropathological changes of the nucleus accumbens in Huntington's chorea, *Acta. Neuropathol.* 55:21–22.

Bowman, M., and Lewis, M.S., 1980, Site of subcortical damage in diseases which resemble schizophrenia, *Neuropsychology* 18:597–601.

Brandt, J., Folstein, S.E., and Folstein, M.F., 1988, Differential cognitive impairment in Alzheimer's disease and Huntington's disease, *Ann Neurol.* 23:555–561.

Brann, M.R., and Emson, P.C., 12980, Microiontophoretic injection of fluorescent tracer combined with simultaneous immunofluorescent histochemistry for the demonstration of efferents from the caudate putamen projecting to the globus pallidus, *Neurosci. Lett.* 16:61–66.

Bridges, R.J., Neito-Sampedro, M., and Cotman, C.W., 1985, Stereospecific binding of L-glutamate to astrocyte membranes, *Soc. Neurosci. Abstr.* 11:110.

Bruyn, G.W., 1968, Huntington's chorea, historical, clinical and laboratory synopsis, *Handbook Clin. Neurol.* 6:298.

Bruyn, G.W., Bots, G., Th, A.M., and Dom, R., 1979, Huntington's chorea: Current neuropathological status, *Adv. Neurol.* 23:83–93.

Buck, S.H., Burks, T.F., Brown, M.R., and Yamamura, H.I., 1981, Reduction in basal ganglia and substantia nigra substance P levels in Huntington's disease, *Brain Res.* 209:464–469.

Buell, U., Costa, D.C., Kirsch, G., Moretti, J.L., Van Royen, E.A., and Schober, O., 1990, The investigation of dementia with single photon emission tomography, *Nucl. Med. Commun.* 11:823–841.

Burns, A., Psych, M.R.C., Folstein, S., Brandt, J., and Folstein, M., 1990, Clinical assessment of irritability, aggression, and apathy in Huntington and Alzheimer disease, *J. Nerv. Ment. Dis.* 178(1):20–26.

Calderon, S.F., Sanberg, P.R., and Norman, A.B., 1988, Quinolinic acid lesions of rat striatum abolish D_1- and D_2-dopamine receptor mediated catalepsy, *Brain Res.* 450:403–407.

Carter, C.J., 1982, Glutamine synthetase activity in Huntington's disease, *Life Sci.* 31:1151–1159.

Castaigne, P., Escourelle, R., and Gray, F., 1976, Huntington's chorea and cerebellar atrophy (a case report with clinical and pathological data), *Rev. Neurol. (Paris)* 132:233.

Chan-Palay, V., 1977, Indolamine neurons and their processes in the normal rat brain and in chronic diet-induced thiamine deficiency demonstrated by uptake of ^{3}H-serotonin, *J. Comp. Neurol.* 176:467–494.

Chase, T.N., 1973, Biochemical and pharmacological studies of monoamines in Huntington's chorea, *Adv. Neurol.* 1:533–542.

Clarke, D.J., Dunnett, S.B., Isacson, O., and Bjorklund, A., 1988a, Striatal grafts in the ibotenic acid lesioned neostriatum: Ultrastructural and immunocytochemical studies, *Prog. Brain Res.* 78:47–53.

Clarke, D.J., Dunnett, S.B., Isacson, O., Sirinathsinghji, D.J.S., and Bjorklund, A., 1988b, Striatal grafts in rats with unilateral neostriatal lesions. I: Ultrastructural evidence of afferent synaptic inputs from the host nigrostriatal pathway, *Neuroscience* 24:791–801.

Cotman, C.W., and Iversen, L.L., 1987, Excitatory amino acids in the brain—focus on NMDA receptors, *Trends Neurosci.* 10:261–265.

Cotman, W.W., and Monaghan, D.T., 1988, Excitatory amino acid neurotransmission: NMDA receptors and Hebb-type synaptic plasticity, *Ann. Rev. Neurosci.* 11:61–80.

Cotman, C.W., Managhan, D.T., Ottersen, O.P., and Storm-Mathisen, J., 1987, Anatomical organization of excitory amino acid receptors and their pathways, *Trends Neurosci.* 10:273–279.

Coyle, J.T., 1979, An animal model for Huntington's disease, *Biol. Psychiatry* 14:251–276.

Coyle, J.T., and Schwarcz, R., 1976, Lesion of striatal neurones with kainic acid provides a model for Huntington's chorea, *Nature* 253:244–246.

Coyle, J.T., Molliver, M.E., and Kuhar, M.J., 1978, In situ injection of kainic acid: A new model for selectively lesioning neural cell bodies and sparing axons of passage, *J. Comp. Neurol.* 180:301–323.

Coyle, J.T., Price, D.L., and Delong, M.R., 1983, Alzheimer's disease: A disorder of cholinergic innervation of cortex, *Science* 219:1184–1190.

Cross, A.J., Slater, P., and Reynolds, G.P., 1986, Reduced high-affinity glutamate uptake sites in the brains of patients with Huntington's disease, *Neurosci. Lett.* 67:198–202.

Cudkowicz, M., and Kowall, N.W., 1990, Degeneration of pyramidal projection neurons in Huntington's disease cortex, *Ann. Neurol.* 27(2):200–204.

Cummings, J.L., and Benson, D.F., 1988, Psychological dysfunction accompanying subcortical dementias, *Ann. Rev. Med.* 39:53–61.

Davies, S.W., and Roberts, P.J., 1987, No evidence for preservation of somatostatin-containing neurons after intrastriatal injections of quinolinic acid, *Nature* 327:326–329.

Davies, S.W., and Roberts, P.J., 1988, Model of Huntington's disease, *Science* 241:474–475.

Davis, A., 1976, Emily—a victim of Huntington's chorea, *Nurs. Times* 72:449.

Deckel, A.W., and Robinson, R.G., 1987, Receptor characteristics and behavioral consequences of kainic acid lesions and fetal transplants of the striatum, *in* "Cell and Tissue Transplantation into Adult Brain," Volume 495 (E.C. Azmitia and A. Bjorklund, eds.), pp. 556–580, Annals of the New York Academy of Sciences, New York.

Deckel, A.W., Robinson, R.G., Coyle, J.T., and Sanberg, P.R., 1983, Reversal of long-term locomotor abnormalities in the kainic acid model of Huntington's disease by day-18 fetal striatal implants, *Eur. J. Pharmacol.* 93:287–288.

Deckel, A.W., Moran, T.H., and Robinson, R.G., 1986a, Behavioral recovery following kainic acid lesions and fetal implants of the striatum occurs independent of dopaminergic mechanisms, *Brain Res.* 363:383–385.

Deckel, A.W., Moran, T.H., Coyle, J.T., Sanberg, P.R., and Robinson, R.G., 1986b, Anatomical predictors of behavioral recovery following fetal striatal transplants, *Brain Res.* 365:249–258.

Deckel, A.W., Moran, T.H., and Robinson, R.G., 1988a, Receptor characteristic and recovery of function following kainic acid lesions and fetal transplants of the striatum. I: Cholinergic systems, *Brain Res.* 474:27–38.

Deckel, A.W., Moran, T.H., and Robinson, R.G., 1988b, Receptor characteristic and recovery function following kainic acid lesions and fetal transplants of the striatum. II: Dopaminergic systems, *Brain Res.* 474:39–47.

Del Fiacco, M., Paxinos, G., and Cuello, A.C., 1982, Neostriatal enkephalin-immunoreactive neurons project to the globus pallidus, *Brain Res.* 231:1–17.

Dewhurst, K., Oliver, J., Trick, K.L.K., and McKnight, A.L., 1979, Neuropsychiatric aspects of Huntington's disease, *Confinia Neurol.* 31:258–268.

DiFiglia, M., Schiff, L., and Deckel, A.W., 1988, Neuronal organization of fetal striatal grafts in kainate and sham-lesioned rat caudate nucleus: Light and electron microscope observations, *J. Neurosci.* 8:1112–1130.

Divac, J., Fonnum, F., and Storm-Mathisen, J., 1977, High-affinity uptake of glutamate in terminals of corticostriatal axons, *Nature* 266:377–378.

Divac, J., Markowitsch, J.J., and Ptrizel, M., 1978, Behavioral and anatomical consequences of small intrastriatal lesions in the rat, *Brain Res.* 151:523–532.

Dix, M.R., 1970, Clinical observations upon the vestibular responses in certain disorders of the central nervous system, *Adv. Otorhinolaryngol.* 17:118.

Dom, R., Malfoid, M., and Baro, F., 1976, Neuropathology of Huntington's chorea, *Neurology* 26:64–68.

Dragunow, M., Williams, M., and Faull, R.L.M., 1990, Haloperidol induces Fos and related molecules in intrastriatal grafts derived from fetal striatal primordia, *Brain Res.* 530:309–311.

Du, F., Okuno, E., Whetsell, W.O., Jr., Kohler, C., and Schwarcz, R., 1991, Immunohistochemical localization of quinolinic acid phosphoribosyltransferase in the human neostriatum, *Neurosci.* 42(2):397–406.

Dulap, C.B., 1927, Pathological changes in Huntington's chorea with special reference to the corpus striatum, *Arch. Neurol. Psychiatry* 18:867–943.

Dunnett, S.B., and Iversen, S.D., 1981, Learning impairments following selective kainic acid induced lesions within the neostriatum of rats, *Behav. Brain Res.* 2:189–209.

Dunnett, S.B., Isacson, O., Siriathsinghji, D.J.S., Clarke, D.J., and Bjorklund, A., 1988, Striatal grafts in rats with unilateral neostriatal lesions. III: Recovery from dopamine-dependent motor asymmetry and deficits in skilled paw reaching, *Neuroscience* 24:813–820.

Dure, L.S. IV, Young, A.B., and Penney, J.B., 1991, Excitatory amino acid binding sites in the caudate nucleus and frontal cortex of Huntington's disease, *Ann. Neurol.* 30(6)1:785–793.

Ellison, D.W., Beal, M.F., Mazurek, M.F., Malloy, J.R., Bird, E.D., and Martin, J.B., 1987, Amino acid neurotransmitter abnormalities in Huntington's disease and the quinolinic acid model of Huntington's disease, *Brain* 110:1657–1673.

Emerich, D.F., Zubrick, E.M., Shipley, M.T., Norman, A.B., and Sanberg, P.R., 1991, Female rats are more sensitive to the locomotor alterations following quinolinic acid-induced striatal lesions: Effects of striatal transplants, *Exp. Neurol.* 111:369–378.

Emson, P.C., Rehfeld, J.F., Langevin, H., and Rosser, M., 1980, Reduction in cholecystokinin-like immunoreactivity in the basal ganglia in Huntington's disease, *Brain Res.* 198:497–500.

Emson, P.C., Dawbarn, D., Rosser, M.N., Rehfeld, J.F., Brundin, P., Isacson, O., and Bjorklund, A., 1985, Cholecystokinin content in the basal ganglia in Huntington's disease: The expression of cholecystokinin immunoreactivity in striatal grafts to ibotenic acid-lesioned rat striatum, *in* "Neuronal Cholecystokinin" (J. Vanderhaegen and J.D. Cawley, eds.), pp. 488–494, The New York Academy of Sciences, New York.

Enna, S.J., Bennett, J.P., Bylund, D.B., Snyder, S.H., Bird, E.D., and Iversen, L.L., 1976a, Alterations of brain neurotransmitter receptor binding in Huntington's chorea, *Brain Res.* 116:531–537.

Enna, S.J., Bird, E.D., Bennett, J.P., Bylund, D.B., Snyder, S.H., and Iversen, L.L., 1976b, Huntington's chorea changes in neurotransmitter receptors in the brain, *N. Engl. J. Med.* 29:1305–1309.

Ferrante, R.J., Beal, M.F., Kowall, N.W., Richardson, E.P., and Martin, J.B., 1987, Sparing of acetylcholinesterase-containing striatal neurons in Huntington's disease, *Brain Res.* 415:178–182.

Ferrante, R.J., Kowall, N.W., and Richardson, E.P. Jr., 1991, Proliferative and degenerative changes in striatal spiny neurons in Huntington's disease: A combined study using the section-Golgi method and calbindin D28k immunocytochemistry, *J. Neurosci.* 11(12):3877–3887.

Fibiger, H.C., Pruditz, R.E., McGeer, P.L., and McGeer, E.G., 1972, Axonal transport in nigrostriatal and nigrothalamic neurons: Effects of medial forebrain bundle lesions and 6-hydroxydopamine, *J. Neurochem.* 19:1697–1708.

Fields, J.Z., Reizine, T.D., and Yamamura, H.I., 1978, Loss of striatal dopaminergic receptors after intrastriatal kainic acid injection, *Life Sci.* 23:569–574.

Filloux, F., Wagster, M.V., Folstein, S., Price, D.L., Hedreen, J.C., Dawson, T.M., and Wamsley, J.K., 1990, Nigral dopamine type-1 receptors are reduced in Huntington's disease: A postmortem autoradiographic study using [^{3}H]SCH 23390 and correlation with [^{3}H]forskolin binding, *Exp. Neurol.* 110:219–227.

Finn, S.F., Hyman, B.T., Storey, E., Miller, J.M., and Beal, M.F., 1991, Effects of aging on quinolinic acid lesions in rat striatum, *Brain Res.* 562:276–280.

Fonnum, F., Storm-Mathisen, J., and Divac, I., 1981, Biochemical evidence for glutamate as neurotransmitter in corticostriatal and corticothalamic fibers in rat brain, *Neuroscience* 6:863–873.

Forno, L.S., and Jose, C., 1973, Huntington's chorea: A pathological study, *Adv. Neurol.* 1:453—470.

Foster, A.C., Collins, J.F., and Schwarcz, R., 1983, On the excitotoxic properties of quinolinic acid, 2,3-piperidine dicarboxylic acids and structurally related compounds, *Neuropharmacology* 22.1331–1342.

Foster, A.C., Vezzani, A., French, E.D., and Schwarcz, R., 1984, Kynurenic acid blocks metabolite quinolinic acid, *Neurosci. Lett.* 48:273–278.

Foster, A.C., Whetsell, W.O., Bird, E.D., and Schwarcz, R., 1985, Quinolinic acid phosphoribosyltransferase in human and rat brain: Activity in Huntington's disease and in quinolate-lesioned rat striatum, *Brain Res.* 336:207–214.

Foster, A.C., and Wong, E.H.F., 1987, The novel anticonvulsant MK-801 binds to the activated site of the N-methyl-D-aspartate receptor in rat brain, *Br. J. Pharmacol.* 91:403–409.

Foster, A.C., Gill, R., Kemp, J.A., and Woodruff, G.N., 1987, Systemic administration of MK-801 prevents N-methyl-D-aspartate–induced neuronal degeneration in rat brain, *Neurosci. Lett.* 76:307–311.

Freed, W.J., Cannon-Spoor, H.E., and Krauthamer, E., 1986, Intrastriatal adrenal medulla grafts in rats: Long-term survival and behavioral effects, *J. Neurosurg.* 65:664–670.

Gage, F.H., and Bjorklund, A., 1986, Cholinergic septal grafts into the hippocampal formation improve spatial learning and memory in aged rats by an atropine-sensitive mechanism, *J. Neurosci.* 6:2837–2847.

Gayle, J.S., Bird, E.D., Spokes, E.G., Iversen, L.L., and Jessell, T., 1978, Human brain substance P distribution in controls in Huntington's chorea, *J. Neurochem.* 30:633–634.

Gebbink, T.B., 1986, Huntington's chorea: Fibre changes in the basal ganglia, *Handbook Clin. Neurol.* 6:399.

Gerfen, C.R., Herkenham, M., and Thibault, J., 1987, The neostriatal mosaic. II: Patch- and matrix-directed mesostriatal dopaminergic system, *J. Neurosci.* 7:3915–3934.

Gill, R., Foster, A.C., and Woodruff, G.N., 1987, Systemic administration of MK-801 protects against ischemia-induced hoppocampal neurodegeneration in the gerbil, *J. Neurosci.* 7:3343–3349.

Gilliam, T.C., Gusella, J.F., and Lehrach, H., 1988, Molecular genetic strategies to investigate Huntington's disease, *Adv. Neurol.* 48:465–469.

Giordano, M., Houser, S.H., and Sanberg, P.R., 1988, Intraparenchymal fetal striatal transplants and recovery in kainic-acid-lesioned rats, *Brain Res.* 446:183–188.

Giordano, M., Ford, L., Shipley, M.T., and Sanberg, P.R., 1990a, Neural grafts and pharmacological intervention in a model of Huntington's disease, *Brain Res. Bull.* 25:453–465.

Giordano, M., Ford, L., Norman, A.B., and Sanberg, P.R., 1990b, MK-801 prevents quinolinic acid-induced behavioral deficits and neurotoxicity in the striatum, *Brain Res. Bull.* 24:433–465.

Giordano, M., Lu, S.Y., Emerich, D.F., Pixley, S.K., Norgren, R.B., Lehman, M.N., and Norman, A.B., 1991, Behavioral effects of transplants on cultured astrocytes on excitotoxin-induced locomotor behavior, *Soc. Neurosci. Abstr.* 17:903.

Girotti, F., Marano, R., Soliver, Geminiani, G., and Scagliano, G., 1988, Relationship between motor and cognitive disorders in Huntington's disease, *J. Neurol.* 235:454–457.

Gonzalez-Heydrich, J., and Peroutka, S.J., 1991, Postsynaptic localization of 5-HT $_{1D}$ receptor binding sites in human caudate, *Exp. Neurol.* 113:28–30.

Gray, P.N., May, P.C., and Elkin, J., 1980, L-glutamate toxicity in Huntington's disease, *Biochem. Biophys. Res. Commun.* 95:707–714.

Graybiel, A.M., 1983, Biochemical anatomy of the striatum, *in* "Chemical Neuroanatomy" (P.C. Emson, ed.), pp. 427–503, Raven Press, New York.

Graybiel, A.M., Ragsdale, C.W., Jr., Yoneoka, E.S., and Elde, R.P., 1981, An immunohistochemical study on enkephalins and other neuropeptides in the striatum of the cat with evidence that the opiate peptides are arranged to form mosaic patterns in register with the striasomal compartments visible by acetylcholinesterase staining, *Neuroscience* 6:377–397.

Graybiel, A.M., Lui, F.C., and Dunnett, S.B., 1989, Intrastriatal grafts derived from fetal striatal primordia. L: Phenotype and modular organization, *J. Neurosci.* 9:3250–3271.

Gusella, J.F., Wexler, N.S., Conneally, P.M., Naylor, S.L., Anderson, M.A., Tanzi, R.E., Shoulson, I., Bonilla, E., and Martin, J.B., 1983, A polymorphic marker genetically linked to Huntington's disease, *Nature* 306:234–238.

Hagenmeyer-Houser, S.H., and Sanberg, P.R., 1987, Locomotor behavior changes induced by E-17 striatal transplants in normal rats, *Pharmacol. Biochem. Behav.* 27:583–586.

Hayden, M.R., Hewitt, J., Wasmuth, J.J., Kastelein, J.J., Langlois, S., Conneally, M., Haines, J., Smith, B., Hilbert, C., and Allard, D., 1988, A polymorphic DNA marker that represents a conserved expression sequence in the region of the Huntington's disease gene, *Am. J. Hum. Genet.* 42:125–131.

Hantraye, P., Riche, D., Maziere, M., and Isacson, O., 1990, A primate model of Huntington's disease: Behavioral and anatomical studies of unilateral excitotoxic lesions of the caudate-putamen in the baboon, *Exp. Neurol.* 108:91–104.

Hays, S.E., Goodwin, F.K., and Paul, S.M., 1981, Cholecystokinin receptors are decreased in basal ganglia and cerebral cortex of Huntington's disease, *Brain Res.* 225:452–456.

Hedreen, J.C., Peyser, C.E., Folstein, S.E., and Ross, C.A., 1991, Neuronal loss in layers V and VI of cerebral cortex in Huntington's disease, *Neurosci. Lett.* 133:257–261.

Hefter, H., Homberg, V., Lange, H.W., Freund, H.J., 1987, Impairment of rapid movement in Huntington's disease, *Brain* 110:585–612.

Heindel, W.C., Butters, N., and Salmon, D.P., 1988, Impaired learning of a motor skill in patients with Huntington's disease, *Behav. Neurosci.* 102:141–147.

Heindel, W.C., Salmon, D.P., and Butters, N., 1990, Pictorial priming and cued recall in Alzheimer's and Huntington's disease, *Brain Cogn.* 13:282–295.

Helm, G.A., Palmer, P.E., and Bennett, J.P., Jr., 1990, Fetal neostriatal transplants in the rat: A light and electron microscope Golgi study, *Neuroscience* 37(3):735–756.

Henke, H., 1979, Kainic acid binding in human caudate nucleus: Effect of Huntington's disease, *Neurosci. Lett.* 14:247–251.

Hiley, R.C., and Bird, E.D., 1974, Decreased muscarinic receptor concentration in post-mortem brain in Huntington's chorea, *Brain Res.* 30:355–358.

Hodges, J.R., Salmon, D.P., and Butters, N., 1991, The nature of the naming deficit in Alzheimer's and Huntington's disease, *Brain* 114:1547–1558.

Hruska, R.E., and Silbergeld, E.K., 1979, Abnormal locomotion in rats after bilateral intrastriatal injection of kainic acid, *Life Sci.* 25:181–194.

Huettner, J.E., and Bean, B.P., 1988, Block of N-methyl-D-aspartate-activated current by the anticonvulsant MK-801: Selective binding to open channels, *Proc. Natl. Acad. Sci. U.S.A.* 85:1307–1311.

Innis, R.B., and Snyder, S.H., 1980, Cholecystokinin receptor binding in brain and pancreas: Regulation of pancreatic binding by cyclic and acyclic quanine nucleotides, *Eur. J. Pharmacol.* 65:123–124.

Innis, R.B., Correa, F.M.A., Uhl, G.R., Schneider, B., and Snyder, S.H., 1979, Cholecystokinin octapeptide-like immunoreactivity: Histochemical localization in rat brain, *Proc. Natl. Acad. Sci. U.S.A.* 76:521–635.

Isacson, O., Brundin, P., Kelly, P., Gage, F.H., and Bjorklund, A., 1984, Functional neuronal replacement by grafted striatal neurons in the ibotenic acid-lesioned rat striatum, *Nature* 311:458–460.

Isacson, O., Brundin, P., Gage, F.H., and Bjorklund, A., 1985, Neural grafting in the rat model of Huntington's disease: Progressive neurochemical changes after neostriatal ibotenate lesions and striatal tissue grafting, *Neuroscience* 16:799–817.

Isacson, O., Dunnett, S.B., and Bjorklund, A., 1986, Graft-induced behavioral recovery in an animal model of Huntington's disease, *Proc. Natl. Acad. Sci. U.S.A.* 83:2728–2732.

Isacson, O., Dawbarn, D., Brundin, P., Gage, F.H., Emson, P.C., and Bjorklund, A., 1987a, Neural grafting in a rat model of Huntington's disease: Striosomal-like organization of striatal grafts as revealed by acetylcholinesterase histochemistry, immunocytochemistry, and receptor autoradiography, *Neuroscience* 22:481–497.

Isacson, O., Pritzel, M., Dawbarn, D., Brundin, P., Kelly, A.T., Wiklund, L., Emson, P.C., Gage, F.H., Dunnett, S.B., and Bjorklund, A., 1987b, Striatal neural transplants in the ibotenic acid-lesioned rat neostriatum: Cellular and functional aspects, *in* "Cell and Tissue Transplantation into the Adult Brain" (E.C. Azmitia and A. Bjorklund, eds.), pp. 537–555, New York Academy of Sciences, New York.

Isacson, O., Riche, D., Hantraye, P., Sofroniew, M.V., and Maziere, M., 1989, A primate model of Huntington's disease; cross-species implantation of striatal precursor cells to the excitotoxically lesioned baboon caudate-putamen, *Exp. Brain Res.* 75:213–220.

Isacson, O., Hantraye, P., Maziere, M., Sofroniew, M.V., and Riche, D., 1990, Apomorphine-induced dyskinesias after excitotoxic caudate-putamen lesions and the effects of neural transplantation in non-human primates, *Prog. Brain Res.* 82:523–533.

James, W.E., Mefferd, R.B., and Kimbell, I., 1969, Early signs of Huntington's chorea, *Dis. Nerv. Syst.* 30:556–559.

Jernigan, T.L., Salmon, D.P., Buters, N., and Hesselink, J.R., 1991, Cerebral structure on MRI. Part II: Specific changes in Alzheimer's and Huntington's diseases, *Biol. Psychiatry* 29:68–81.

Jervis, G.A., 1963, Huntington's chorea in childhood, *Arch. Neurol.* 9:244–257.

Jessell, T.M., Emson, P.S., Paxinos, G., and Cuello, A.C., 1978, Topographic projections of substance P and GABA pathways in the atriato- and pallido-nigral system: A biochemical and immunohistochemical study, *Brain Res.* 152:487–498.

Johnson, J.W., and Ascher, P., 1987, Glycine potentiates the NMDA response in cultured mouse brain neurons, *Nature* 325:529–531.

Joyce, J.N., Sapp, D.W., and Marshall, J.F., 1986, Human striatal dopamine receptors are organized in compartments, *Proc. Natl. Acad. Sci. U.S.A.* 83:8002–8006.

Kanazawa, I., Bird, E.D., Gale, J.S., Iversen, L.L., Jessell, T.M., Muramoto, O., Spokes, E.G., and Sutoo, D., 1979, Substance P: Decrease in substantia nigra and globus pallidus in Huntington's disease, *Adv. Neurol.* 23:495–504.

Kemp, J.A., Foster, A.C., and Wong, E.H.K., 1987, Non-competitive antagonists of excitatory amino acid receptors, *Trends Neurosci.* 10:294–298.n

Kemp, J.A., Foster, A.C., Leeson, P.D., Priestly, T., Tidgett, R., Iversen, L.L., and Woodruff, G.N., 1988, 7-Chlorokynurenic acid is a selective antagonist at the glycine modulatory site of the N-methyl-D-aspartate receptor complex, *Proc. Natl. Acad. Sci. U.S.A.* 85:6547–6550.

Kesslak, J.P., Neito-Sampedro, M., Globus, J., and Cotman, C.W., 1986, Transplants of purified astrocytes promote behavioral recovery after frontal cortex ablation, *Exp. Neurol.* 92:377–390.

Kim, J.S., Kornhuber, H.H., Holtzmuller, B., Schmid-Burgk, W., Mergner, T., and Krzepinski, G., 1980, Reduction of cerebrospinal fluid glutamic acid in Huntington's chorea and in schizophrenic patients, *Arch. Psychiatr. Nervenkr.* 228:7.

Kish, S.J., Shannack, K., and Hornykiewicz, O., 1987, Elevated serotonin and reduced dopamine in subregionally divided Huntington's disease striatum, *Ann. Neurol.* 22:386–389.

Klawans, H.L., Goetz, C.G., and Westheimer, R., 1972, Pathophysiology of schizophrenia and the striatum, *Dis. Nerv. Syst.* 33:711–719.

Klechner, N.W., and Dingledine, R., 1988, Requirements for glycine in activation of NMDA receptors expressed in *Xenopus* oocytes, *Science* 241:835–838.

Kloog, Y., Nadler, V., and Sokolovsky, M., 1988, Mode of binding of [^{3}H]dibenzocycloalkenimine (MK-801) to the N-methyl-D-aspartate (NMDA) receptor and its therapeutic implication, *FEBS Lett.* 230:167–170.

Knopman, D., and Nissen, M.J., 1991, Procedural learning is impaired in Huntington's disease: Evidence from the serial reaction time task, *Neuropsychologia* 29:245–254.

Kochhar, A., Zivin, J.A., Lyden, P.D., and Mazzarella, V., 1988, Glutamate antagonist therapy reduces neurologic deficits produced by focal central nervous system ischemia, *Arch. Neurol.* 45:148–153.

Koh, J., Peters, S., and Choi, D.W., 1986, Neurons containing NADPH-diaphorase are selectively resistant to quinolinate toxicity, *Science* 234:73–76.

Koh, J.Y., and Choi, D.W., 1988, Cultured striatal neurons containing NADPH-diaphorase or acetylcholinesterase are selectively resistant to injury by NMDA agonists, *Brain Res.* 446:374–378.

Korenyi, C., and Whittier, J.R., 1973, The juvenile form of Huntington's chorea: Its prevalence and other observations, *Adv. Neurol.* 1:75–85.

Kremer, H.P.H., Roos, R.A.C., Dingjan, G., Marani, E., Bots, G., Th., A.M., 1990, Atrophy of the hypothalamic lateral tuberal nucleus in Huntington's disease, *J. Neuropathol. Exp. Neurol.* 49:371–382.

Kromer, L.F., and Cornbrooks, L.S., 1985, Transplants of Schwann cell cultures promote axonal regeneration in adult mammalian brain, *Proc. Natl. Acad. Sci. U.S.A.* 82:6330–6334.

Kuwert, T., Lange, H.W., Langen, K-J., Herzog, H., Aulich, A., and Feinendegen, L.E., 1989,

Cerebral glucose consumption measured by PET in patients with and without psychiatric symptoms of Huntington's disease, *Psychiatry Res.* 29:361–362.

Kuwert, T., Lange, G.W., Langen, K-J., Herzog, H., Aulich, A., and Feinendegen, L.E., 1990, Cortical and subcortical glucose consumption measured by PET in patients with Huntington's disease, *Brain* 113:1405–1423.

Labandeira-Garcia, J.L., Wictorin, K., Cunningham, E.T., and Bjorklund, A., 1991, Development of intrastriatal striatal grafts and their afferent innervation from the host, *Neuroscience* 42(2):407–426.

Lange, H., Thorner, G., Hopf, A., and Schroeder, K.F., 1976, Morphometric studies of the neuropathological changes in choreatic diseases, *J. Neurol. Sci.* 28:401–425.

Lange, H.W., 1981, Quantitative changes of telencephalon, diencephalon, and mesencephalon in Huntington's chorea, postencephalitic and idiopathic parkinsonism, *Verh. Anat. Ges.* 75:923–925.

Lublhuber, F., Hoell, K., Reisecker, F., Gebetsberger, B., Puerhringer, W., Trenkler, E., and Deisenhammer, E., 1989, Single photon emission computed tomography in Huntington's chorea, *Psychiatry Res.* 29:337–339.

London, E.D., Yamamura, H.I., Bird, E.D., and Coyle, J.T., 1981, Decreased receptor-binding sites for kainic acid in brains of patients with Huntington's disease, *Biol. Psychiatry* 16:155–162.

Lu, S.Y., Giordano, M., Norman, A.B., Shipley, M.T., and Sanberg, P.R., 1990, Behavioral effects of neural transplants into the intact striatum, *Pharmacol. Biochem. Behav.* 37:135–148.

Lu, S.Y., Shipley, M.T., Norman, A.B., and Sanberg, P.R., 1991, Striatal, ventral mesenchephalic and cortical transplants into the intact striatum: A neuroanatomical study, *Exp. Neurol.* 113:109–130.

Mann, J.J., Stanley, M., Gershon, S., and Rosser, M., 1980, Mental symptoms in Huntington's disease and a possible primary aminergic neuron lesion, *Science* 210:1369–1371.

Manyam, B.V., Giacobini, E., and Colliver, J.A., 1990, Cerebrospinal fluid acetylcholinesterase and choline measurements in Huntington's disease, *J. Neurol.* 237:281–284.

Martin, J.B., 1989, Molecular genetic studies in the neuropsychiatric disorders, *Trends Neurosci.* 12:130–136.

Mason, S.T., and Fibiger, H.C., 1979, Kainic acid lesions of the striatum in rats mimic the spontaneous motor abnormalities of Huntington's disease. *Neuropharmacology* 18:403–407.

Mason, S.T., Sanberg, P.R., and Fibiger, H.C., 1978, Kainic acid lesions of the striatum dissociate amphetamine and apomorphine stereotypy: Similarities to Huntington's chorea, *Science* 201:352–355.

Massman, P.J., Delis, D.C., and Butters, N., 1990, Are all subcortical dementias alike? Verbal learning and memory in Parkinson's and Huntington's disease patients, *J. Clin. Exp. Neuropsychol.* 12(5):729–744.

Mattson, B., Gottfries, C.E., Roos, B.E., and Winblad, B., 1974, Huntington's chorea: Pathology and brain amines, *Acta Psychiatr. Scand.* 255:269–277.

McAllister, J.P., 1987, Tritiated thymidine identification of embryonic neostriatal transplants, *in* "Cell and Tissue Transplantation into the Adult Brain", Volume 495 (E.C., Azmitia and A. Bjorklund, eds.), pp. 745–748, Annals of the N.Y. Academy of Sciences, New York.

McAllister, J.P., Kaplan, L., and Reynolds, M.C., 1984, Morphology and connectivity of fetal neostriatal tissue transplanted into the neostriatum of adult hosts, *Anat. Rec.* 208:107A.

McAllister, J.P., Walker, P.D., Zemanick, M.C., Weber, A.B., Kaplan, L.I., and Reynolds, M.C., 1985, Morphology of embryonic neostriatal cell suspensions transplanted into adult neostriata, *Develop. Brain Res.* 23:282–286.

McBean, C.J., and Roberts, P.J., 1985, Neurotoxicity of L-glutamate and DL-threo-3-hydroxyaspartate in the rat striatum, *J. Neurochem.* 44:246–254.

McGeer, P.L., and McGeer, E.G., 1976a, Enzymes associated with the metabolism of cate-

cholamines, acetylcholine and GABA in human controls and patients with Parkinson's disease and Huntington's chorea, *J. Neurochem.* 26:65–76.
McGeer, E.G., and McGeer, P.L., 1976b, Duplication of biochemical changes of Huntington's chorea by intrastriatal injections of glutamic and kainic acids, *Nature* 263:517–519.
McGeer, P.L., and McGeer, E.G., 1982, Kainic acid: The neurotoxic breakthrough, *CRC Crit. Rev. Toxicol.* 10:1–26.
McGeer, P.L., McGeer, E.G., and Fibiger, H.C., 1973, Choline acetylase and glutamic acid decarboxylase in Huntington's chorea, *Neurology* 23:912–917.
McGeer, E.G., McGeer, E.G., Hattori, T., and Vincent, S.R., 1979, Kainic acid neurotoxicity and Huntington's disease, *Adv. Neurol.* 23:577–591.
McGeer, P.L., Kimura, H., and McGeer, E.G. 1984, Transplantation of newborn brain tissue into adult kainic-acid-lesioned neostriatum, *in* "Neural Transplants" (J.R. Sladek and D.J. Gash, eds.), pp. 361–371, Plenum Press, New York.
McGeer, P.L., Eccles, J.C., and McGeer, E.G., (eds.), 1987, "Molecular Neurobiology of the Mammalian Brain, Plenum Press, New York.
McHugh, P.R., and Folstein, M.F., 1975, Psychiatric syndromes of Huntington's chorea: A clinical and phenomenologic study, *in* "Psychiatric Aspects of Neurologic Disease" (F. Benson and D. Blumer, eds.), pp. 267–281, Grune and Stratton, New York.
McIntosh, G.E., Jameson, D., and Markesbury, W.R., 1978, Huntington's disease associated with Alzheimer's disease, *Ann. Neurol.* 3:545–548.
Melamed, E., Hefti, F., and Bird, E.D., 1982, Huntington's chorea is not associated with hyperactivity of dopaminergic neurons: Studies in post-mortem tissues and rats with kainic acid lesions, *Neurology* 33:640–644.
Meyer, D.K., Beinfeld, M.C., Oertel, W.H., and Brownstein, M.J., 1982, Origin of the cholecystokinin-containing fibers in the caudatoputamen, *Science* 215:187–188.
Moore, R.Y., and Card, J.P., 1984, Noradrenaline-containing neuron systems, *in* "Handbook of Chemical Neuroanatomy, Classical Transmitters in the CNS," Part I. (A. Bjorklund and T. Hokfelt, eds.), pp. 123–156, Elsevier, Amsterdam.
Mroz, E.A., Brownstein, M.J., and Leeman, S.E., 1977, Evidence for substance P in the striatonigral tract, *Brain Res.* 125:305–311.
Muller, H.W., and Seifert, W.A., 1982, A neurotrophic factor (NTF) released from primary glial cultures supports survival and fiber outgrowth of cultured hippocampal neurons, *J. Neurosci. Res.* 8:195–204.
Neito-Sampedro, M., and Cotman, C.W., 1985, Growth factor induction and temporal order in CNS repair, *in* "Synaptic Plasticity" (C.W. Cotman, ed.), pp. 407–455, Guilford Press, New York.
Norman, A.B., McGowan, T., Calderon, S.F., Giordano, M., and Sanberg, P.R., 1987, Attenuation of apomorphine-induced rotational behavior by fetal striatal tissue transplants in rats with unilateralstriatal kainic acid lesions, *Soc. Neurosci. Abstr.* 13:785.
Norman, A.B., Calderon, S.F., Giordano, M., and Sanberg, P.R., 1988a, Striatal tissue transplants attenuate apomorphine-induced rotational behavior in rats with unilateral kainic acid lesions, *Neuropharmacology* 27:333–336.
Norman, A.B., Calderon, S.F., Giordano, M., and Sanberg, P.R., 1988b, A novel rotational behavior model for assessing the restructuring of striatal dopamine effector systems: Are transplants sensitive to peripherally acting drugs?, *in* "Transplantation into the Mammalian CNS" (D.M. Gash and J.R. Sladek, eds.), pp. 61–67, Elsevier, Amsterdam.
Norman, A.B., Giordano, M., and Sanberg, P.R., 1989a, Fetal striatal tissue grafts into excitotoxin-lesioned striatum: Pharmacological and behavioral aspects, *Pharmacol. Biochem. Behav.* 34:139–147.
Norman, A.B., Lehman, M., and Sanberg, P.R., 1989b, Functional effects of fetal striatal transplants, *Brain Res. Bull.* 22:163–172.

Norman, A.B., Ford, L.M., Kolmonpunporn, M., and Sanberg, P.R., 1990, Chronic treatment with MK-801 increases the quinolinic acid-induced loss of D-$_1$ dopamine receptors in rat striatum, *Eur. J. Pharmacol.* 176:363–366.

Oka, H., 1980, Organization of the cortico-caudate projections, *Exp. Brain Res.* 40:203–208.

Okuno, E., Kohler, C., and Schwarcz, R., 1987, Rat-3-hydroxyanthranlic acid oxygenase: purification from the liver and immunocytochemical localization in the brain, *J. Neurochem.* 49:771–780.

Olney, J.W., and de Gubareff, T., 1978, Glutamate neurotoxicity and Huntington's chorea, *Nature* 271:557–559.

Oyanagi, K., and Ikuta, F.A., 1987, A morphometric reevaluation of Huntington's chorea with special reference to the large neurons of the neostriatum, *Clin. Neuropathol.* 6:71–79.

Oyanagi, K., Takeda, S., Takahashi, J., Ohama, E., and Ikuta, F., 1989, A quantitative investigation of the substantia nigra in Huntington's disease, *Ann. Neurol.* 26:13–19.

Palacios, J.M., Rigo, M., Chinaglia, G., and Probst, A., 1990, Reduced density of striatal somatostatin receptors in Huntington's chorea, *Brain Res.* 522:342–346.

Palacios, J.M., Chinaglia, G., Rigo, M., Ulrich, J., and Probst, A., 1991, Neurotensin receptor binding levels in basal ganglia are not altered in Huntington's chorea or schizophrenia, *Synapse* 7:114–122.

Perry, T.L., Hansen, S., and Kloster, M., 1973, Huntington's chorea: Deficiency of GABA in brain, *N. Engl. J. Med.* 288:337–342.

Pillon, B., Dubois, B., Ploska, A., and Agid, Y., 1991, Severity and specificity of cognitive impairment in Alzheimer's, Huntington's, and Parkinson's diseases and progressive supranuclear palsy, *Neurology* 41:634–643.

Pinel, C., 1976, Huntington's chorea, *Nurs. Times* 72:447.

Pisa, M., Sanberg, P.R., and Fibiger, H.C., 1980, Locomotor activity, exploration and spatial alternation learning in rats with striatal injections of kainic acid, *Physiol. Behav.* 24:1120.

Pisa, M., Sanberg, P.R., and Fibiger, H.C., 1981, Striatal injections of kainic acid selectively impair serial memory performance in the rat, *Exp. Neurol.* 74:633–653.

Podoll, K., Caspary, P., Lange, H.W., and Noth, J., 1988, Language functions in Huntington's disease, *Brain* 111:1475–1503.

Pritzel, M., Isacson, O., Brundin, P., Wiklund, L., and Bjorklund, A., 1986, Afferent and efferent connections of striatal grafts implanted into the ibotenic lesioned neostriatum in adult rats, *Exp. Brain Res.* 65:112–126.

Randolph, C., 1991, Implicit, explicit, and semantic memory functions in Alzheimer's disease and Huntington's disease, *J. Clin. and Exp. Neuropsychol.* 13(4):479–494.

Reed, T.E., and Chandler, J.H., 1958, Huntington's chorea in Michigan. I: Demography and genetics, *Am. J. Hum. Genet.* 10:201—225.

Reiner, A., Albin, D.L., Anderson, K.D., D'Amato, C.J., Penny, J.B., and Young, A.B., 1988, Differential loss of striatal projection neurons in Huntington's disease, *Proc. Natl. Acad. Sci. U.S.A.* 85:5733–5737.

Reisine, T.D., Fields, J.Z., Stern, L.Z., Johnson, P.C., Bird, E.D., and Yamamura, H.I., 1977, Alterations in dopaminergic receptors in Huntington's disease, *Life Sci.* 21:1123–1128.

Reynolds, G.P., Pearson, S.J., Halket, J., and Sandler, M., 1988, Brain quinolinic acid in Huntington's disease, *J. Neurochem.* 50:1959–1960.

Reynolds, G.P., Pearson, S.J., and Heathfield, K.W.G., 1990, Dementia in Huntington's disease is associated with neurochemical deficits in the caudate nucleus, not the cerebral cortex, *Neurosci. Lett.* 113:95–100.

Rhodes, K.J., Joyce, J.N., Sapp, D.W., and Marshall, J.F., 1987, [^{3}H]hemicholinium-3 binding in rabbit striatum: Correspondence with patchy acetylcholinesterase staining and a method for quantifying striatal compartments, *Brain Res.* 412:400–404.

Roberts, R.C., and DiFiglia, M., 1989, Short- and long-term survival of large neurons in the excitotoxic lesioned rat caudate nucleus: A light and electron microscopic study, *Synapse* 3:363–371.

Roberts, R.C., and DiFiglia, M., 1990, Long-term survival of GABA-, enkephalin-, NADPH-diaphorase- and calbindin-d28k-containing neurons in fetal striatal grafts, *Brain Res.* 532:151–159.

Rodda, R.A., 1981, Cerebellar atrophy in Huntington's disease, *J. Neural. Sci.* 50:147.

Roizen, L., Kaufman, M.A., Wilson, W., Stellar, S., and Liu, J.C., 1976, Neuropathologic observations in Huntington's chorea, *Prog. Neuropathol.* 3:447.

Rutherford, A., Garcia-Munoz, Dunnett, S.B., and Arbuthott, G.W., 1987, Electrophysiological demonstration of host cortical inputs to striatal grafts, *Neurosci. Lett.* 83:275–281.

Saint-Cyr, J.A., Taylor, A.E., and Lang, A.E., 1988, Procedural learning and neostriatal dysfunction in man, *Brain* 111:941–959.

Salmon, D.P., Kwo-on-Yeun, P.F., Heindel, W.C., Butters, N., and Thal, L.J., 1989, Differentiation of Alzheimer's disease and Huntington's disease with the dementia rating scale, *Arch. Neurol.* 46:1204–1208.

Sanberg, P.R., 1980, Haloperidol-induced catalepsy is mediated by postsynaptic dopamine receptors, *Nature* 284:472–473.

Sanberg, P.R., and Coyle, J.T., 1984, Scientific approaches to Huntington's disease, *CRC Crit. Rev. Clin. Neurobiol.* 1:1–44.

Sanberg, P.R., and Creese, I., 1981, Dopamine receptor stimulation and striatal kainic acid neurotoxicity, *J. Pharm. Pharmacol.* 33:674–675.

Sanberg, P.R., and Fibiger, H.C., 1979, Body weight, feeding and drinking behaviors in rats with kainic acid lesions of striatal neurons, with a note on body weight symptomatology in Huntington's disease, *Exp. Neurol.* 66:444–466.

Sanberg, P.R., and Johnston, G.A., 1981, Glutamate and Huntington's disease, *Med. J. Aust.* 2:460–465.

Sanberg, P.R., Lehmann, J., and Fibiger, H.C., 1978, Impaired learning and memory after kainic acid lesions of the striatum: A behavioral model of Huntington's disease, *Brain Res.* 149:546–551.

Sanberg, P.R., Pisa, M., and Fibiger, H.C., 1979, Avoidance operant and locomotor behavior in rats with neostriatal injections of kainic acid, *Pharmacol. Biochem. Behav.* 10:137–144.

Sanberg, P.R., Fibiger, H.C., and Mark, R.R., 1981a, Body weight and dietary factors in Huntington's disease patients compared with matched controls, *Med. J. Aust.* 1:407–409.

Sanberg, P.R., Pisa, M., and Fibiger, H.C., 1981b, Kainic acid injections in the striatum alter the cataleptic and locomotor effects of drugs influencing the dopaminergic and cholinergic systems, *Eur. J. Pharmacol.* 74:347–357.

Sanberg, P.R., Hagenmeyer, S.H., and Henault, M.A., 1985, Automated measurement of multivariate locomotor behavior in rodents, *Neurobehav. Toxicol. Teratol.* 7:87–94.

Sanberg, P.R., Henault, and M.A., Deckel, A.W., 1986, Locomotor hyperactivity: Effects of multiple striatal transplants in an animal model of Huntington's disease, *Pharmacol. Biochem. Behav.* 25:297–300.

Sanberg, P.R., Calderon, S.F., Garver, D.L., and Norman, A.B., 1987a, Brain tissue transplants in an animal model of Huntington's disease, *Psychopharmacol. Bull.* 23:476–482.

Sanberg, P.R., Henault, M.A., Hagenmeyer-Houser, S.H., and Russel, R.H., 1987b, The topography of amphetamine and scoplamine-induced hyperactivity: Toward an activity print, *Behav. Neurosci.* 101:131–133.

Sanberg, P.R., Henault, M.A., Hagenmeyer-Houser, S.H., Giordano, M., and Russell, K.H., 1987c, Multiple transplants of fetal striatal tissue in the kainic acid model of Huntington's disease: Behavioral recovery may not be related to acetylcholinesterase, *in* "Cell and Tissue

Transplantation into the Adult Brain" (E.C. Azmitia and A. Bjorklund, eds.), pp. 781–785, Annals of the N.Y. Academy of Science New York.

Sanberg, P.R., Nash, D.R., Calderon, S.F., Giordano, M., Shipley, M.T., and Norman, A.B., 1988, Neural transplants disrupt the blood–brain barrier and allow peripherally acting drugs to exert a centrally mediated behavioral effect, *Exp. Neurol.* 102:149–152.

Sanberg, P.R., Calderon, S.F., Giordano, M., Tew, J.M., and Norman, A.B., 1989a, The quinolinic acid model of Huntington's disease: Locomotor abnormalities, *Exp. Neurol.* 105:45–53.

Sanberg, P.R., Giordano, M., Henault, M.A., Nash, D.R., Ragozzino, M.E., and Hagenmeyer-Houser, S.H., 1989b, Intraparenchymal striatal transplants required for maintenance of behavioral recovery in an animal model of Huntington's disease, *J. Neurotransplant.* 1:23–31.

Sanberg, P.R., Zubrycki, E.M., Ragozzino, M.E., Lu, S.Y., Norman, A.B., and Shipley, M.T., 1990, NADPH-diaphorase-containing neurons and cytochrome oxidase activity following striatal quinolinic acid lesions and fetal striatal transplants, *Prog. Brain Res.* 82:427–431.

Schmidt, R.H., Bjorklund, A., and Stenevi, U., 1981, Intracerebral grafting of dissociated CNS tissue suspensions: A new approach for neuronal transplantation to deep brain sites, *Brain Res.* 218:347–356.

Scholz, O.B., and Berlemann, C., 1987, Memory performances in Huntington's disease, *Int. J. Neurosci.* 35:155–162.

Schwarcz, R., and Kohler, C., 1983, Differential vulnerability of central neurons of the rat to quinolinic acid, *Neurosci. Lett.* 38:85–90.

Schwarcz, R., and Shoulson, I., 1987, Excitotoxins and Huntington's disease, *in* "Animal Models of Dementia: A Synaptic Neurochemical Perspective" (J.T. Coyle, ed.), pp. 39–68, Alan R. Liss, New York.

Schwarcz, R., Ruxe, K., Agnati, L.F., Hokfelt, T., and Coyle, J.T., 1979, Rotational behavior in rats with unilateral striatal kainic acid lesions: A behavioral model for studies of intact dopamine receptors, *Brain Res.* 170:485–495.

Schwarcz, R., Whetsell, W.O., Jr., and Mangano, R.M., 1983, Quinolinic acid: An endogenous metabolite that produces axon-sparing lesions in rat brain, *Science* 219:316–318.

Schwarcz, R., Foster, A.C., French, E.D., Whetsell, W.O., and Kohler, C., 1984, Excitotoxic models for neurodegenerative disorders, *Life Sci.* 35:19–32.

Schwarcz, R., Okono, E., and White, R.J., 1989, Basal ganglia lesions in the rat: effects on quinolinic acid metabolism, *Brain Res.* 490:103–109.

Segovia, J., Meloni, R., and Gale, K., 1989, Effect of dopaminergic denervation and transplant-derived reinnervation on a marker of striatal GABAergic function, *Brain Res.* 493:185–189.

Shoulson, I., 1986, Huntington's disease, *in* "Disease of the Nervous System" (A. Asbury, G.M. McKhann, and I. McDonald, eds.), pp. 1258–1267, Ardmore Medical Books (W.B. Saunders), Philadelphia.

Sirinathsinghji, D.J.S., Dunnett, S.B., Isacson, O., Clarke, D.J., Kendrick, K., and Bjorklund, A., 1988, Striatal grafts in rats with unilateral neostriatal lesions. II: In vivo monitoring of GABA release in globus pallidus and substantia nigra, *Neuroscience* 24:803–811.

Sladek, J.R., Redmond, D.E., Collier, T.J., Bount, J.P., Elsworth, J.R., Taylor, J.R., and Roth, R.H., 1988, Fetal dopamine neural grafts: Extended reversal of methylphenyltetrahydropyridine-induced parkinsonism in monkeys, *Prog. Brain Res.* 78:497–506.

Sortrel, A., Paskevich, P.A., Kiely, D.K., Bird, E.D., Williams, R.S., and Myers, R.H., 1991, Morphometric analysis of the prefrontal cortex in Huntington's disease, *Neurology* 41:1117–1123.

Spokes, E.G.S., 1980, Neurochemical alterations in Huntington's chorea: A study of post-mortem brain tissue, *Brain* 103:179–210.

Stahl, W.L., and Swanson, P.D., 1974, Biochemical abnormalities in Huntington's chorea brains, *Neurology* 24:813–819.

Staines, W.A., Nagy, J.I., Vincent, S.R., and Fibiger, H.C., 1980, Neurotransmitters contained in the striatum, *Brain Res.* 194:391–402.

Starr, A., 1967, A disorder of rapid eye movements in Huntington's chorea, *Brain* 90:545–564.

Steinbusch, H.W.M., 1984, Serotonin-immunoreactive neurons and their projections in the CNS, *in* "Handbook of Chemical Neuroanatomy, Classical Transmitters and Transmitter Receptors in CNS," Part II, Volume 3 (A Bjorklund, T. Hokfelt, and M.J. Kuhar, eds.), pp. 68–125, Elsevier, Amsterdam.

Stromberg, I., Herrera-Marschitz, M., Ungerstedt, U., Ebendal, T., and Olson, L., 1985, Chronic implants of chromaffin tissue into the dopamine-denervated striatum: Effects of NGF on graft survival fiber growth and rotational behavior, *Exp. Brain Res.* 60:335–349.

Susel, Z., Engber, T.M., Kuo, S., and Chase, T.N., 1991, Prolonged infusion of quinolinic acid into rat striatum as an excitotoxic model of neurodegenerative disease, *Neurosci. Lett.* 121:234–238.

Van Putten, T., and Menkes, J.H., 1973, Huntington's disease masquerading as chronic schizophrenia, *Dis. Nerv. Syst.* 34:54—56.

Vincent, S.R., Hokfelt, T., Christensson, I., and Terenius, L., 1982, Immunohistochemical evidence for a dynorphin immunoreactive striatonigral pathway, *Eur. J. Pharmacol.* 85:251–252.

Vonsattel, J-P., Myers, R.H., and Stevens, T.J., 1985, Neuropathological classification of Huntington's disease, *J. Neuropathol. Exp. Neurol.* 44:559–577.

Waeber, C., and Palacios, J.M., 1989, Serotonin-1 receptor binding sites in the human basal ganglia are decreased in Huntington's chorea but not in Parkinson's disease: A quantitative *in vitro* autoradiography study, *Neuroscience* 32(2):337–347.

Walker, P.D., and McAllister, J.P., 1987, Minimal connectivity between neostriatal transplants and the host brain, *Brain Res.* 425:34–44.

Walker, P.D., Chovanes, G.I., and McAllister, J.P., 1987, Identification of acetyl cholinesterase-reactive neurons and neuropil in neostriatal transplants, *J. Comp. Neurol.* 259:1–12.

Walsh, J.P., Zhou, F.C., Hull, C.D., Fisher, R.S., Levine, M.S., and Buchwald, N.A., 1988, Physiological and morphological characterization of striatal neurons transplanted into adult rat striatum, *Synapse* 2:37–44.

Wastek, G.J., Stern, L.Z., Johnson, O.C., and Yamamura, H.L., 1976, Huntington's disease: Regional alterations in muscarinic cholinergic receptor binding in human brain, *Life Sci.* 19:1033–1040.

Wastek, G.J., and Yamamura, H.I., 1978, Biochemical characterization of the muscarinic cholinergic receptor in human brain: Alterations in Huntington's disease, *Mol. Pharmacol.* 14:768–774.

Whetsell, W.O., and Schwarcz, R., 1989, Prolonged exposure to submicromolar concentration of quinolinic acid causes excitotoxic damage in organotypic cultures of rat corticostriatal system, *Neurosci. Lett.* 97:271–275.

Wictorin, K., and Bjorklund, A., 1989, Connectivity of striatal grafts implanted into the ibotenic acid-lesioned striatum. II: Cortical afferents, *Neuroscience* 30:297–311.

Wictorin, K., Isacson, O., Fischer, W., Nothias, F., Peschanski, M., and Bjorklund, A., 1988a, Studies on host afferent inputs to fetal striatal transplants in excitotoxically lesioned striatum, *in* "Transplantations into the Mammalian CNS: Progress in Brain Research," Volume 78 (D.M. Gash and J.R. Sladek, eds), pp. 55–60, Elsevier, Amsterdam.

Wictorin, K., Isacson, O., Fischer, W., Nothias, F., Peschanski, M., and Bjorklund, A., 1988b, Connectivity of striatal grafts implanted into the ibotenic acid lesioned striatum. I: Subcortical afferents, *Neuroscience* 27:547–562.

Wictorin, K., Quimet, C.C., and Bjorklund, A., 1989a, Intrinsic organization and connectivity of intrastriatal transplants in rats as revealed by DARPP-32 immunohistochemistry: Specificity of connections with the lesioned host brain, *Eur. J. Neurosci.* 1:690–701.

Wictorin, K., Clarke, D.J., Bolam, J.P., and Bjorklund, A., 1989b, Host corticostriatal fibres establish synaptic connections with grafted striatal neurons in the ibotenic acid lesioned striatum, *Eur. J. Neurosci.* 1:189–195.

Wictorin, K., Simerly, R.B., Isacson, O., Swanson, L.W., and Bjorklund, A., 1989c, Connectivity of striatal grafts implanted into the ibotenic acid lesioned striatum. III: Efferent projecting graft neurons and their relation to host afferents within the grafts, *Neuroscience* 30:313–330.

Wictorin, K., Lagenaur, C.F., Lund, R.D., and Bjorklund, A., 1990, Efferent projections to the host brain from intrastriatal striatal mouse-to-rat grafts: Time-course and tissue-type specificity as revealed by a mouse-specific neuronal marker, *Eur. J. Neurosci.* 3:86–101.

Wilson, C.J., Emson, P., and Feles, C., 1987, Electrophysiological evidence for the formation of corticostriatal pathway in neostriatal tissue grafts, *Soc. Neurosci. Abstr.* 13:79.

Wong, E.H.F., Kemp, J.A., Priestly, T., Knight, A.R., Woodruff, G.N., and Iversen, L.L., 1986, The anticonvulsant MK-801 is a potent N-methyl-D-aspartated antagonist, *Proc. Natl. Acad. Sci. U.S.A.* 83:7104–7108.

Wong, P.T.-H., Singh, U.K., and McGeer, E.G., 1982, Ornithine aminotransferase in Huntington's disease, *Brain Res.* 231:466–471.

Xu, Z.C., Wilson, C.J., and Emson, P.C., 1991a, Restoration of thalamostriatal projections in rat neostriatal grafts: An electron microscopic analysis, *J. Comp. Neurol.* 303:22–34.

Xu, Z.C., Wilson, C.J., and Emson, P.C., 1991b, Synaptic potentials evoked in spiny neurons in rat neostriatal grafts by cortical and thalamic stimulation, *J. Neurophysiol.* 65(3):477–493.

Yamada, K., Fuji, K., Nsbeshima, T., and Kameyama, T., 1990, Neurotoxicity induced by continuous infusion of quinolinic acid into the lateral ventricle in rats, *Neurosci. Lett.* 118:128–131.

Yang, H.-Y.T., Panula, P., Tang, J., and Costa, E., 1983, Characterization and localization of Met^5-enkephalin-Arg^6-Phe stored in various rat brain regions, *J. Neurochem.* 40:969–976.

Young, A.B., Shoulson, I., Penny, J.B., Starosta-Rubenstein, S., Gomez, F., Travers, H., Ramos, M., Snodgrass, S.R., Bonilla, A., Moreno, H., and Wexler, N., 1986, Huntington's disease in Venezuela: Neurological features and functional decline, *Neurology* 36:244–249.

Young, A.B., Greenamyre, J.T., Hollingsworth, Z., Albin, R., D'Amato, C., Shoulson, I., and Penny, J.B., 1988, NMDA receptor losses in putamen from patients with Huntington's disease, *Science* 241:981–983.

Zaczek, R., Schwarcz, R., and Coyle, J.T., 1978, Long-term seqeulae of striatal kainate lesion, *Brain Res.* 152:626–632.

Zemanick, M.D., Walker, P.D., and McAllister, J.P., 1987, Quantitative analysis of dendrites from transplanted neostriatal neurons, *Brain Res.* 414:149–152.

Zhou, F.C., and Buchwald, N., 1989, Connectivities of the striatal grafts in adult rat brain: A rich afference and scant striatonigral efference, *Brain Res.* 504:15–30.

Zhou, F.C., Hull, C.D., Levine, M.S., and Buchwald, N.A., 1986, Host nigral fibers may grow into striatal neurons, *Soc. Neurosci. Abstr.* 12:1479.

Zubrycki, E.W., Emerich, D.F., and Sanberg, P.R., 1990, Sex differences in regulatory changes following quinolinic acid-induced striatal lesions, *Brain Res. Bull.* 25:633–637.

9

The Use of AF64A (Ethylcholine Aziridinium Ion) to Model Alzheimer's Disease

THOMAS J. WALSH and KAREN D. OPELLO

1. Neurotoxins as Tools to Model Neurodegenerative Disorders

Natural and synthetic toxins have been valuable tools for exploring the morphological, biochemical, and molecular correlates of brain function. Compounds that disrupt specific ion channels, axoplasmic transport processes, neuronal populations, or neurotransmitter systems have helped to characterize the fundamental neurobiology of brain cells and the principles of synaptic transmission (McGeer *et al.*, 1987). For example, neurotoxins have been used to (1) identify the steps involved in the synthesis, storage, and release of neurotransmitters, (2) determine how a transmitter interacts with specific receptors and ion channels, (3) examine plasticity of a given brain system and the nature and extent of neural reorganization following its injury, (4) examine the covariation between altered neurotransmitter dynamics and behavior, and finally (5) develop animal models of neurological disorders such as Parkinson's disease, Huntington's disease, and temporal lobe epilepsy (Sanberg and Coyle, 1984).

The purpose of this chapter is to discuss how neurotoxins have been used to generate animal models of Alzheimer's disease (AD). We will address (1) the rationale and goals of developing such models, (2) the considerations that guide the development of animal models, and both (3) the successes and (4) failures of current models of AD. We will concentrate on how the cholinotoxin ethylcholine aziridinium ion (AF64A) has been used as a tool to model various aspects of the disease.

THOMAS J. WALSH and KAREN D. OPELLO • Department of Psychology, Rutgers University, New Brunswick, New Jersey 08903.

Toxin-Induced Models of Neurological Disorders, edited by Michael L. Woodruff and Arthur J. Nonneman. Plenum Press, New York, 1994.

2. Alzheimer's Disease

2.1. Characteristics and Demographics

Alzheimer's disease is the most prevalent of the age-related cognitive disorders (Cooper, 1991). Incidence increases with age, and approximately 3%, 14%, and 34% of people in the age brackets 60–69, 70–79, and 80+, are affected (Cooper, 1991). Prevalence varies across geographical regions, reflecting regional differences in diagnostic criteria, medical care, and demographic structure (Amaducci and Lippi, 1990). In the United States, the disease affects approximately 15% of those people over the age of 65, the most rapidly expanding segment of the population (Evans *et al.*, 1989). The etiology of the disorder is unknown and a variety of factors including advanced age, genetic predispositions, exposure to environmental toxins, neuroactive viruses, and prior brain injury might contribute to the onset of the disease.

The AD brain exhibits distinct morphological changes that are region-specific. While there is an extensive loss of neurons in the association cortices and the hippocampus (HPC) there is a relative preservation of sensory and motor areas of cortex. The cellular changes that are a hallmark of the disease—neurofibrillary tangles, granulovacular degeneration, and amyloid placques—are also most concentrated in the HPC, association cortex, and several other areas of the limbic system. One of the earliest alterations involves the anatomical isolation of the HPC from its cortical input due to loss of neurons in the entorhinal cortex and a decrease in the output of the HPC to cortical and subcortical areas owing to loss of neurons in the subiculum (Hyman *et al.*, 1984). There is also loss of cholinergic neurons in the basal forebrain, which provide the primary cholinergic input to the cortex and the HPC (Whitehouse *et al.*, 1985). Other brain areas that can be affected include noradrenergic neurons in the locus ceruleus and the serotonergic neurons in the raphe (see Hardy *et al.*, 1985). Although these changes are evident in subpopulations of patients or late in the course of the disease, they have not been shown to relate to the prevailing memory disorder (Mann *et al.*, 1981).

2.2. The Cholinergic System in Alzheimer's Disease

A major focus for the study of AD has been the cholinergic system (Bartus *et al.*, 1982). The initial neurochemical studies of AD revealed extensive loss of choline acetyltransferase (ChAT) activity in the cortex and HPC (Davies and Maloney, 1976). ChAT is an enzyme that synthesizes acetylcholine (ACh) from acetyl-CoA and choline. Subsequent studies have shown that all biochemical markers of presynaptic cholinergic function including ACh synthesis, high-affinity choline transport (HAChT), and ACh release are decreased in the HPC and neocortex in AD (Coyle *et al.*, 1983; Rylett *et al.*, 1983). There is also a pronounced loss of cholinergic neurons in the basal forebrain, which innervate these regions (Whitehouse *et al.*, 1985). The functional relevance of these alterations is supported by the significant correlation between the severity of dementia, the number of senile plaques, and the loss of cholinergic markers (Perry *et al.*, 1978; Wilcock *et al.*, 1982). Furthermore, a wealth of evidence supports a role for cholinergic systems in learning and memory processes. For example, there are alterations in the activity of basal forebrain cholinergic neurons associated with learning, and both pharmacological antagonism of the cholinergic system and damage to cholinergic neurons produce memory impairments (reviewed in Walsh and Chrobak, 1991).

Cholinergic neurons in the medial septum and vertical limb of the diagonal band innervate the entire septotemporal extent of the HPC and parahippocampal gyrus (see Frotscher and Leranth, 1985, 1986). Cholinergic fibers synapse on pyramidal cells in CA1-4, granule cells in the dentate gyrus, and interneurons in the hilus. While cholinergic synapses might account for only 10% of all the synapses in the HPC, the cholinergic component of the septohippocampal pathway exerts a profound influence on this structure by inducing the theta rhythm (see Bland, 1986, for review). Lynch and colleagues have hypothesized that "neuronal activity at this frequency (theta) is optimal for the induction of synaptic plasticity in the HPC" (Larson *et al.*, 1986). Stimulation of the Schaeffer collateral projections to CA1 at theta frequency (5 Hz) produces an optimal condition for the induction of long-term potentiation (Larson *et al.*, 1986; Greenstein *et al.*, 1988). The cholinergic innervation of the HPC modulates the neural activity of this structure and in particular those neural mechanisms that might subserve memory processes. Furthermore, damage to the septohippocampal pathway should compromise those memory processes that depend on the integrity of the HPC.

2.3. Memory Impairments in Alzheimer's Disease

Patients with AD exhibit a complex spectrum of cognitive alterations that prominently affect memory and language. The most debilitating deficit early in the disease is a loss of memory for recent events (working/episodic memory). This is followed by a progressive deterioration of cognitive function, which is accompanied by additional behavioral changes including emotional volatility, aphasia, and social withdrawal (Terry and Katzman, 1983).

The profile of cognitive changes observed in AD has provided indirect support for the hypothesis that at least two memory processes are responsible for certain types of cognitive operations (Cohen and Squire, 1980; Nissen *et al.*, 1987). One process, referred to as *reference memory,* seems to support the performance of skills or habits, a class of operations dependent upon rules or procedures. The other, referred to as *working memory,* allows the maintenance of representations of previous events or episodes and is governed by declarative information. Based upon the human and animal literature, Sherry and Schacter (1987, p. 446) have suggested a "distinction between a memory system that supports gradual or incremental learning and is involved in the acquisition of habits and skills and a system that supports one trial learning and is necessary for forming memories that represent specific situations and episodes." These authors further suggest that the functions of these memory systems might be incompatible and thus must depend upon unique neural representations.

Early in the course of AD there is a profile of select cognitive deficits in which some types of memory processes are disrupted and others are not. For example, patients are impaired in the acquisition and retention of recently presented information but are not impaired in the acquisition of cognitive and perceptual skills (Cohen and Squire, 1980). This dissociation of memory processes is most evident in those situations where a subject can perform a learned perceptual or motor skill in the absence of any recollection of the specific episodes in which it was acquired (Knopman and Nissen, 1987). Eslinger and Damasio (1986) reported that patients with AD exhibited profound deficits in the recall of a word list and in the performance of a visual delayed match-to-sample task but that their ability to learn and retain the perceptual motor skills necessary to perform a rotary-pursuit task was not impaired.

The cognitive impairments observed in AD patients have been analyzed using Baddeley's Working Memory Model as a theoretical framework (Baddeley, 1986). Baddeley postulates a tripartite model of working memory that includes a Central Executive System (CES), which coordinates and directs two peripheral working memory systems: (1) the Articulatory Loop System, which retains verbal information, and (2) the Visual-Spatial System, which retains nonverbal information. Early in the course of AD there is a selective impairment in the central executive mechanism evidenced by deficits in dual tasks requiring the coordination of concurrent cognitive operations (e.g., retention of verbal materials and tracking of sporadically presented auditory stimuli). These deficits are evident even though the Articulatory Loop System is not affected. While it not possible to relate these impairments to a specific pathological change observed in AD, pharmacological evidence indicates that antagonism of brain cholinergic systems can produce a similar profile of cognitive changes. For example, Nissen and colleagues (1987) reported that the muscarinic antagonist scopolamine selectively disrupted free recall of a word list but did not alter the acquisition and retention of a visual reaction-time task in adult volunteers. Furthermore, there was no disruption of general semantic knowledge (i.e., reference memory) such as the generation of city names in alphabetical order. Scopolamine selectively compromised working memory without disrupting reference memory. Recent work by Rusted and Warburton (1988) further confirmed that scopolamine selectively disrupted working memory processes in young adults. The behavioral changes observed across a number of tasks designed to assess differentially these subcomponents of working memory indicated that scopolamine selectively compromised the activity of the CES. Taken together, these observations suggest that an early cognitive change in AD is an impairment of the CES that coordinates working memory processes and that this change might relate to the loss of cholinergic function.

3. Animal Models of Alzheimer's Disease

An important approach for understanding the neurobiology of memory and its disorders is through the development of animal models (Olton, 1985). While models are unable to reproduce the complex multifactorial etiology and symptomatology of neurodegenerative diseases they do provide a viable way to explore specific questions concerning structure and function (Kornetsky, 1977). A major effort of our laboratory has been to develop a model of cholinergic hypofunction that shares some of the cardinal features of AD (Walsh and Chrobak, 1991). Therefore, it is useful to introduce this topic with a consideration of the factors that contribute to the evolution of such a model.

The development of a viable animal model of any neurological or psychiatric disorder must be predicated upon a set of preestablished criteria. In one of the first attempts to formulate meaningful criteria for animal models, McKinney and Bunney (1969) proposed that models should recapitulate the disease in four dimensions: etiology, symptomatology, treatment efficacy, and underlying pathophysiology. A consideration of each of these factors contributes to the understanding of the utility of the model and helps to bridge the gap between the model and the human situation. However, this strategy is too exacting for models of disorders where etiology is obscure and probably a combination of factors. A variety of interacting influences might lead to a physiological condition that is conducive to the development of brain changes associated with neurodegenerative diseases. Rather than looking to specific etiological factors that predispose to these disorders it might be more useful to examine changes in the brain that induce a similar profile of symptoms (see Walsh and Opello, 1992).

Willner (1991) suggests that animal models should be evaluated in relation to their face validity, predictive validity, and construct validity. *Face validity* refers to the phenomenological similarity between the model and the disease state. Are some of the essential and characteristic behavioral and neurobiological symptoms of the disorder reproduced with acceptable fidelity in the model? Alzheimer's disease is associated with the degeneration of cholinergic neurons in the basal forebrain that innervate structures involved in the coordination of memory processes, most notably the HPC and cortex. The functional consequence is a loss of presynaptic cholinergic indices and the impairment of memory processes dependent upon these brain systems. To have face validity a model of AD must produce a persistent cholinergic hypofunction and associated memory impairments.

A second consideration is *predictive validity,* which is based upon the correspondence between efficacy of therapeutic strategies that are effective in the model and the disease state. This is particularly problematic for models of AD owing to the lack of established treatment regimens (i.e., positive controls) that consistently improve symptoms in patients with the disease. In fact, a conspicuous disparity exists between the effectiveness of drugs in animal models of AD and in patients with the disease. In an extensive review of the literature, Sarter and colleagues (1992a,b) have illustrated that, whereas various drugs (cholinergic agonists, cholinesterase inhibitors, nootropics) are effective in reversing cognitive deficits in currently available models of AD, there is a total lack of correspondence between these effects and their clinical efficacy. While they are efficacious in several animal models they have not provided significant benefit in patients with AD. These authors argue that better strategies are needed to evaluate the behavioral properties of cognitive enhancers and their potential usefulness in the treatment of cognitive disorders.

The final criteria to consider is the *construct validity* of an animal model. What are the theoretical considerations and the rationale for the development of a specific animal model? Alzheimer's disease is associated with changes in brain systems, neurotransmitters, metabolic processes, cytoskeletal integrity, protein processing, and gene expression. Which of these changes, alone or in combination, should serve as a focus for the development of a model of the disease? Which pathological event most contributes to the behavioral pathology? Because the etiology of AD remains obscure, and in fact might be combinatorial in nature, an animal model must focus on relevant behavioral and neurobiological target symptoms. Alzheimer's disease is characterized by the degeneration of cholinergic neurons and a specific constellation of memory impairments.

4. Cholinergic Neurotoxins

Compelling evidence suggests that the memory impairments in AD are dependent upon the degree of cholinergic hypofunction. Therefore, a logical strategy for developing an animal model of AD might be to focus on the "target" symptoms of working memory deficits and cholinergic hypofunction. An ideal model of AD would exhibit "the same biochemical, behavioral, and histopathologic abnormalities as the human state" (Fisher and Hanin, 1986, p. 163). However, the development of a reliable model of AD has proved to be challenging since the etiology of the disease is obscure and is likely to involve the complex interplay of a variety of factors. Therefore, while it is possible to induce some of the behavioral and neurobiological alterations associated with AD, the proximal cause of these changes is probably not analogous to the multitude of etiological factors contributing to the human condition. Currently, no one model mimics all the major characteristics of the human disease state.

Numerous laboratories have attempted to develop animal models of AD based upon the selective pharmacological compromise of brain cholinergic systems. A variety of agents can temporarily modify the synthesis or release of ACh, or the interaction of ACh with either muscarinic or nicotinic receptors. These pharmacological manipulations alter performance in a variety of appetitive and aversive tasks and provide support for the involvement of the cholinergic system in cognitive function (see Walsh and Chrobak, 1991, for review). However, a persistent compromise of the cholinergic system has been difficult to produce owing to the rapid synthesis and turnover of ACh in the brain (Hanin *et al.*, 1987). This property renders cholinergic systems resistent to lasting insult. Numerous problems are inherent in the pharmacological analysis of memory processes because anticholinergic drugs (1) have limited central nervous system specificity, (2) affect multiple muscarinic receptor types, (3) exhibit a high degree of toxicity, (4) have a short duration of action, and (5) indiscriminately affect all central cholinergic systems. Furthermore, the action that a given muscarinic receptor will have depends on its cellular location (presynaptic or postsynaptic), its second messenger system, and the ion channel linked to that system. Therefore, cholinergic antagonists lack the specificity and the time-course to reproduce the chronic cholinergic hypofunction observed in AD.

4.1. Ethylcholine Aziridinium Ion (AF64A)

The terminal of the cholinergic neuron contains a high-affinity choline transport (HAChT) system that acquires extracellular choline and serves as the rate-limiting step in ACh synthesis (Jope, 1979). This system is specific to cholinergic terminals and is functionally coupled to the activity of the neuron. Treatments that either increase or decrease the activity of cholinergic neurons produce correlative changes in HAChT (Simon *et al.*, 1976). A low-affinity choline transport system is widely distributed and contributes to the synthesis of membrane phospholipids in all cell types.

The unique presence of the HAChT system on cholinergic terminals has prompted investigators to develop compounds that selectively target this site. Most notably, various analogs of choline have been examined as potential cholinergic-specific neurotoxins that interact with the HAChT system (Hanin *et al.*, 1987; Rylett and Colhoun, 1980; Tagari *et al.*, 1986).

Fisher and Hanin (1980) explored a series of choline analogs and found that one of these compounds, ethylcholine aziridinium ion (AF64A), a nitrogen mustard analog of choline that contains a highly reactive aziridinium ring, has

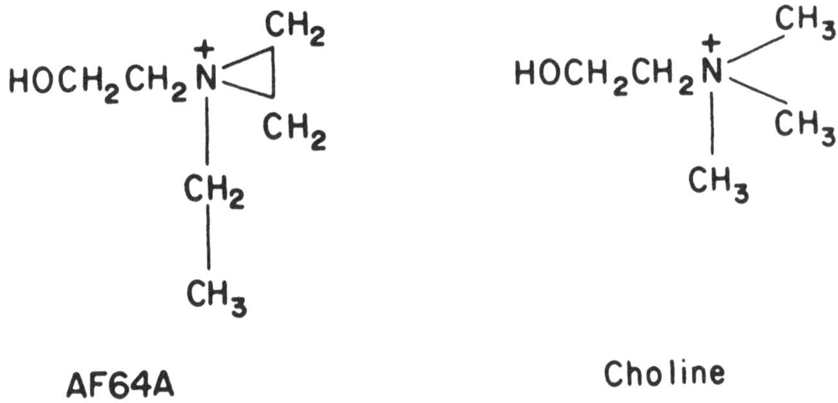

Figure 1 Chemical structures of AF64A (ethylcholine aziridinium ion) and choline.

cholinotoxic properties that make it useful for both *in vivo* and *in vitro* studies (Figure 1).

Most studies have examined the neurobiological and behavioral effects induced by intracerebroventricular (icv) injection of AF64A. With this route of administration AF64A produces a 30% to 60% reduction in all indices of presynaptic cholinergic function including (1) regional concentrations of ACh, (2) the activity of ChAT, (3) HAChT, (4) K^+-stimulated release of ACh from slices of HPC, and (5) intra-axonal transport of acetylcholinesterase (see Hanin, 1990, for review of AF64A's neurochemical effects and for relevant references). These effects develop over time with maximal decreases evident approximately 7 days following treatment and persisting up to 1 year following injection (Leventer *et al.*, 1987). Despite the decrease in presynaptic parameters there are no changes in the number of muscarinic or nicotinic binding sites following AF64A treatment (Vickroy *et al.*, 1985). The AF64A-induced decreases in cholinergic function are not accompanied by persistent changes in the regional concentrations of catecholamines, serotonin, or glutamate. While AF64A does not produce lasting changes in other transmitter systems it can produce temporally limited changes that might be secondary to a disruption of normal cholinergic function. These changes parallel the decrease in cholinergic markers, and they can occur at loci that are distant from the injection site (Hortnagl *et al.*, 1987a,b). In addition, it has been shown that administration of hemicholinium-3 or its analog A-4 prior to AF64A attenuates the loss of cholinergic and serotonergic markers in the HPC (Potter *et al.*, 1987). These compounds are both potent inhibitors of HAChT with inhibitory rate constants 30–100 times greater than that of AF64A and they prevent AF64A from gaining access to the HAChT system (Barker and Mittag, 1975). These data indicate that the effects of AF64A on neurotransmitter systems other than ACh are probably secondary to the cholinergic hypofunction.

Following intraventricular injection, AF64A produces a regionally specific compromise of cholinergic function in the HPC and the septohippocampal pathway. This conspicuous regional specificity could reflect the preferential distribution of the compound to structures adjacent to the ventricles and/or to the unique vulnerability of hippocampal cholinergic nerve terminals. For example, metabolic rates might vary for different cholinergic systems and render them more or less vulnerable to a decrease in choline availability. This important issue awaits resolution.

4.2. Mechanisms of Toxicity

The effects of AF64A on cholinergic neurons might be conceptually divided into three distinct phases involving (1) interaction with the HAChT system and accumulation in the nerve terminal, (2) inhibition of choline-specific enzymes and a disruption of cytoskeletal processes, and (3) damage to DNA resulting in cell death. The temporal dynamics of these stages and their interdependence are not yet understood and could involve both sequential and parallel processes.

The first stage in the toxicity of AF64A involves the compound being taken into cholinergic terminals by the HAChT system owing to its structural similarity to choline. Depending upon its local concentration, AF64A can irreversibly inhibit HAChT by alkylating nucleophilic sites on this system's carrier protein as it is transported into the cell. The initial cholinotoxic effects of AF64A are dependent upon its interaction with the HAChT system since they can be prevented by compounds that compete for the HAChT system such as high levels of choline or hemicholinium-3 (Barlow and Marchbanks, 1984; Curti and Marchbanks, 1984). More recently, Chrobak and colleagues (1989) reported that pretreatment of rats with hemicholinium-3 prevented AF64A-induced cholinotoxicity as well as the behavioral impairments induced by the toxin.

Once inside the terminal AF64A can be acetylated by ChAT to yield acetyl-AF64A, a putative toxic, false-transmitter. Both AF64A and acetyl-AF64A then initiate a second phase of toxicity in which they inhibit the activity of several choline-specific enzymes including ChAT, choline kinase, choline dehydrogenase, and AChE (Barlow and Marchbanks, 1984; Rylett and Colhoun, 1980; Sandberg *et al.*, 1985). This results in a persistent presynaptic cholinergic hypofunction in which all measures of ACh synthesis and release are affected.

Axoplasmic transport processes are also compromised by AF64A. Kasa and Hanin (1985) reported that AF64A produced a dose-related blockade of axoplasmic transport in both the septohippocampal pathway and in sciatic nerve fibers. These changes suggest a disruption of intra-axonal transport of newly synthesized neurofilaments and other cellular components to the nerve terminal that would contribute to the degeneration of cholinergic axon terminals (Kasa and Hanin, 1985). The relevant evidence suggests that the second phase of toxicity involves a disruption of both the synthetic capacity of cholinergic neurons and of cytoskeletal processes that contribute to intracellular and extracellular transport and cell viability.

Although the effects of AF64A on cholinergic parameters have been extensively studied there has been little effort to understand how AF64A kills cells. AF64A does produce a dose-related decrease in the number of cholinergic neurons in the medial septum, with different cell groups being uniquely vulnerable (Chrobak *et al.*, 1988; Lorens *et al.*, 1991). Lorens and colleagues reported the following order of susceptibility among the different cholinergic cell groups in the septum: intermediate >> dorsal > midline >> ventral. Defining the factors that underlie this regional sensitivity might provide important clues about the compound's mechanism of toxicity. Is there an inherent selective vulnerability or resistance (i.e., pathoclisis) of distinct cholinergic neurons? Understanding this phenomenon might help to delineate the factors regulating degeneration of cholinergic neurons in a variety of neurological disorders.

The final phase of AF64A toxicity (i.e., cell death) might relate to the effects of this compound on nucleic acids and their essential intracellular functions. AF64A is structurally similar to several antitumor agents, such as nitrogen mustard, that produce cytotoxic effects through their interactions with DNA, RNA, and other cellular macromolecules. Futscher and colleagues (1992) recently demonstrated that AF64A does produce dose-dependent DNA strand breaks and the premature termination of RNA transcription. Therefore, the final and terminal phase of toxicity might be due to a disruption of fundamental cellular processes involving the translation and transcription of genes that are necessary for neuronal survival.

4.3. Behavioral Effects

A number of laboratories have examined whether AF64A induces a profile of cognitive changes resembling the memory deficits seen in AD. The central question is whether there is a dissociation of memory following AF64A that is similar to the one observed in AD. Is working memory impaired while reference memory is intact? Various experimental paradigms have been used to explore this problem, but we will concentrate on studies using spatial learning tasks since they share formal similarities with the most common tests of human cognition and can be manipulated to examine the subcomponents of cognitive behavior (Olton, 1983; Walsh and Chrobak, 1987).

AF64A Effects on Working and Reference Memory

A consistent observation is that AF64A impairs performance in a variety of spatial tasks such as the radial arm maze (RAM). A problem that arises in the interpretation of these data is that rats can often adopt response strategies or algorithms that are compatible with completion of the task and that reflect refer-

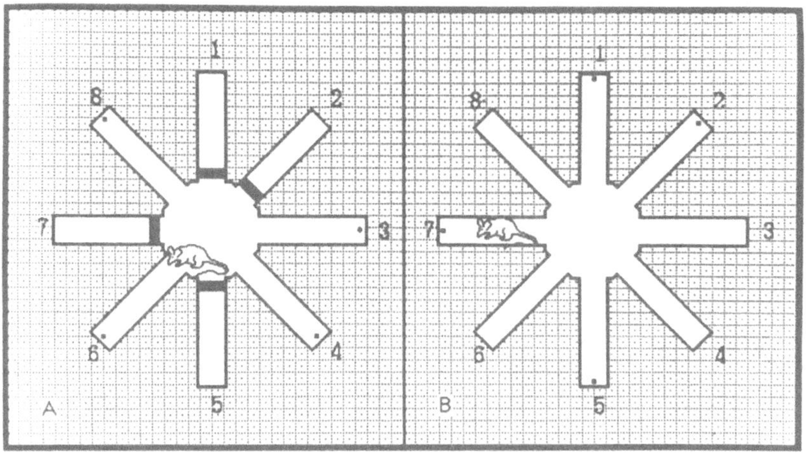

Figure 2 Schematic representation of the delayed nonmatch-to-sample radial-arm maze task. See text for procedural details.

ence memory. A rat might use the same response pattern and enter arms 1,2,3. . . 8 in sequential order without error for each trial. This would clearly minimize or eliminate the working memory aspect of the task. In an effort to minimize this problem we have developed a version of the RAM task that requires an animal to remember a specific training episode to perform the task correctly. In this task, barriers are used to force the animals to select a predetermined set of four arms. Four arms are open and baited with food and four arms are blocked (Figure 2). After entering all four open arms, rats are returned to their home cages for a specified delay period. Following this period they are returned to the maze and all eight arms are now open, but only those arms that were blocked during the predelay sessions now contain food. This represents a delayed-nonmatch to sample (DNMTS) task in which the rat must use a win-shift strategy to complete the task. The pattern of open and closed arms varies randomly from day to day. Therefore, the animals are presented with a new working memory problem each day.

To examine the effects of AF64A on working and reference memory, rats were tested in a series of cognitive tasks (Chrobak *et al.*, 1988). Prior to surgery, rats acquired the DNMTS task and then were bilaterally injected icv with either CSF or 3.0 nmole of AF64A and given a 14-day recovery period before being retested.

Animals were tested for 30 postoperative trials with delays of 0, 0.5, 1, 5, 15, or 30 min randomly imposed between the fourth and fifth arm choices. Animals injected with CSF exhibited excellent performance (80% to 85%) at each of the delay intervals. However, the AF64A-injected group was performing

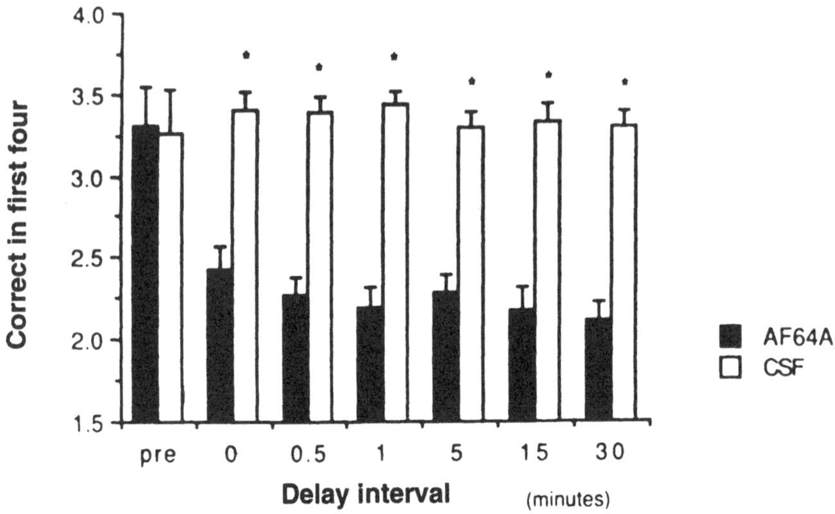

Figure 3 The effects of AF64A on the performance of the delayed-nonmatch-to-sample radial-arm maze task. Delays of 0.0, 0.5, 1, 5, 15, or 30 min were imposed between the fourth and fifth arm choices. The figure presents the mean (± SEM) number of correct choices made in the first four postdelay arm selections.* $p < .001$ vs. CSF controls, Spjotvall-Stoline *t*-test. (From Chrobak *et al.*, 1988; Copyright by Elsevier; used with permission.)

at approximately chance levels (50% to 60%) of accuracy regardless of delay condition (see Figure 3).

Despite the impairments in the DNMTS task the AF64A group was able to acquire a position discrimination in which they learned to perform an invariant trial-independent response of always going to the same spatial location in a T-maze. All animals required a comparable number of trials to achieve a performance criterion of 70%, 80%, or 90% correct in any 10-trial block.

Further, AF64A reduced ChAT activity in the HPC by 50% but did not alter the activity of this enzyme in any other region examined. Moreover, there were no alterations in the regional concentrations of norepinephrine, dopamine, DOPAC, serotonin or 5-HIAA in the HPC, the frontal cortex, the parietal cortex, the cingulate, the amygdala, or the striatum.

A similar dissociation between working and reference memory was also observed in a T-maze task that independently assessed both types of memory within the same trial (Chrobak *et al.*, 1986). In this task each trial was initiated by placing the rat into either the left or the right start/goal box of the T-maze for 15 sec. The food-deprived rat was then rewarded with a single chocolate-flavored pellet for turning into and traversing the stem of the maze and for subsequently returning to the alternate start/goal location. This task involves (1) a trial-

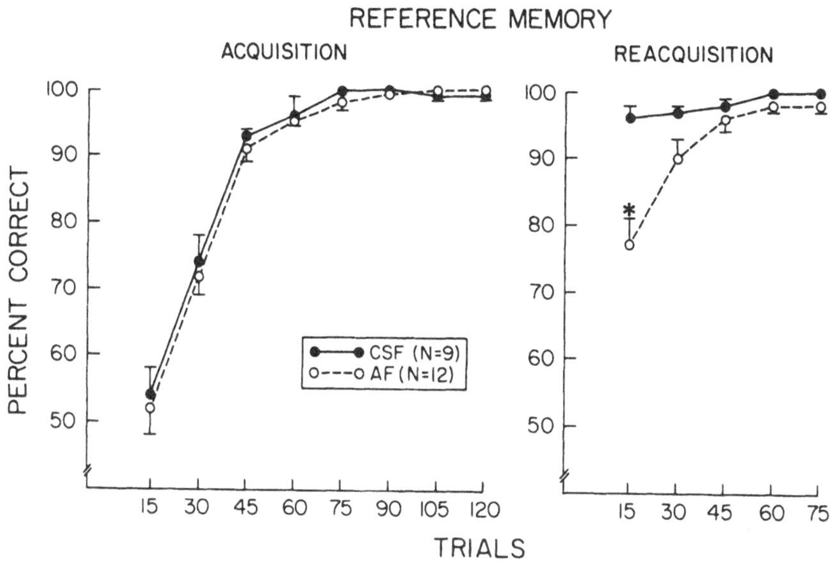

Figure 4 The effects of AF64A on the reacquisition of the reference memory component of the T-maze task. Following 120 acquisition trials, rats were injected with AF64A (3 nmole/side, icv) or CSF, allowed 14 days to recover from surgery, and were then given 75 reacquisition trials. This figure presents mean (± SEM) percentage of correct responses summarized over each 3-day block of sessions in which there were 5 trials per day. *$p < .01$ vs. CSF controls, Spjotvall-Stoline *t*-test. (From Chrobak *et al.*, 1987; Copyright by Elsevier; used with permission.)

independent component (reference memory) involving an invariant response of running down the stem of the maze and (2) a trial-dependent response (working memory) in which the animal must maintain a representation of the start location in order to perform accurately (Chrobak *et al.*, 1986)

Rats who were injected icv with CSF readily reacquired both components of the task. The AF64A group, however, exhibited a transient decrease in reference memory that recovered within 30 trials and a persistent decrease in working memory that was evident throughout the 75 postoperative trials (Figure 4). Working memory was impaired in this group even when their reference memory performance was comparable to controls and close to 100% accurate (Figure 5). The AF64A-treated rats navigated through the maze, consumed the pellets after a correct response, completed each trial in the same amount of time as the CSF-injected controls, and were able to perform the appropriate response to complete the reference memory component of the task. These behavioral deficits were associated with a 42% decrease in ChAT activity in the HPC with no alterations in striatum or frontal cortex.

Subsequent studies found that the working memory impairments induced by

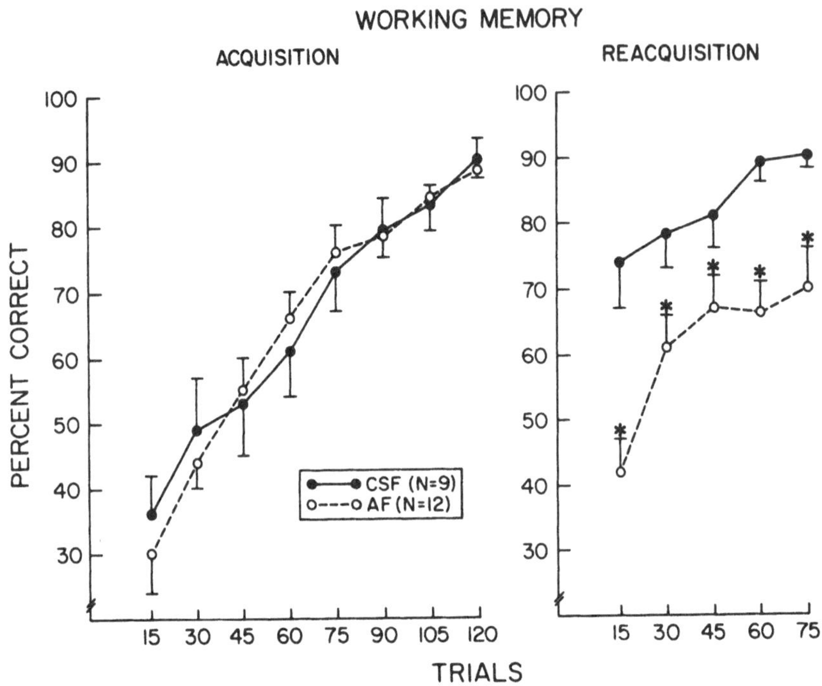

Figure 5 The effects of AF64A on the reacquisition of the working memory component of the T-maze task described in Figure 1. Following 120 acquisition trials, rats were injected with AF64A (3 nmole/side, icv) or CSF, allowed 14 days to recover from surgery, and were then given 75 reacquisition trials. This figure presents mean (± SEM) percentage of correct responses summarized over each 3-day block of sessions in which there were 5 trials per day. $*p < .01$ vs. CSF controls, Spjotvall-Stoline *t*-test. (From Chrobak *et al.*, 1987; Copyright by Elsevier; used with permission.)

AF64A were dose-dependent and related to the cognitive demands of the task (Chrobak and Walsh, 1991). Rats were bilaterally injected icv with 0, 0.75, or 1.5 nmole AF64A and tested with either 0-, 1-, or 2-hr delays in the DNMTS task. While none of the groups were impaired at the 0 delay condition, the 1.5-nmole group was impaired at both the 1- and the 2-hr delay. However, the 0.75-nmole group was only impaired at the 2-hr delay (Figure 6). Therefore, the memory impairments were related to the dose of AF64A *and* to the demand placed upon the working memory system by the task. This is particularly interesting due to the delay-dependent memory impairments observed in patients with AD (see Hart *et al.*, 1988).

The working memory deficits were also found to depend upon the cholinergic hypofunction induced by AF64A. Treating rats with hemi-

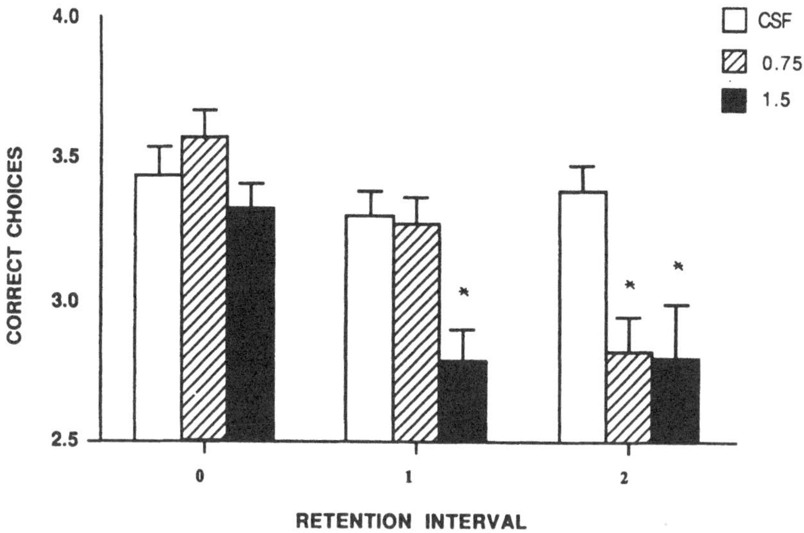

Figure 6 Dose- and delay-dependent effects of AF64A on the delayed nonmatch-to-sample radial-arm maze task. Data are presented as the mean (± SEM) number of correct choices made in the first 4 postdelay-arm entries. $*p < .05$ vs. CSF group at the same delay, Spjotvall-Stoline *t*-test. (From Chrobak and Walsh, 1991; Copyright by Academic Press; used with permission.)

cholinium-3, a more potent short-acting inhibitor of HAChT (Barker and Mittag, 1975), prevented the AF64A-induced deficits in the DNMTS task and the decreases in HAChT (Chrobak *et al.*, 1989). Since hemicholinium-3 prevents AF64A from interacting with the HAChT system it follows that the working memory impairments induced by the AF64A depend upon the select cholinergic hypofunction.

The Morris water-maze (MWM) task is also a commonly used test of spatial learning and memory. This task requires a rat to escape from a pool of water onto a hidden (submerged) platform that remains in the same location across all trials. After a certain number of trials, the platform is removed from the pool and a probe trial is given to determine whether rats concentrate their swimming in the quadrant that previously contained the platform. This serves as an index of retention of specific spatial information, namely location of the target platform.

Gower and colleagues (1989) reported that AF64A produced dose-related impairments in the performance of this task. Rats injected icv with 0.3, 1.0, and 3.0 nmole/side exhibited significantly longer escape latencies even at 6 weeks following surgery. These same investigators found that AF64A had no effect on such reference memory tasks as a light-dark discrimination in a Y-maze. Other reports have found more transient deficits in the water-maze task following AF64A (Brandeis *et al.*, 1986).

A potential problem with these data is that the latency to find the submerged platform was the only index of learning in the task. No probe trials were presented, so the effects of AF64A on retention of specific spatial information (i.e., location of the platform) was not assessed. In an effort to explore this question we used a MWM task to (1) further characterize the cognitive aspects of the task, and (2) to determine whether AF64A produces a dissociation of memory in this task (Opello *et al.*, 1993).

Rats injected icv with CSF quickly learned to swim directly to the platform and spent about 45% of the probe trial searching the target quadrant. Rats bilaterally injected icv with 2.5 nmole AF64A exhibited significantly longer swimpaths and latencies to the platform and failed to show any preference for the target quadrant during the probe trial. However, the AF64A group did exhibit a significant decrease in latency to find the platform over the 40 trials. This appeared to depend on the acquisition of an efficient search strategy (i.e., skill or reference memory). However, AF64A produced a persistent impairment of retention; they spent a random amount of time in the target quadrant during the probe trial. Rats treated with AF64A exhibited a significant decrease in HAChT in the HPC, but not the amygdala, striatum, or cortex. Furthermore, HAChT in the HPC was a highly significant predictor of retention during the probe trial, whereas reference memory measures (i.e., latency) were not related to this variable.

These data further illustrate a dissociation between memory processes following AF64A, and illustrate the need for more qualitative analyses of behavior in spatial tasks to understand the influence of any treatment on cognitive function. Assessing the degree of impairment does not necessarily reveal the type of memory function that is compromised or the compensatory use of surviving memory processes.

5. Summary and Caveats

A central cholinergic hypofunction is believed to underlie the cognitive impairments observed in AD. Alzheimer patients exhibit decreases in presynaptic cholinergic indices and a loss of cholinergic neurons in the basal forebrain, both of which are correlated with the degree of cognitive impairment. These observations have inspired investigators to develop animal models of AD by inducing a chronic central cholinergic hypofunction. Because a disruption of cholinergic afferents to the HPC may be the locus for the cognitive impairments observed in AD (Rossor *et al.*, 1984), AF64A may be used to model this aspect of the disease. In fact, AF64A does produce a profile of neurobiological and behavioral changes in the rat that closely resemble the changes observed in AD; it produces a model with both face validity and construct validity. Whether these changes are homologous or analogous with the pathophysiology manifest in AD requires a more complete understanding of the disease process. At present the AF64A model lacks predictive validity, as do all animal models of AD. There are no "positive control" therapies that work in AD and that should work in the AF64A model. The question of predictive validity remains unanswered. The AF64A model, however, can be used to evaluate the efficacy and assist in the development of treatment modalities that may prove useful in the prevention, remediation, or management of the disorder (Emerich and Walsh, 1990; Emerich *et al.*, 1992). A variety of laboratories have examined whether cholinomimetic drugs, neurotrophic factors, neurotransplants, or anti-oxidants can attenuate the behavioral and/or neurobiological effects of AF64A. In this case predictive validity might depend upon the establishment of effective therapies in this or other models and their subsequent validation in patients with AD.

A final consideration is that AF64A appears to be a selective and powerful tool for examining the relationship between the activity of the cholinergic septohippocampal pathway and cognitive function. However, it is clear that the specificity of AF64A depends upon a multitude of interdependent variables including (1) the purity of the starting compound, (2) the dose and concentration, (3) the rate and total volume of injection, and (4) the site of administration. Injection of AF64A into cerebroventricles seems to be an appropriate way to produce a selective compromise of the cholinergic septohippocampal pathway. However, injection of AF64A into discrete sites can produce extensive nonspecific damage (see McGurk *et al.*, 1987). A very high concentration of AF64A in a restricted area will probably result in the compound being taken up into a variety of noncholinergic cells by the ubiquitous low-affinity choline uptake carrier. The result would be a general cytotoxicity evidenced by nonspecific morphological damage.

The use of AF64A, like any neurotoxin, is appropriate for addressing a restricted set of specific questions. In particular, based on the relevant literature and our own experience we believe that AF64A is an extremely powerful tool to selectively compromise the cholinergic innervation of the HPC following its icv injection in appropriate doses. This can be exploited to (1) explore the biological and behavioral properties of this brain system, (2) examine strategies to limit, attenuate, or reverse the functional consequences of damage to this system, (3) determine the plasticity of this and interacting brain systems following select insult, and finally (4) model the functional deficits that occur following damage to this system in neurological disorders such as Alzheimer's disease.

ACKNOWLEDGMENTS. The authors would like to thank Robert Stackman and Jennifer Buckman for their helpful comments on this manuscript. The AF64A studies from our laboratory were supported by BRSG grants (PHS RR 07058), NSF grant BNS-9222097, a Charles and Johanna Busch Bequest grant, and a gift in memory of Colonel Norman C. Kalmar, to T.J.W.

6. References

Amaducci, L., and Lippi, A., 1990, Diagnosis and care of Alzheimer's disease patients: A European perspective, *in* "Alzheimer's Disease: Advances in Neurology, Vol. 51" (R.J. Wurtman, S. Corkin, J.H. Growdon, and E. Ritter-Walker, eds.), pp. 7–14, Raven Press, New York.

Baddeley, A.D., 1986, "Working Memory," Clarendon Press, Oxford.

Barker, L.A., and Mittag, T.W., 1975, Comparative studies of substrates and inhibitors of choline transport and choline acetyltransferase, *J. Pharmacol. Ex. Ther.* 192:86–94.

Barlow, P., and Marchbanks, R.M., 1984, Effect of ethylcholine mustard on choline dehydrogenase and other enzymes of choline metabolism, *J. Neurochem.* 43:1568–1573.

Bartus, R.T., Dean, R.L., Beer, B., and Lippa, A.S., 1982, The cholinergic hypothesis of geriatric memory dysfunction, *Science* 217:408–417.

Bland, B.H., 1986, The physiology and pharmacology of the hippocampal formation theta rhythms, *Prog. Neurobiol.* 26:1–54.

Brandeis, R., Pittel, Z., Lachman, C., Heldman, E., Luz, S., Dachir, S., Levy, A., Hanin, I., and Fisher, A., 1986, AF64A-induced cholinotoxicity: Behavioral and biochemical correlates, *in* "Alzheimer's and Parkinson's Disease: Strategies for Research and Development" (A. Fisher, I. Hanin, and C. Lachman, eds.), pp. 469–478, Plenum Press, New York.

Chrobak, J.J., and Walsh, T.J., 1991, Dose and delay dependent working/episodic memory impairments following intraventricular administration of ethylcholine aziridinium ion (AF64A, *Behav. Neural Biol.* 56:200–212.

Chrobak, J.J., Hanin, I., and Walsh, T.J., 1986, AF64A (ethylcholine mustard aziridinium ion), a cholinergic neurotoxin, selectively impairs working memory in a multiple component T-maze task, *Brain Res.* 414:15–21.

Chrobak, J.J., Hanin, I., Schmechel, D.E., and Walsh, T.J., 1988, AF64A-induced working memory impairment: Behavioral, neurochemical and histological correlates, *Brain Res.* 463:107–117.

Chrobak, J.J., Spates, M.J., Stackman, R.W., and Walsh, T.J., 1989, Hemicholinium-3 prevents the working memory impairments and the cholinergic hypofunction induced by ethylcholine aziridinium ion (AF64A), *Brain Res.* **504:**269–275.

Cohen, N.J., and Squire, L.R., 1980, Preserved learning and retention of pattern analyzing skill in amnesia: Dissociation of knowing how and knowing that, *Science* 210:207–209.

Cooper, B., 1991, The epidemiology of primary degenerative dementia and related neurological disorders, *Eur. Arch. Psychiatry Clin. Neurosci.* 240:223–233.

Coyle, J.T., Price, D.L., and DeLong, M.R., 1983, Alzheimer's disease: A disorder of cortical cholinergic innervation, *Science* 219:1184–1190.

Curti, D., and Marchbanks, R.M., 1984, Kinetics of irreversible inhibition of choline transport in synaptosomes by ethylcholine mustard aziridinium, *J. Membr. Biol.* 82:259–268.

Davies, P., and Maloney, A.F.J., 1976, Selective loss of cerebral cholinergic neurones in Alzheimer's disease, *Lancet* 2:1403.

Emerich, D.F., and Walsh, T.J., 1990, Ganglioside AGF2 promotes task-dependent behavioral recovery and attenuates the cholinergic hypofunction induced by AF64A, *Brain Res.* 527:299–307.

Emerich, D.F., Black, B., Kesslak, P., Cottman, C., and Walsh, T.J., 1992, Transplantation of fetal cholinergic neurons into hippocampus attenuates the cognitive and neurochemical deficits induced by AF64A, *Brain Res. Bull.* 28:219–226.

Eslinger, P.J., and Damasio, A.R., 1986, Preserved motor learning in Alzheimer's disease: Implications for anatomy and behavior, *J. Neurosci.* 10:3006–3009.

Evans, D.A., Funkenstein, H.H., Albert, M.S., Scherr, P.A., Cook, N.R., Chown, M.J., Herbert, L.E., Hennekens, C.H., and Taylor, J.O., 1989, Prevalence of Alzheimer's disease in a community population of older persons, JAMA 262:2551–2556.

Fisher, A., and Hanin, I., 1980, Minireview: Choline analogs as potential tools in developing selective animal models of central cholinergic hypofunction, *Life Sci* 27:1615–1643.
Fisher, A., and Hanin, I., 1986, Potential animal models for senile dementia of Alzheimer's type, with emphasis on AF64A-induced cholinotoxicity, *Ann. Rev. Pharmacol. Toxicol.* 26:161–181.
Frotscher, M., and Leranth, C., 1985, Cholinergic innervation of the rat hippocampus as revealed by choline acetyltransferase immunocytochemistry: A combined light and electron microscopic study, *J. Comp. Neurol.* 239:237–246.
Frotscher, M., and Leranth, C., 1986, The cholinergic innervation of the rat fascia dentata: Identification of target structures on granule cells by combining choline acetyltransferase immunocytochemistry and golgi impregnation, *J. Comp. Neurol.* 243:58–70.
Futscher, B.W., Pieper, R.O., Barnes, D.M., Hanin, I., and Erickson, L.C., 1992, DNA-damaging and transcription terminating lesions induced by AF64A in vitro, *J. Neurochem.* 58:1504–1509.
Greenstein, Y.J., Pavlides, C., and Winson, J., 1988, Long-term potentiation in the dentate gyrus is preferentially induced at theta rhythm periodicity, *Brain Res.* 438:331–334.
Gower, A.J., Rousseau, D., Jamsin, P., Gobert, J., Hanin, I., and Wulfert, E., 1989, Behavioral and histological effects of low concentrations of intraventricular AF64A, *Eur. J. Pharmacol.* 166:271–281.
Hanin, I., Fisher, A., Hortnagl, H., Leventer, S.M., Potter, P.E., and Walsh, T.J., 1987, Ethylcholine mustard aziridinium (AF64A; ECMA) and other potential cholinergic neuron-specific neurotoxins, *in* "Psychopharmacology: The Third Generation of Progress" (H.Y. Meltzer, ed.), pp. 341–349, Raven Press, New York.
Hardy, J., Adolfsson, R., Alafuzoff, I., Bucht, G., Marcusson, J., Nyberg, P., Perdahl, E., Wester, P., and Winblad, B., 1985, Transmitter deficits in Alzheimer's disease, *Neurochem. Int.* 7:545–563.
Hart, R.P., Kwentus, J.A., Harkins, S.W., and Taylor, J.R., 1988, Rate of forgetting in mild Alzheimers-type dementia, *Brain Cogn.* 7:31–38.
Hortnagl, H., Potter, P.E., and Hanin, I., 1987a, Effect of cholinergic deficit induced by ethylcholine aziridinium on serotonergic parameters in rat brain, *Neuroscience* 22:203–213.
Hortnagl, H., Potter, P.E., and Hanin, I., 1987b, Effect of cholinergic deficit induced by ethylcholine aziridinium (AF64A) on noradrenergic and dopaminergic parameters in rat brain, *Brain Res.* 421:75–84.
Hyman, B.T., Damasio, A.R., Van Hoesen, G.W., and Barnes, C.L., 1984, Alzheimer's disease: Cell-specific pathology isolates the hippocampal formation, *Science* 225:1168–1170.
Jope, R.S., 1979, High-affinity choline transport and acetyl CoA production in brain and their roles in the regulation of acetylcholine synthesis, *Brain Res. Rev.* 180:313–344.
Kasa, P., and Hanin, I., 1985, Ethylcholine mustard aziridinium blocks the axoplasmic transport of acetylcholinesterase in cholinergic nerve fibers of the rat, *Histochemistry* 83:343–345.
Knopman, D.S., and Nissen, M.J., 1987, Implicit learning in patients with probable Alzheimer's disease, *Neurology* 37:784–787.
Kornetsky, C., 1977, Animal models: Promises and problems, in "Animal Models in Psychiatry and Neurology" (I. Hanin and E. Usdin, eds.), pp. 1–7, Pergamon Press, New York.
Larson, J., Wong, D., and Lynch, G., 1986, Pattern stimulation at the theta frequency is optimal for the induction of hippocampal long-term potentiation, *Brain Res.* 363:347–350.
Leventer, S.M., Wulfert, E., and Hanin, I., 1987, Time course of ethylcholine aziridinium ion (AF64A)-induced cholinotoxicity in vivo, *Neuropharmacology* 26:361–365.
Lorens, S.K., Kindel, G., Dong, X.W., Lee, J.M., and Hanin, I., 1991, Septal choline acetyltransferase immunoreactive neurons: Dose-dependent effects of AF64A, *Brain Res. Bull.* 26:965–971.
Mann, J.J., Stanley, M., Neophitides, A., de Leon, M., Ferris, S.H., and Gershon, S., 1981, Central amine metabolism in Alzheimer's disease: In vivo relationship to cognitive deficit, *Neurobiol. Aging* 2:57–60.

McGeer, P.L., Eccles, J.C., and McGeer, E.G., 1987, "Molecular Neurobiology of the Mammalian Brain," Plenum Press, New York.

McGurk, S.R., Hartgraves, S.L., Kelly, P.H., Gordons, M.N., and Butcher, L.L., 1987, Is ethylcholine aziridinium ion a specific cholinergic neurotoxin? *Neuroscience* 22:215–224.

McKinney, W.T., and Bunney, W.E., 1969, Animal model of depression: Review of evidence and implications for research, *Arch. Gen. Psychiatry* 21:240–248.

Nissen, M.J., Knopman, D.S., and Schacter, D.L., 1987, Neurochemical dissociation of memory systems, *Neurology* 37:789–794.

Olton, D.S., 1983, Memory functions and the hippocampus, *in* "Neurobiology of the Hippocampus" (W. Seifert, ed.), pp. 335–373, Academic Press, New York.

Olton, D.S., 1985, Strategies for the development of animal models of human memory impairments, *Ann. N Y Acad. Sci.* 444:113–121.

Opello, K.D., Stackman, R.W., Ackerman, S. and Walsh, T.J., 1993, AF64A (ethylcholine mustard aziridinium) impairs aquisition and performance of a spatial but not a cued water maze task: Relation to cholinergic hypofunction, *Physiol. Behav.* 54:1227–1233.

Perry, E.K., Tomlinson, B.E., Blessed, G., Bergmann, K., Gibson, P., and Perry, R.H., 1978, Correlation of cholinergic abnormalities with senile plaques and mental test scores in senile dementia, *Brit. Med. J.* 2:1457–1459.

Potter, P.E., Tedford, C.E., Kindel, G.H., & Hanin, I., 1987, Inhibition of high affinity choline transport attenuates both cholinergic and noncholinergic effects of ethylcholine aziridinium (AF64A), *Brain Res.* 13:238–244.

Rossor, M.N., Iversen, L.L., Reynolds, G.P., Mountjoy, C.Q., and Roth, M., 1984, Neurochemical characteristics of early and late onset types of Alzheimer's disease, *Brit. Med. J.* 283:961–964.

Rusted, J.M., and Warburton, D.M., 1988, The effects of scopolamine on working memory in healthy young volunteers, *Psychopharmacology* 96:145–152.

Rylett, R.J., and Colhoun, E.H., 1980, Kinetic data on the inhibition of high affinity choline transport into rat forebrain synaptosomes by choline-like compounds and nitrogen mustard analogues, *J. Neurochem.* 34:713–719.

Rylett, R.J., Ball, M.J., and Colhoun, E.H., 1983, Evidence for high affinity choline transport in synaptosomes prepared from hippocampus and neocortex of patients with Alzheimer's disease, *Brain Res.* 289:169–175.

Sanberg, P.R., and Coyle, J.T., 1984, Scientific approaches to Huntington's disease, *CRC Crit. Rev. Neurobiol.* 1:1–44.

Sandberg, K., Schnaar, R.L., McKinney, M., Hanin, I., Fisher, A., and Coyle, J.T., 1985, AF64A: An active site-directed irreversible inhibitor of choline acetyltransferase, *J. Neurochem.* 44:439–445.

Sarter, M., Hagan, J., and Dudchenko, P., 1992a, Behavioral screening for cognition enhancers: From indiscriminate to valid testing: Part I, *Psychopharmacology* 107:144–150.

Sarter, M., Hagan, J., and Dudchenko, P., 1992b, Behavioral screening for cognition enhancers: From indiscriminate to valid testing: Part II, *Psychopharmacology* 107:461–473.

Sherry, D.F., and Schacter, D.L., 1987, The evolution of multiple memory systems, *Psychol. Rev.* 98:439–454.

Simon, J.R., Atweh, S., and Kuhar, M.J., 1976, Sodium-dependent high affinity choline uptake: A regulatory step in the synthesis of acetylcholine, *J. Neurochem.* 26:909–922.

Tagari, P.C., Maysinger, D., and Cuello, A.C., 1986, Hemicholinium mustard derivatives: Preliminary assessment of cholinergic neurotoxicity, *Neurochem. Res.* 11:1091–1102.

Terry, R.D., and Katzman, R., 1983, Senile dementia of Alzheimer's disease. *Ann. Neurol.* 14:497–506.

Vickroy, T.W., Watson, M., Leventer, S.M., Roeske, W.R., Hanin, I., and Yamamura, H.I., 1985, Regional differences in ethylcholine mustard aziridinium ion (AF64A)-induced deficits in presynaptic cholinergic markers for the rat central nervous system, *J. Pharmacol. Exp. Ther.* 235:577–582.

Walsh, T.J., and Chrobak, J.J., 1987, The use of the radial arm maze in neurotoxicology, *Physiol. Behav.* 40:799–803.

Walsh, T.J., and Chrobak, J.J., 1991, Animal models of Alzheimer's disease: Role of hippocampal cholinergic systems in working memory, *in* "Current Topics in Animal Learning: Brain, Emotion, and Cognition" (L. Dachowski and C. Flaherty, eds.), pp 347–379, Lawrence Erlbaum, Hillsdale, NJ.

Walsh, T.J., and Opello, K.D., 1992, Neuroplasticity, the aging brain, and Alzheimer's disease, *Neurotoxicology* 13:101–110.

Whitehouse, P.J., Struble, R.G., Hedreen, J.C., Clark, A.W., and Price, D.L., 1985, Alzheimer's disease and related dementias: Selective involvement of specific neuronal systems, *CRC Crit. Rev. Clin. Neurobiol.* 1:319–339.

Wilcock, G.K., Esiri, M.M., Bowen, D.M., and Smith, C.C.T., 1982, Alzheimer's disease: Correlation of cortical choline acetyltransferase activity with the severity of dementia and histological abnormalities, *J. Neurol. Sci.* 57:407–417.

Willner, P., 1991, Behavioural models in psychopharmacology, *in* "Behavioural Models in Psychopharmacology" (P. Willner, ed.), pp. 3–18, Cambridge University Press, Cambridge.

10

Perinatal Intake of Polychlorinated Biphenyl (PCB)

Alteration of Endocrine and Neurochemical Parameters Related to Alzheimer's Disease

LAURA M. JUÁREZ DE KU and LEE A. MESERVE

1. Introduction

1.1. Polychlorinated Biphenyls (PCBs)

Polychlorinated biphenyls (PCBs) are a family of toxic chemicals that have been used commercially for more than 60 years as plasticizers, dielectric fluids, flame retardants, hydraulic and heat transfer fluids, and as coolants and lubricants in electrical equipment (Brandt, 1977). Improper disposal of PCBs by the industries that used them has resulted in environmental contamination and the ubiquitous presence of this material in aquatic environments, both freshwater and marine (Wassermann *et al.*, 1979). Because they are persistent, lipophilic, and become concentrated through food webs, PCBs are present in the tissues of fish, cattle, poultry, wildlife, and humans (Hammond, 1972; Swain, 1983; Jensen, 1989).

The effects of PCBs on humans have been evaluated from accidental consumption by several different populations and by exposure of industrial workers. In the United States, humans have been exposed to PCBs via consumption of fish from the PCB-polluted Great Lakes (Veith *et al.*, 1981; Humphrey, 1983). Thousands of Japanese and Taiwanese suffered from PCB poisoning (*yusho* and *yucheng*, Japanese and Taiwanese words for "oil disease," respectively) when

LAURA M. JUÁREZ DE KU and LEE A. MESERVE • Department of Biological Sciences, Bowling Green State University, Bowling Green, Ohio 43402.

Toxin-Induced Models of Neurological Disorders, edited by Michael L. Woodruff and Arthur J. Nonneman. Plenum Press, New York, 1994.

contaminated rice oil was consumed. Some of the symptoms of this poisoning may have been the result of highly toxic dibenzofurans that were also present in the oil (Yoshimura, 1974; Masuda *et al.*, 1978). As a result of these exposures, human PCB poisoning symptoms have been described as fatigue, nausea, vomiting, skin rashes, alterations in liver enzyme activity, impotence in men, abnormal menstrual cycles in women, and decreased growth rate in children (Kimbrough, 1980; Jacobson *et al.*, 1990). Hypothyroidism (a deficiency of the two primary thyroid hormones, thyroxine [T_4] and triiodothyronine [T_3]) is also thought to be a symptom of PCB poisoning (McKinney and Pedersen, 1987). In contrast, Murai *et al.* (1987) reported hyperthyroxinemia in *yusho* patients 16 years after exposure to the poisoned oil, but the patients did not show clinical signs of hyperthyroidism, and the hyperthyroidism was partially attributed to the presence of dibenzofurans and quaterphenyls in the oil.

Rats administered PCB experimentally suffer symptoms similar to those of accidentally-exposed humans, such as epidermal abnormalities, altered liver enzyme activity, depressed growth rate, reproductive dysfunctions, and hypothyroidism (Brezner *et al.*, 1984; Bruckner *et al.*, 1973; Kunita *et al.*, 1985; Byrne *et al.*, 1987; Collins and Capen, 1980; Meserve *et al.*, 1992; Sharma *et al.*, 1991). The offspring of humans and rats are affected by the passage of PCB through the placenta and milk (Masuda *et al.*, 1978; Takagi *et al.*, 1976).

1.2. Characteristics of Valid Animal Models of Neurological Disorders

An animal model for a neurological disorder must possess characteristics that not only provide the opportunity to investigate the mechanisms involved in the neurological disorder but also to develop an effective treatment and/or better understanding of the mechanisms associated with the disorder (Wenk, 1990). Given this description, it is reasonable to assume that any model system for a neurological disorder will display not only the associated neurological symptoms but also a collection of other peripheral physiological alterations inherent in the disorder since such effects may be interrelated. Additionally, both the utility and the relevance of any model system are enhanced by the facility with which the disorder may be induced experimentally and the possibility of similar conditions resulting in the clinical symptoms.

Among the methods used to induce neurological disorders are manipulation of genetic control mechanisms; administration of neurotoxic chemicals by infusion, injection, or ingestion; and ablation of selected brain areas by surgical or electrolytic lesions. While these methods have been applied to different experimental rodents, the success in developing a model for Alzheimer's disease (AD) is equivocal. The recent excitement regarding genetically engineered mice displaying Alzheimer's neuropathology has faded (Marx, 1992). Institution of physical damage by lesioning compromises the potential for developing treatment mechanisms in the clinical setting. As described in the next section, maternal dietary incorporation of PCB results in birth and development of young rats that display several of the deficits of AD, making them a model system in which development of the pathology and possible peripheral treatment strategies can be investigated. An additional attraction of this model is the widespread human exposure to environmental PCB, making its clinical relevance an intriguing possibility.

1.3. PCB-Exposed Rat as Animal Model for Alzheimer's Disease

There are several characteristic neuropathologies in the brains of patients with AD. One of these is a marked depression in the activity of the biosynthetic enzyme *choline acetyltransferase* (ChAT) in two areas of the brain important in learning and memory—the hippocampus and basal forebrain (Davies and Maloney, 1976; White *et al.*, 1977; Araujo *et al.*, 1988; Perry *et al.*, 1978).

Some cases of AD are considered to be the direct result of inheritance (St George Hyslop *et al.*, 1987), but the majority of them are not considered to result from inheritance alone. Thus, the possibility that a virus or environmental toxin could trigger the onset of AD has become an important issue to consider (Perl and Brody, 1980; Wisniewski *et al.*, 1981). Alzheimer's disease is not directly transmissible to other humans or laboratory animals, and neither viral DNA nor viral antigen has been reproducibly found in tissues affected with the disease. There are, however, several examples of exogenous compounds that lead directly to death of specific neurons resulting in neurological disease (Henneberry and Spatz, 1991). A parallel may be drawn with Parkinson's disease for which no environmental correlate had been identified until an accidental experiment of the drug culture resulted in discovery of an exogenous material (methylphenyltetrahydropyridine, MPTP) that causes a Parkinson-like disorder (Davis *et al.*, 1979; Langston *et al.*, 1983). Subsequent studies found a natural constituent of cycad seeds to cause Parkinson's in natives of Guam who had eaten these seeds (Spencer *et al.*, 1987). Additionally, exposure to other environmental toxins early in life has been hypothesized to activate mechanisms resulting in the later development of AD (Spencer *et al.*, 1987; Krishnan *et al.*, 1988; Bjorklund, 1991).

If a connection can be made between environmental exposure and deficits of neural function, then a stronger association between exogenous factors and AD may be made. It has been possible to construct such a parallel by virtue of the hypothyroidism induced by PCB and that induced by other chemical means (e.g., propylthiouracil, PTU). Young rats made hypothyroid with PTU display depressed ChAT activity in the hippocampus and basal forebrain (Patel *et al.*, 1987; Gould and Butcher, 1989). We hypothesized that incorporation of PCB in the maternal diet would, in the young, (1) induce hypothyroidism and (2) cause a depression of ChAT activity. Indeed, if PCB were to induce a depression of ChAT activity selectively in the same brain areas of young rats where such depression occurs in Alzheimer's patients, these animals could serve as a model for the provision of information relevant to Alzheimer's disease. Because PCB is globally present in the environment (Tanabe, 1988), it is likely that a sizable portion of the human population has ingested some PCB.

In the experiments reported in this chapter, circulating thyroid hormone status and ChAT activity were determined in 15-day-old rat pups whose mothers ingested a diet containing 250 ppm PCB, from the first day of pregnancy and through gestation and lactation. Additionally, the efficacy of exogenous thyroid hormone administration as a replacement procedure was investigated.

2. Materials and Methods

2.1. Animals

Female and male Sprague-Dawley rats (*Rattus norvegicus;* Harlan Sprague-Dawley, Indianapolis, IN) were housed together on a 12-hour light/dark cycle and fed standard laboratory diet *ad libitum*. Vaginal smears were performed daily to detect pregnancies. Once smears were sperm-positive, the females were housed individually and fed either standard rat chow (Teklad-Wayne Lab Blox Mash; Continental Grain Co., Chicago) or chow mixed with PCB (Aroclor 1254®, 250 ppm by weight) throughout pregnancy and lactation. An anonymous donor provided the PCB, which was originally produced by the Monsanto Company (St. Louis, MO). At 3 days of age, litters were standardized to 8 pups, with 4 males and 4 females. No differences with respect to sex were found; therefore, analyses did not differentiate between males and females. A minimum of 10 litters was used for each treatment group.

On postnatal day 15, animals were decapitated, trunk blood was collected for later determination of circulating T_4 and T_3 levels, and brain areas were removed for determination of ChAT activity.

To determine whether PCB-induced depression of ChAT was a result of hypothyroidism, some of the pups were injected daily with physiological replacement doses of either T_3 (10 ng/g body wt.) or T_4 (50 ng/g body wt.), or with 0.9% saline vehicle on postnatal days 4–14.

2.2. Assays

Activity of ChAT in the hippocampus and basal forebrain as well as in a third control area (cerebral cortex) was assessed by a method measuring rate of incorporation of a radiolabeled precursor into acetylcholine (Fonnum, 1975). Briefly, the appropriate brain area was homogenized in water to lyse cell membranes and release the enzyme. Then a volume of homogenate equivalent to 40 μg of brain protein (Lowry *et al.*, 1951) was added to assay buffer (300 mM NaCl; 50 mM $NaPO_4$, pH 7.4; 8 mM choline bromide; 20 mM EDTA; 0.1 mM eserine; 0.5% v/v Triton X-100) and incubated at 37°C for 15 min with 2.25 μCi/ml (55.9 mCi/mmole) of [^{14}C]-acetyl CoA (New England Nuclear, Boston, resulting in the production of [^{14}C]-labeled acetylcholine in proportion to ChAT activity. Ice-cold phosphate buffer (pH 7.4, 50 mM) was used to stop the reaction. The [^{14}C]-acetylcholine was extracted with 0.5% tetraphenylboron in acetonitrite, and immiscible phases were generated by mixing with toluene scintillation fluid (PPO, POPOP, toluene). Unreacted labeled acetyl CoA remaining in the aqueous phase produced no scintillation. The radioactivity of each sample was determined for 10 min using a Beckman LS 7000 scintillation counter.

Circulating T_4 and T_3 levels of all animals were determined by radioimmunoassay (RIA) using commercially available kits (ICN Biomedicals, Inc., Carson, CA).

2.3. Statistics

The significance of PCB influence on hormone concentrations and ChAT activity was determined by analysis of variance with multiple comparisons of the means performed using Tukey's test. Statistical significance was ascribed at $p <$ .05.

3. Results

Addition of 250 ppm Aroclor 1254 to the maternal diet depressed circulating T_4 in 15-day-old pups to less than 10% of normal (Figure 1). While mean values for T_3 in PCB-exposed pups were slightly less than those of controls, they were not significantly below normal ($p >$ .05). Activity of ChAT in hippocampus and basal forebrain was markedly depressed by PCB to 63% and 44% of normal, respectively (Figure 2), but was not significantly altered in cerebral cortex. What appears to be a PCB-related increase in mean cerebral cortical ChAT activity is not significant, and is likely the combined result of a small number of measures performed on this tissue, and normal enzyme activity that is considerably less than that in the hippocampus and basal forebrain.

In the thyroid hormone replacement experiments, injection of saline vehicle had no influence on ChAT activity in any brain areas observed (data not shown). Provision of a replacement regime for T_4 by daily injection significantly increased ChAT activity of PCB rats to 86% of normal in hippocampus and 65% of normal in basal forebrain (Figure 3). Injection of T_3 had no ameliorative influence on ChAT activity.

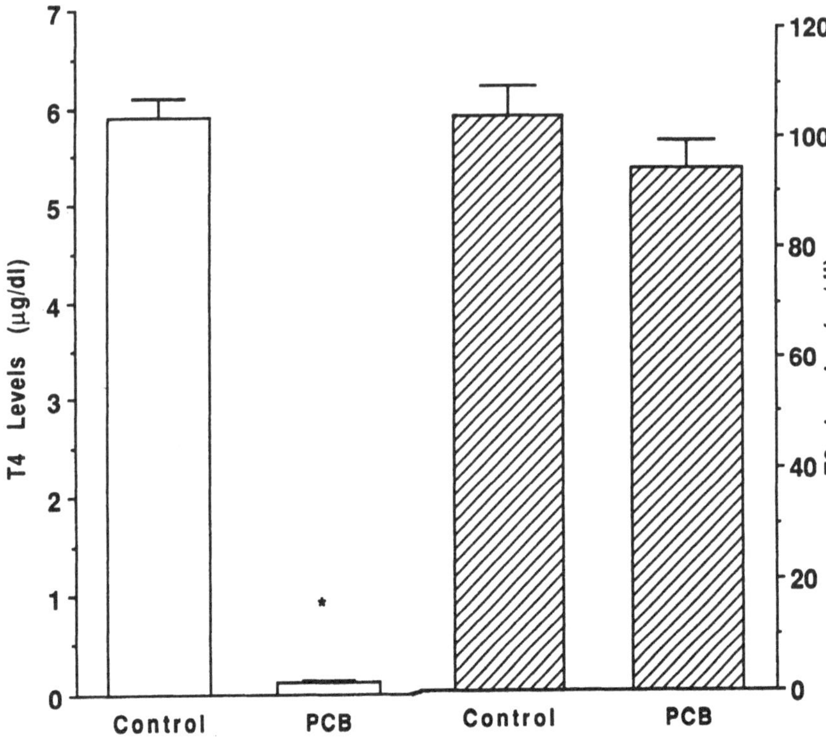

Figure 1 Mean circulating thyroxine (T_4) and triiodothyronine (T_3) levels ± SEM in 15-day-old rats exposed *in utero* and through lactation to 0 (Control) or 250 ppm PCB (PCB). * $p < .05$ vs. Control rats.

4. Discussion

4.1. Use of PCB-Exposed Rats as a Model for Alzheimer's Disease

Results from the present experiments support the use of PCB-exposed rats as animal models for studing neuropathologies involved in AD. Hippocampal and basal forebrain ChAT activity was suppressed in 15-day-old rats exposed *in utero* and during lactation to 250 ppm PCB. The influence of PCB on ChAT activity in these two areas and the absence of influence in the cerebral cortex are indicative of some degree of localization of either the organohalide itself, or at least its effects, at sites where Alzheimer's patients are also known to have depressed ChAT activity (Davies and Maloney, 1976; Perry *et al.*, 1978; White *et al.*, 1977).

At least 20 different paradigms are being used to model AD (Karczmar,

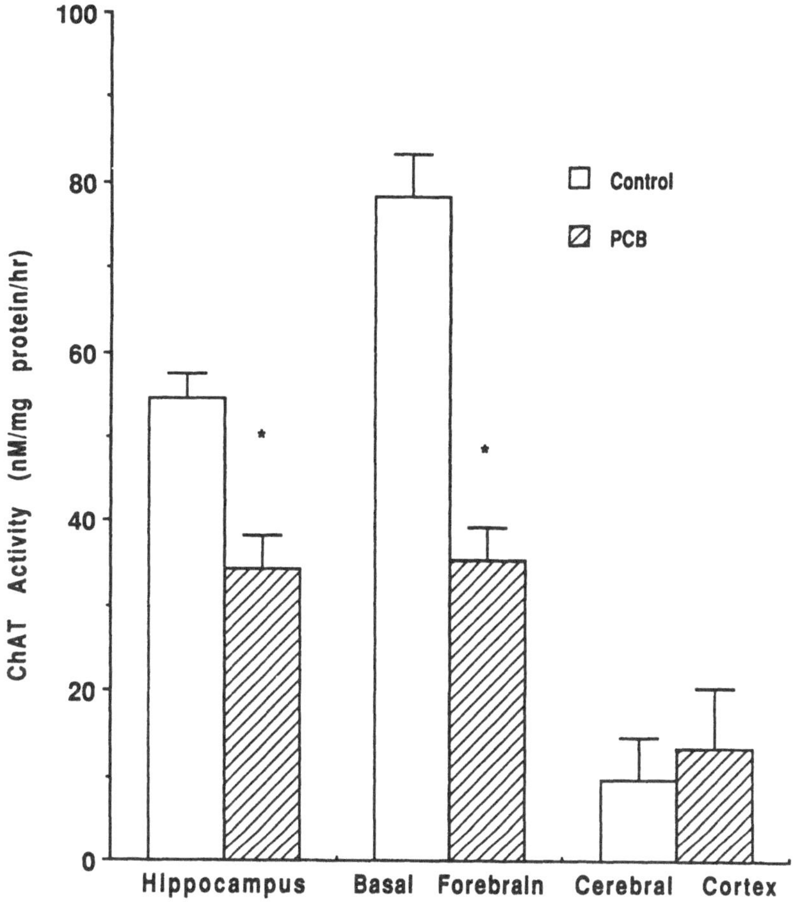

Figure 2 Mean ChAT activity ± SEM in homogenates of hippocampus, basal forebrain, and cerebral cortex of 15-day old rats exposed *in utero* and through lactation to 0 (Control) or 250 ppm PCB (PCB). * $p < .05$ vs. Control rats.

1991). However, it has been difficult to find a model that has several of the major symptoms present in AD (cholinergic deficits, plaques and/or neurofibrillary tangles, and memory impairment). Animals exposed pre- and/or postnatally to PCB have cholinergic deficits and behavioral deficits including impaired learning and memory, altered activity levels, and delayed ontogeny of reflexes. For example, Pantaleoni *et al.* (1988) found that PCB-exposed rats displayed impaired acquisition of an active avoidance task. Neonatal rats whose mothers were exposed to 26 ppm PCB showed delayed ontogeny of negative geotaxis, auditory startle, and air righting reflex (Overmann *et al.*, 1987). Prenatally exposed mice

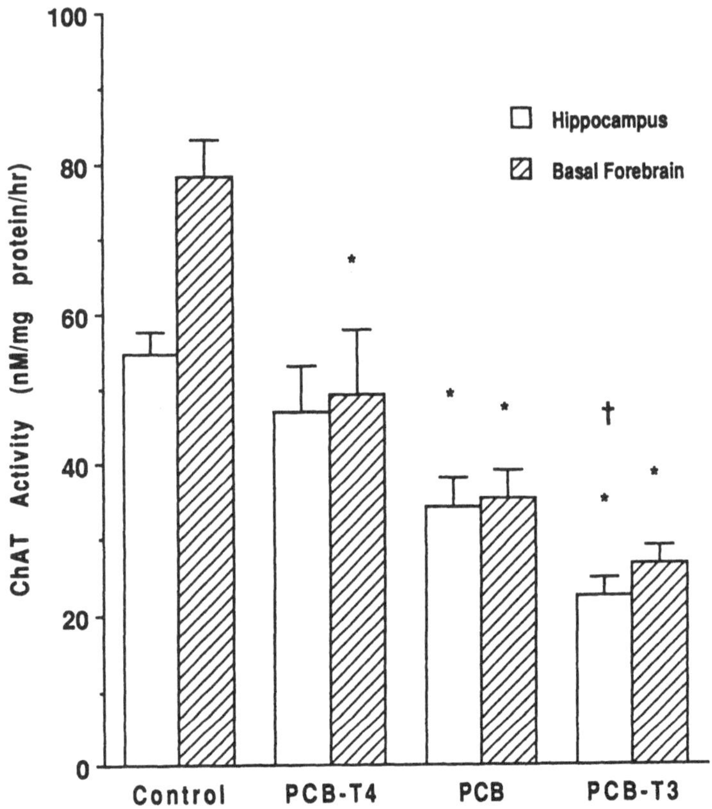

Figure 3 Mean ChAT activity ± SEM in homogenates of hippocampus and basal forebrain of 15-day-old rats exposed *in utero* and through lactation to 0 (Control) or 250 ppm PCB (PCB) (both uninjected) and PCB rats injected with T_4 (PCB-T4) or T_3 (PCB-T3) on postnatal days 4–14. Control and PCB values are repeated from Figure 2 to facilitate comparison. * $p < .05$ vs. Control rats. † $p < .05$ vs. PCB-T4.

exhibited spinning, hyperactivity, and impaired acquisition of a one-way, shock-motivated avoidance response (Tilson *et al.*, 1979). Eriksson *et al.* (1991) have demonstrated depressed spontaneous motor behavior in 4-month-old mice exposed at 10 days of age. Levin *et al.* (1988) showed PCB-exposed monkeys were retarded in learning progressive probability shifts and object alternation tasks.

The results from the animal studies correlate well with poor visual recognition memory, hypotonia, and hyporeflexia in transplacentally exposed children (J.L. Jacobson *et al.*, 1984; W.W. Jacobson *et al.*, 1985; J.L. Jacobson *et al.*, 1990; Rogan *et al.*, 1986; Schantz *et al.*, 1990). Although the proposed model displays two of the three major AD symptoms, it has not yet been determined whether plaques and/or tangles occur in PCB-exposed rats.

Alzheimer's disease has been reported to influence a number of neurotransmitter systems and neuropeptides, including the generation, brain content, and/or degradation of dopamine, serotonin, neuropeptide Y, somatostatin, and corticotropin-releasing factor (Perry *et al.*, 1984; Allen *et al.*, 1984; Davies and Terry, 1981; Rossor *et al.*, 1980; Bissette *et al.*, 1985). A direct connection between modification of these systems and the collection of neurological symptoms (especially memory loss) attendent with AD remains to be made (Allen *et al.*, 1984; Arai *et al.*, 1984). On the other hand, deficits in mechanisms and neuroanatomical sites that are primarily cholinergic have been linked with the memory loss of AD with sufficient consistency to be proposed as important causative factors in this loss (Bartus *et al.*, 1982; Coyle *et al.*, 1983). Thus, an animal model that displays cholinergic deficits concomitant with behavioral deficiency would be particularly attractive. Because 15-day-old normal rats are a bit young for use in standard animal behavioral learning and memory procedures, we are in the process of carrying rats exposed to PCB past weaning and into adulthood to perform behavioral measures. Additionally, it is our intent to determine whether removal of PCB from the diet at some point during development allows recovery of thyroid and ChAT deficits.

The present and previous studies (Collins and Capen, 1980; Sharma *et al.*, 1991; Meserve *et al.*, 1992) have shown that exposing neonatal rats to PCB induces hypothyroidism. Rats made hypothyroid by means other than PCB ingestion also exhibit depressed ChAT activity (Patel *et al.*, 1987; Gould and Butcher, 1989). Because replacement of thyroxine elevated ChAT activity in rats made hypothyroid by ingestion of PTU (Patel *et al.*, 1988), it appears that thyroid hormones are involved in the regulation of ChAT in the hippocampus and basal forebrain. The present study indicates that the influence of PCB on depressed ChAT activity may also be partially a result of the accompanying hypothyroidism rather than, or in addition to, a direct effect on these brain areas because thyroxine injections in PCB-exposed rats significantly increased ChAT activity in both the hippocampus and basal forebrain.

Some of the neurological symptoms of clinical hypothyroidism resemble those of Alzheimer's disease (e.g., fatigue, mental apathy, memory deficits) (Eayrs, 1960). Thus, several studies have been conducted to develop an association between hypothyroidism and AD (Heyman *et al.*, 1984; Lopez *et al.*, 1989; Percy *et al.*, 1990; Ewins *et al.*, 1991; Yoshimasu *et al.*, 1991) but the connection remains controversial. There are specific intriguing findings that may relate hypothyroidism to AD. For example, AD patients display elevated circulating levels of antithyroglobulin antibodies (Percy, 1990; McRae-Degueurce *et al.*, 1988). This condition could result in clinical or subclinical autoimmune thyroiditis, interfering with adequate release of thyroid hormones into circulation. Riiosen (1988) has reported AD patients to have reduced levels of prealbumin, a normal thyroid hormone-binding protein, in cerebrospinal fluid. Subnormal concentrations of this protein could compromise normal delivery of thyroid hormones to target cells, especially in the central nervous system. The mRNA transcribed from the gene cERB Aα is subnormal in AD patients (Kung Sutherland *et al.*, 1992). Because the gene product of cERB Aα is one of the types of intracellular receptors for thyroid hormone (Sap *et al.*, 1986), depressed expression of the gene could result in hyporesponsivity of thyroid hormone target cells. As previously mentioned, an important characteristic of animal models for AD is that they display symptoms similar to those of the AD patient (Karczmar, 1991). It would seem that the hypothyroidism displayed by PCB rats (Figure 1) provides symptoms analogous to those thyroid-related effects of AD. There do not appear to have been previous reports of an increased incidence of AD in human cretins.

Two subtypes of Alzheimer's disease have been suggested: Type 1 (late onset) and Type 2 (early onset) (Roth, 1986). This classification appears to be important to treatment strategies, with cholinergic therapy likely to be more effective in treating Type 1 patients. Because thyroid hormone replacement elevates depressed ChAT activity in the present model (Figure 3), the system may more closely resemble Type 1 than Type 2 AD. A clinical correlate of the thyroid effect in PCB-exposed rats is that humans with the hyperthyroidism of Graves' disease are thought to have a "protective effect" against Alzheimer's disease (Yoshimasu *et al.*, 1991).

One strategy for reducing the severity of AD symptoms with cholinergic mediation is to augment the function of compromised neurons by administration of precursors or agonists to acetylcholine, or to increase availability by inhibiting acetylcholinesterase (Johns *et al.*, 1985). Another strategy, given early detection of the deficit, might be to rescue or "reinvigorate" cholinergic neurons. Because Patel *et al.* (1988) and the results of the present study (Figure 3) have found thyroxine to be palliative when depression of ChAT activity accompanies either PTU- or PCB-induced hypothyroidism, thyroid hormone therapy may be an appropriate intervention.

In the present study, thyroxine did not completely return ChAT activity to normal (Figure 3), suggesting the requirement of other neurotropic substances to achieve complete repletion. A possible candidate substance is the polypeptide nerve growth factor (NGF), which has been co-localized in neurons with acetylcholine (Hefti *et al.*, 1985; Woolf *et al.*, 1989) and has been proposed as a functional regulator of cholinergic neurons in the forebrain (Gnahn *et al.*, 1983; Fusco *et al.*, 1989). Patel *et al.* (1988) have found intracerebroventricular administration of NGF to replace ChAT activity in PTU-hypothyroid rats, with additive effects of a NGF-thyroxine combination. Indeed, Hefti and Weiner (1986) have proposed the possibility of developing NGF therapy for AD and have described several preclinical evaluations necessary for implementation of such treatment. Given the possible efficacy of NGF, it would be of interest to determine whether the combination of thyroxine and NGF would completely restore ChAT activity in the PCB model.

Among the advantages of further investigating thyroid hormone therapy are:

thyroid hormones pass the blood-brain barrier, whereas other substances, including NGF, do not; thyroid hormone augmentation may result in enhanced production of endogenous neurotropic factors, such as NGF, because of the "permissive" nature of this class of hormones; thyroid hormone treatment may prevent the compromised activity of the cholinergic neurons, as appears to be the case in the PCB rat model. Further studies to determine the mechanisms whereby thyroid hormone spares ChAT activity in PCB rats will allow a better estimation of whether cholinergic neurons are rescued by this process. Some consideration of the relative efficacy of T_4 and T_3 is necessary. Given that PCB depressed levels of T_4 more severely than those of T_3 in the present study (Figure 1), it follows intuitively that replacement of T_4 would be more effective in normalizing ChAT than would injection of T_3. Counterintuitive to this notion is the conventional wisdom that T_3 is the biologically active thyroid hormone and T_4 is primarily a prohormone (Braverman *et al.*, 1970). However, the absolute nature of this relation continues to be debated (Silva and Larsen, 1977). In brain neurons, T_4 is the thyroid hormone primarily removed from circulation, where it is then converted to T_3 locally (Crantz *et al.*, 1982).

The foregoing discussion has primarily considered a comparison of AD and PCB exposure with regard to compromised potential to generate the neurotransmitter acetylcholine. However, similarities in modification of transmitter-receptor interaction should not be ignored. For example, a subnormal population of cholinergic muscarinic receptors has been reported for neurons from AD patients (Weinberger *et al.*, 1991; Mash *et al.*, 1985) and from 10-day-old mice given a single dose of PCB (Eriksson *et al.*, 1991). The receptor deficit in mice in the latter study persists into adulthood. We have not determined the density of muscarinic receptors in our PCB rat model, nor has the influence of thyroid hormone on this parameter been investigated.

4.2. Mechanisms by Which PCB-Induced Hypothyroidism May Depress ChAT Activity

Although the molecular mechanisms by which PCB induces its effects, including hypothyroidism, are still not known, several have been proposed including PCB parent compounds binding to a receptor (arylhydrocarbon receptor, Ah) and PCB metabolites generated from the parent compounds binding to thyroxine/vitamin A-binding proteins (Poland *et al.*, 1976; Knutson and Poland, 1984; Brouwer and Van den Berg, 1986; Brouwer, 1991). The Ah receptor has been found in several cell types, including neurons (Carlstedt-Duke, 1979), and the receptor is thought to be activated when PCB or other ligands bind to it. The activated ligand-receptor complex is then translocated to the nucleus where it binds to a responsive element within the genome, which increases expression of specific mRNA and protein.

The PCB metabolites are thought to induce effects related to vitamin A and thyroid hormone metabolism. The metabolites bind several thyroxine- and vitamin A-binding proteins such as the major transport protein complex in rodents, thyroxine-transthyretin (TTR), which is also called thyroxine binding prealbumin (TBPA)-retinol binding protein (RBP) complex (Brouwer and Van den Berg, 1986). There appears to be a direct competition between T_4 and PCB metabolites for the T_4 binding site on TTR, which results in a loss of free T_4 from the circulation. Metabolites of PCB also bind to thyroxine receptors in liver nuclei (McKinney *et al.*, 1987), and to Type 1 iodothyronine monodeiodinase (Rickenbacher *et al.*, 1989). Iodothyronine deiodonase catalyzes the conversion of T_4 to T_3 in target cells and is responsible for approximately 70% of the circulating T_3 in the euthyroid rat (Kaplan, 1984), so its modification could drastically influence thyroid status. There is less information about the effects of PCB on thyroid-binding globulin (TBG).

Thus, the PCB-induced hypothyroidism and depression of ChAT may be a result of PCB and its metabolites binding to both the thyroxine-binding proteins and to the Ah receptor as well. The depression in ChAT activity appears to be partly the result of T_4 deficiency, since replacement increases ChAT activity (Figure 3), and partly the result of as yet unidentified mechanisms.

4.3. Limitations of PCB-Exposed Rat as Animal Model for Alzheimer's Disease

Perhaps the greatest weakness of the proposed model at this stage of its development is that it uses very young rats to study neurological damage pertinent to a clinical condition that occurs most frequently in the elderly. One of the drawbacks of previous models, which impose cholinergic deficits, is that they become spontaneously corrected with time (Bartus *et al.*, 1985). Indeed, in the PTU-hypothyroid rats of Patel *et al.* (1987), some brain areas in which ChAT activity was subnormal neonatally recovered as the animals reached adulthood. Sufficient data have not been collected with the PCB model to firmly support continued cholinergic deficits with the passage of time. Very preliminary studies suggest a persistence of depressed ChAT activity, at least in hippocampus. To correlate the PCB rat to clinical AD with any degree of assuredness requires that early exposure to the organohalide results in deficits later in life. Schantz *et al.* (1991) have shown that perinatal exposure to PCB at levels typically reported in the human population produces cognitive deficits in subhuman primates (as determined by two-choice discrimination-reversal learning and delayed spatial alternation tests) that persist into young adulthood and quite possibly may be permanent.

A second weakness is the frailty of connection between depressed ChAT activity in this and other models, and alteration of learning and memory or other behavioral deficits. Some studies have reported PCB to produce deficient behavior, and others have not, depending on the behavioral measure (see Tilson *et al.*, 1990, for review). Indeed, this may be part of the larger question of whether cholinergic deficits are of primary importance with regard to memory loss in AD. For example, the glutamate analog quisqualic acid destroys 80%–90% of the neurons in the basal forebrain of rats but produces only slight memory impairment (Dunnett *et al.*, 1987). This led Wenk (1990) to suggest that the 10% to 20% of neurons not destroyed by quisqualate are noncholinergic and are the forebrain elements important in consolidation of memory.

In any event, PCB exposure compromises cholinergic parameters (Figure 2), although the reason is presently undetermined. Additionally, there is no direct connection between loss of cholinergic neurons in AD and thyroid hormone insufficiency or hyporesponsivity of thyroid hormone target cells to sufficiently delivered hormone. In true AD, there is always the possibility that a collection of insults such as selectively toxic chemicals, limiting nutrients, modified neurotropic factors, and many others may be causative (Hefti and Weiner, 1986).

It is anticipated that further study using the PCB rat model will allow generation of considerable information regarding the neurotoxicity of this material. This information is potentially valuable in the understanding and treatment of Alzheimer's disease.

4.4. Summary

A possible animal model has been generated by incorporating PCB (250 ppm Aroclor 1254) into the diet of pregnant and lactating rats, which results in 15-day-old offspring with neurological modifications characteristic of Alzheimer's disease. Specifically, ChAT activity in hippocampus and basal forebrain are markedly depressed. Such depression is concomitant with subnormal circulating T_4 levels, and a replacement regime of T_4 injections normalizes ChAT activity.

At this stage of development of the model, a number of questions remain to be answered with regard to the molecular mechanism of action of PCB and how this material produces symptoms similar to those of AD. Furthermore, it is of primary importance to determine whether cholinergic deficits in young PCB-exposed rats persist as they reach an age with greater chronological relevance to AD, and whether these deficits correlate with appropriate behavioral modifications. Future studies should also determine whether other neuropathological modifications in structure and function, characteristic of Alzheimer's disease, are present in the PCB rat model.

ACKNOWLEDGMENTS. Thanks to P. Luke of ICN Biomedicals, Inc., for generous gift of T_3 and T_4 RIAs, and to R. Amiri, M. Baratian, and J. Ku for lab assistance. Thanks also to V. Bingman, L. Dokas, E. Russ, and S. Schantz for reviewing the manuscript. Funding provided by Research Challenge Grant OBOR/BGSU.

5. References

Allen, J.M., Ferrier, I.N., Roberts, G.W., Cross, A.J., Adrian, T.E., Crow, T.J., and Bloom, S.R., 1984, Elevation of neuropeptide Y (NPY) in substantia innominata in Alzheimer's-type dementia, *J. Neurol. Sci.* 64:325–331.

Arai, H., Kosaka, K., and Iizuka, R., 1984, Changes of biogenic amines and their metabolites in postmortem brains from patients with Alzheimer-type dementia, *J. Neurochem.* 43:388–393.

Araujo, D.M., Lapchak, P.A., Robitaille, Y., Gauthier, S., and Quirion, R., 1988, Differential alteration of various cholinergic markers in cortical and subcortical regions of human brain in Alzheimer's disease, *J. Neurochem.* 50:1914–1923.

Bartus, R.T., Dean, R.L., III, Beer, B., and Lippa, A.S., 1982, The cholinergic hypothesis of geriatric memory dysfunction, *Science* 217:408–417.

Bartus, R.T., Glicker, C., Dean, R.L.I., Pontecorvo, M., Figueiredo, J.C., and Fisher, S.K., 1985, Selective memory loss following nucleus basalis lesions: Long-term behavioral recovery despite persistent cholinergic deficiencies, *Pharmacol. Biochem. Behav.* 23:125–135.

Bissette, G., Reynolds, G.P., Kilts, C.D., Widerlov, E., and Nemeroff, C.B., 1985, Corticotropin-releasing factor-like immunoreactivity in senile dementia of the Alzheimer type: Reduced cortical and striatal concentrations, *JAMA* 254:3067–3069.

Bjorklund, G., 1991, Mercury as a potential source for the etiology of Alzheimer's Disease, *Trans. Elem. Med.* 8(4):208.

Brandt, T., 1977, Tissue localization of PCB's chemical structure related to pattern of distribution, *Acta. Pharmacol. Toxicol. Suppl.* 40(2):1–108.

Braverman, L.E., Ingbar, S.H., and Sterling, K., 1970, Conversion of thyroxine to triiodothyronine in athryreotic human subjects, *J. Clin. Invest.* 49:855–864.

Brezner, E., Terkel, J., and Perry, A.A., 1984, The effect of Aroclor 1254 (PCB) on the physiology of reproduction in the female rat, *J. Comp. Biochem. Physiol.* 77C:65–70.

Brouwer, A., 1991, The role of enzymes in regulating the toxicity of xenobiotics, *Biochem. Soc. Trans.* 19:731–737.

Brouwer, A., and Van den Berg, K.J., 1986, Binding of a metabolite of 3,4,3′,4′-tetrachlorobiphenyl to transthyretin reduces serum vitamin A transport by inhibiting the formation of the protein complex carrying both retinol and thyroxine, *Toxicol. Appl. Pharmacol.* 85:301–312.

Bruckner, J.V., Khanna, K.L., and Cornish, H.H., 1973, Biological responses of the rat to polychlorinated biphenyls, *Toxicol. Appl. Pharmacol.* 24:434–448.

Byrne, J.J., Carbone, J.P., and Hanson, E.A., 1987, Hypothyroidism and abnormalities in the kinetics of thyroid hormone metabolism in rats treated chronically with polychlorinated biphenyl and polybrominated biphenyl, *Endocrinology* 121:520–527.

Carlstedt-Duke, J.M.B., 1979, Tissue distribution of the receptor for 2,3,7,8,-tetrachlorodibenzo-p-dioxin in the rat, *Cancer Res.* 39:3172–3176.

Collins, W.T., and Capen, C.C., 1980, Fine structural lesions and alterations in the thyroid glands of perinatal rats exposed *in utero* and by the milk to PCBs, *Amer. J. Pathol.* 99:125–142.

Coyle, J.T., Price, D.L., and DeLong, M.R., 1983, Alzheimer's disease: A disease of cortical cholinergic innervation, *Science* 219:184–1190.

Crantz, F.R., Silva, J.E., and Larsen, P.R., 1982, An analysis of the sources and quantity of 3,5,3′-triiodothyronine specifically bound to nuclear receptors in rat cerebral cortex and cerebellum, *Endocrinology* 100:367–375.

Davies, P., and Maloney, A.J.F., 1976, Selective loss of central cholinergic neurons in Alzheimer's disease, *Lancet* 2:1403.

Davies, P., and Terry, R.D., 1981, Cortical somatostatin-like immunoreactivity in cases of Alzheimer's disease and senile dementia of the Alzheimer type, *Neurobiol. Aging*. 2:9–14.

Davis, G.C., Williams, A.C., and Markey, S.P., 1979, Chronic parkinsonism secondary to intravenous injection of meperidine analogues, *Psychiatry Res.* 1:249–254.

Dunnett, S.B., Whishaw, I.Q., Jones, G.H., and Bunch, S.T., 1987, Behavioral, biochemical and histochemical effects of different neurotoxic amino acids injected into nucleus basalis magnocellularis of rat, *Neuroscience* 20:653–669.

Eayrs, J.T., 1960, Influence of the thyroid on the central nervous system, *Br. Med. Bull.* 16:122–127.

Eriksson, P., Lundkvist, U., and Fredriksson, A., 1991, Neonatal exposure to 3,3′,4,4′-tetrachlorobiphenyl: Changes in spontaneous behaviour and cholinergic muscarinic receptors in the adult mouse, *Toxicology* 69:27–34.

Ewins, D.L., Rossor, M.N., Butler, J., Roques, P.K., Mullan, M.J., and McGregor, A.M., 1991, Association between autoimmune thyroid disease and familial Alzheimer's disease, *Clin. Endocrinol.* 35:93–96.

Fonnum, F., 1975, A rapid radiochemical method for the determination of choline acetyltransferase, *J. Neurochem.* 24:407–409.

Fusco, M., Oderfeld-Nowak, B., Vantini, G., Schiavo, N., Gradkowska, M., Zaremba, M., and Leon, A., 1989, NGF affects uninjured adult rat septohippocampal cholinergic neurons, *Neuroscience* 33:47–52.

Gnahn, H., Hefti, F., Heumann, R., Schwab, M., and Thoenen, H., 1983, NGF-mediated increase in choline-acetyltransferase (ChAT) in the neonatal forebrain: Evidence for a physiological role of NGF in the brain?, *Dev. Brain Res.* 9:45–52.

Gould, E., and Butcher, L.L., 1989, Developing cholinergic basal forebrain neurons are sensitive to thyroid hormone, *J. Neurosci.* 9(9):3347–3358.

Hammond, A.L., 1972, Chemical pollution: Polychlorinated biphenyls, *Science* 175:155–156.

Hefti, F., and Weiner, W.J., 1986, Nerve growth factor and Alzheimer's disease, *Ann. Neurol.* 20:275–281.

Hefti, F., Haritkka, J., Salvaterra, A., Weiner, W.J., and Mash, D.C., 1985, Localization of nerve growth factor receptor is associated with cholinergic neurons of the human basal forebrain. *Neurosci. Lett.* 69:37–41.

Henneberry, R., and Spatz, L., 1991, The role of environmental factors in neurodegenerative disorders, *Neurobiol. Aging* 12:75–79.

Heyman, A., Wilkinson, W.E., Staffor, J.A., Helms, M.J., Sigmon, A.H., and Weinberg, T., 1984, Alzheimer's disease: A study of epidemiological aspects, *Ann. Neurol.* 15:335–341.

Humphrey, H.E.B., 1983, Population studies of PCBs in Michigan residents, *in* "PCBs: Human and Environmental Hazards" (F.M. D'Itri and M.A. Kamrin, eds.), pp. 299–311, Ann Arbor Science Publishers, Ann Arbor, MI.

Jacobson, J.L., Jacobson, S.W., Fein, G.G., Schwartz, P.M., and Dowler, J.K., 1984, Prenatal exposure to an environmental toxin: A test of the multiple effects model., *Dev. Psychol.* 20:523–532.

Jacobson, J.L., Jacobson, S.W., and Humphrey, H.E.B., 1990, Effects of exposure to PCBs and related compounds on growth and activity in children, *Neurotoxicol. Teratol.* 12:319–326.

Jacobson, W.W., Fein, G.G., Jacobson, J.L., Schwartz, P.M., and Dowler, J.K., 1985, The effect of intrauterine PCB exposure on visual recognition memory, *Child Dev.* 56:853–860.

Jensen, A.A., 1989, Background levels in humans, *in* "Halogenated Biphenyls, Terphenyls, Napthalenes, Dibenzodioxins, and Related Products" (2nd ed.), (R.D. Kimbrough and A.A. Jensen, eds.), pp. 345–380, Elsevier, Amsterdam.

Johns, C.A., Haroutunian, V., Greenwald, B.S., Mohs, R.C., Davis, B.M., Kanof, P., Horvath, T.B., and Davis, K.L., 1985, Development of cholinergic drugs for the treatment of Alzheimer's disease, *Drug Dev. Res.* 5:77–96.

Kaplan, M.M., 1984, The role of thyroid hormone deiodination in the regulation of hypothalamo-pituitary function, *Neuroendocrinology* 38:254–260.

Karczmar, A.G., 1991. SDAT models and their dynamics, *in* "Cholinergic Basis for Alzheimer Therapy" (R. Becker and E. Giacobini, eds.), pp. 142–152, Birkhäuser, Boston.

Kimbrough, R.D., 1980, Occupational exposure, *in* "Halogenated Biphenyls, Terphenyls, Naphthalenes, Dibenzodioxin and Related Compounds" (R.D. Kimbrough, ed.), pp. 373–397, Elsevier, Amsterdam.

Knutson, J.C., and Poland, A., 1984, "Molecular and Cellular Approaches to Understanding Mechanisms of Toxicity" (TA.H. Tashian, ed.), pp. 88–101, Harvard School of Public Health, Boston.

Krishnan, S.S., McLachlan, D.R., Krishnan, B., Fenton, S.S.A., and Harrison, J.E., 1988, Aluminum toxicity to brain, *Sci. Total Environ.* 71:59–64.

Kung Sutherland, M., Wong, L., Somerville, M.J., Handley, P., Yoong, L., Bergeron, C., and Crapper McLachlan, D.R., 1992, Reduction of thyroid hormone receptor c-ERB Aα mRNA levels in the hippocampus of Alzheimer as compared to Huntington brain, *Neurobiol. Aging* 13:301–312.

Kunita, N., Hori, S., Obana, H., Otake, T., Nishimura, H., Kashimoto, T., and Ikegami, N., 1985, Biological effects of PCBs, PCQs, PCDFs present in the oil causing Yusho and Yu-cheng, *Environ. Health Perspect.* 59:79–84.

Langston, J.W., Ballard, P.A., Tetrud, J.W., and Irwin, I., 1983, Chronic parkinsonism in humans due to a product of meperidine-analog synthesis, *Science* 219:979–980.

Levin, E.D., Schantz, S.L., and Bowman, R.E., 1988, Delayed spatial alternation deficits resulting from perinatal PCB exposure of monkeys, *Arch. Toxicol.* 62:267–273.

Lopez, O., Huff, F.J., Martinez, A.J., and Bedetti, C.D., 1989, Prevalence of thyroid abnormalities is not increased in Alzheimer's disease, *Neurobiol. Aging* 10:247–251.

Lowry, O.H., Rosebrough, N.J., Farr, A.L., and Randall, R.J., 1951, Protein measurement with the Folin phenol reagent, *J. Biol. Chem.* 193:265–275.

Marx, J., 1992, Major setback for Alzheimer's models, *Science* 255:1200–1202.

Mash, D.C., Flynn, D.D., and Potter, L.T., 1985, Loss of M2 muscarine receptors in the cerebral cortex in Alzheimer's disease and experimental cholinergic denervation, *Science* 228:1115–1117.

Masuda, Y., Kagawa, R., Kuroki, H., Taki, I., Kusuda, M., Yamashita, F., Hayashi, M., Kuratsune, M., and Yoshimura, T., 1978, Transfer of polychlorinated biphenyls from mothers to foetuses and infants, *Food. Cosmet. Toxicol.* 16:543–546.

McKinney, J.D., Fannin, R., Jordan, S., Chal, K., Rickenbacher, U., and Pedersen, L. 1987, Polychlorinated biphenyls and related compound interactions with specific binding sites for thyroxine in rat liver nuclear extracts, *J. Med. Chem.* 30:79–86.

McKinney, J.D., and Pedersen, L.G., 1987, Do residue levels of polychlorinated biphenyls (PCBs) in human blood produce mild hypothyroidism?, *J. Theor. Biol.* 129(2):1173–1176.

McRae-Degueurce, A., Hagid, K., Rosengren, L., Wallin, A., Blennow, K., Gottfries, C.G., and Dahlström, A., 1988, Antibodies recognizing cholinergic neurons and the thyroglobulin are found in the cerebrospinal fluid of a subgroup of patients with Alzheimer's disease, *Drug Develop. Res.* 15:153–163.

Meserve, L.A., Murray, B.A., and Landis, J.A., 1992, Influence of maternal ingestion of Aroclor 1254® (PCB) or FireMaster BP-6® (PBB) on unstimulated and stimulated corticosterone levels in young rats, *Bull. Environ. Contam. Toxicol.* 48:715–720.

Murai, K., Okamura, K., Tsuji, H., Kajiwara, E., Watanabe, H., Akagi, K., and Fujishima, M., 1987, Thyroid function in "Yusho" patients exposed to polychlorinated biphenyls (PCB), *Environ. Res.* 44:179–187.

Overmann, S.R., Kostas, J., Wilson, L.R., Shain, W., and Bush, B., 1987, Neurobehavioral and somatic effects of perinatal PCB exposure to rats, *Environ. Res.* 44:56–70.

Pantaleoni, G., Fanini, D., Sponta, A.M., Palumbo, G., Giorgi, R., and Adams, P.M., 1988, Effects of maternal exposure to polychlorobiphenyls (PCBs) on F1 generation behavior in the rat, *Fund. Appl. Toxicol.* 11:440–449.

Patel, A.J., Hayashi, M., and Hunt, A., 1987, Selective persistent reduction in choline acetyltransferase activity in basal forebrain of the rat after thyroid deficiency during early life, *Brain Res.* 422:182–185.

Patel, A.J., Hayashi, M., and Hunt, A., 1988, Role of thyroid hormone and nerve growth factor in the development of choline acetyltransferase and other cell-specific marker enzymes in the basal forebrain of the rat, *J. Neurochem.* 50:803–811.

Percy, M.E., Dalton, A.J., Markovic, V.D., Crapper McLachlan, D.R., Gera, E., Hummel, J.T., Rusk, A.C.M., Somerville, M.J., Andrews, D.R., and Walfish, P.G., 1990, Autoimmune thyroiditis associated with mild "subclinical" hypothyroidism in adults with Down syndrome: A comparison of patients with and without manifestations of Alzheimer disease, *Am. J. Med. Genet.* 36:148–154.

Percy, M.E., Dalton, A.J., Markovic, V.D., Crapper McLachlan, D.R., Gera, E., Hummel, J.T., Rusk, A.C.M., Sommerville, M.J., Andrews, D.F., and Wallfish, P.G., 1990, Autoimmune thyroiditis associated with "subclinical" hypothyroidism in adults with Down Syndrome: a comparison of patients with and without manifestations of Alzheimer's Disease, *Amer. J. Med. Genet.* 36:148.

Perl, D.P., and Brody, A.R., 1980, Alzheimer's disease: X-ray spectrometric evidence of aluminum accumulation in neurofibrillary tangle-bearning neurons, *Science* 208:297–299.

Perry, E.K., Tomlinson, B.E., Blessed, G., Bergmann, K., Gibson, P.H., and Perry, R.H., 1978, Correlation of cholinergic abnormalities with senile plaques and mental test scores in senile dementia, *Br. Med. J.* 2:1457–1459.

Perry, E.K., Atack, J.R., Perry, R.H., Hardy, J.A., Dodd, P.R., Edwardson, J.A., Blessed, G., Tomlinson, B.E., and Fairbairn, A.F., 1984, Intralaminar neurochemical distributions in human midtemporal cortex: Comparison between Alzheimer's disease and the normal, *Neurochemistry* 42:1402–1410.

Poland, A., Glover, E., and Kende, A.S., 1976, Stereospecific, high affinity binding of 2,3,7,8-tetrachlorodibenzo-p-dioxin by hepatic cytosol. Evidence that the binding species is receptor for induction of aryl hydrocarbon hydroxylase, *J. Biol. Chem.* 251:4936–4946.

Rickenbacher, U., Jordan, S., and McKinney, J.D., 1989, Structurally specific interaction of halogenated dioxin and biphenyl derivatives with iodothyronine-5'-deiodinase in rat liver, *ACS Symp. S.* 413:354–363.

Rickenbacher, U., McKinney, J.D., Oatley, S.J., and Blake, C.C.F., 1986, Structurally specific binding of halogenated biphenyls to thyroxine transport protein, *J. Med. Chem.* 29:641–648.

Riiosen, H., 1988, Reduced pre-albumin (transthyretin) in CSF of severely demented patients with Alzheimer's disease, *Acta Neurol. Scand.* 78:455–459.

Rogan, W.J., Gladen, B.C., McKinney, J.D., Carreras, N., Hardy, P., Thullen, J., Tinglestad, J., and Tully, M., 1986, Neonatal effects of transplacental exposure to PCBs and DDE, *J. Pediatr.* 109:335–341.

Rossor, M.N., Emson, P.C., Mountjoy, C.Q., Roth, M., and Iversen, L.L., 1980, Reduced amounts of immunoreactive somatostatin in the temporal cortex in senile dementia of Alzheimer type, *Neurosci. Lett.* 20:373.

Roth, M., 1986, The association of clinical and neurobiological findings and its bearing on the classification and aetiology of Alzheimer's disease, *Br. Med. Bull.* 42:42–50.

Sap, J., Muñoz, A., Damm, K., Goldberg, Y., Ghysdael, J., Leutz, A., Beug, H., and Vennström, B., 1986, The *c-erbA* protein is a high-affinity receptor for thyroid hormone, *Nature* 324:635–660.

Schantz, S.L., Jacobson, J.L., Jacobson, S.W., and Humphrey, H.E.B., 1990, Behjavioral correlates of polychlorinated biphenyl (PCB) body burden in school-aged children, *Toxicologist* 10:303.

Schantz, S.L., Levin, E.D., and Bowman, R.E., 1991, Long-term neurobehavioral effects of perinatal polychlorinated biphenyl (PCB) exposure in monkeys, *Environ. Toxicol. Chem.* 10:747–756.

Sharma, M., Juárez de Ku, L.M., Baratian, M.A., Amiri, R.B., and Meserve, L.A., 1991, Effects of various PCB doses on thyroid status and choline acetyltransferase (ChAT) activity in fifteen-day-old rats, *Soc. Neurosci. Abstr.* 17(2):1464.

Silva, J.E., and Larsen, P.R., 1977, Pituitary nuclear 3,5,3′-triiodothyronine and thyrotropin secretion: An explanation of the effect of thyroxine, *Science* 198:617–620.

Spencer, P.S., Nunn, P.B., Hugon, J., Ludolph, A.C., Ross, S.M., Roy, D.N., and Robertson, R.C., 1987, Guam amyotrophic lateral sclerosis-parkinsonism-dementia linked to a plant exitant neurotoxin, *Science* 237:517–522.

St George Hyslop, P., Tanzi, E.E., Polinsky, R.J., Haines, J.L., Nee, L., Watkins, P.C., Myers, R.H., and Feldman, R.S., 1987, The genetic defect causing familial Alzheimer's disease maps on chromosome 21, *Science* 235:885–889.

Swain, W.R., 1983, An overview of the scientific basis for concern with polychlorinated biphenyls in the Great Lakes, *in* "PCBs: Human and Environmental Hazards" (F.M. D'Itri and M.A. Kamrin, eds.), pp. 11–48, Butterworths, Boston.

Takagi, Y., Aburada, S., Hashimoto, K., and Kitaura, T., 1976, Transfer and distribution of accumulated ^{14}C polychlorinated biphenyls from maternal to fetal and suckling rats, *Arch. Environ. Contam. Toxicol.* 15:709–715.

Tanabe, S., 1988, PCB problems in the future: Foresight from current knowledge, *Environ. Pollut.* 60:5–28.

Tilson, H.A., Davis, G.J., McLachlan, J.A., and Lucier, G.W., 1979, The effects of polychlorinated biphenyls given prenatally on the neurobehavioral development of mice, *Environ. Res.* 18:466–474.

Tilson, H.A., Jacobson, J.L., and Rogan, W.J., 1990, Polychlorinated biphenyls and the developing nervous system: Cross-species comparison, *Neurotoxicol. Teratol.* 12:239–248.

Veith, G.C., Kuehl, D.W., Leonard, E.N., Welch, K., and Pratt, G., 1981, Polychlorinated biphenyls and other organic chemical residues in fish from major watersheds near the Great Lakes, *Pestic. Monit. J.* 5:1–8.

Wassermann, M., Wassermann, D., Cucos, S., and Miller, H.J., 1979, World PCBs map: Storage and effects in man and his biologic environment in the 1970s, *in* "Health Effects of Halogenated Aromatic Hydrocarbons" (W.J. Nicholson and J.A. Moore, eds.), pp. 69–124, New York Academy of Sciences, New York.

Weinberger, D.R., Jones, D.W., Coppola, R., Berman, K.F., Gibson, R., and Reba, R.C., 1991, Concerning the distribution of cerebral muscarinic acetylcholine receptors in Alzheimer's disease, *Arch. Neurol.* 48 (11):1118.

Wenk, G.L., 1990, Animal models of Alzheimer's disease: Are they valid and useful?, *Acta Neurobiol. Exp.* 50:219–223.

White, P., Goodhart, M.J., Keet, J.P., Hiley, C.R., Carrasco, L.H., Williams, I.E., and Bowen, D.M., 1977, Neocortical cholinergic neurons in elderly people, *Lancet* 1:668–671.

Wilcock, G.K., Esiri, M.M., Bowen, D.M., and Smith, C.C.T., 1985, The nucleus basalis in Alzheimer's disease: Cell counts and cortical biochemistry, *Neuropathol. Appl. Neurobiol.* 9:175–179.

Wisniewski, H.M., Moretz, R.C., and Lossinsky, A.S., 1981, Evidence for induction of localized amyloid deposits and neuritic plaques by an infectious agent, *Ann. Neurol.* 10:517–522.

Woolf, N.J., Gould, E., and Butcher, L.L., 1989, Nerve growth factor receptor is associated with cholinergic neurons of the basal forebrain but not the pontomesencephalon, *Neuroscience* 30:143–152.

Yoshimasu, F., Kokmen, E., Hay, I.D., Beard, C.M., Offord, K.P., and Kurland, L.T., 1991, The association between Alzheimer's disease and thyroid disease in Rochester, Minnesota, *Neurology* 41:1745–1747.

Yoshimura, T., 1974, Epidemiological study on Yusho babies born to mothers who had consumed oil contaminated by PCB, *Fukuoka Acta Med.* 65:74–78.

11

Aluminum Exposure Produces Learning and Memory Deficits

A Model of Alzheimer's Disease

ROBERT A. YOKEL

1. Evidence Suggesting a Relationship between Aluminum (Al) and Neurodegenerative Disorders

1.1. Occupational Aluminum Exposure

Aluminum is an ubiquitous element. Its exposure is unavoidable. Exposure that greatly exceeds that of the general population is experienced owing to occupation and medical treatments. Neurobehavioral toxicity attributed to occupational exposure to aluminum (Al) has been reported. Memory loss, tremor, jerking movements, and impaired coordination in one person were attributed to the practice of dipping red-hot metal articles, in an aluminum holder, into nitric acid (Spofforth, 1921). A large amount of Al was found in that individual's urine. Gradual mental deterioration and generalized convulsions were seen in a 49-year-old who worked for 13.5 years as a ball mill operator in an Al flake powder factory (McLaughlin *et al.*, 1962). Postmortem autopsy revealed no evidence of Alzheimer's disease. Brain and other organ Al concentrations were equal or greater than those seen in dialyzed uremics (see below). For 25 years, underground miners were exposed to McIntyre powder (finely ground elemental Al and Al oxide) by inhalation prior to mine entry as prophylaxis against silicotic lung disease. The number of years of McIntyre powder exposure correlated with impaired cognitive function, as measured by the Mini-Mental State Examination (MMSE) and

ROBERT A. YOKEL • Division of Pharmacology and Experimental Therapeutics, College of Pharmacy and Graduate Center for Toxicology, University of Kentucky, Lexington, Kentucky 40536.
Toxin-Induced Models of Neurological Disorders, edited by Michael L. Woodruff and Arthur J. Nonneman. Plenum Press, New York, 1994.

two additional cognitive tests (Rifat *et al.*, 1990). Electrical welding, polishing, and shaping of Al increased urinary Al output, suggesting absorption of inhaled Al (Mussi *et al.*, 1984). Increased duration of exposure to Al welding fumes positively correlated with questionnaire answers suggesting disturbances of memory and concentration, affective changes, and symptoms from the autonomic nervous system (Sjögren *et al.*, 1990). Cognitive impairment (particularly memory impairment), motor weakness, and incoordination were seen in ≥ 80% of 25 Al smelting plant employees, although MMSE scores were normal in most (White *et al.*, 1992). Abnormal evoked potentials were seen in three of these employees (Longstreth *et al.*, 1985).

1.2. Environmental Aluminum Exposure

A syndrome of amyotrophic lateral sclerosis parkinsonism dementia (ALS-PD) has been reported in the indigenous residents of three western Pacific islands. One theory attributes this syndrome to the high Al and Mn and low Ca concentrations in the soil, and presumably food and water of these people (Yasui *et al.*, 1991). The syndrome produces early death, with neurofibrillary tangles (NFTs) that mimic Alzheimer's disease (AD), but not the senile (neuritic; amyloid) plaques or cerebrovascular amyloid seen in AD.

Numerous epidemiological studies (reviewed by Crapper McLachlan *et al.*, 1991a) have shown a positive correlation between water Al concentration and AD incidence. Several other studies (Anonymous, 1992) failed to find an association. Visual-motor performance in children negatively correlated with hair Al concentration (Marlowe *et al.*, 1985). Similar studies in adult humans or animals have not been reported. The suggestion that Al can distribute into the brain via the olfactory system (Perl and Good, 1991) has not been well evaluated.

1.3. Pharmaceutical and Iatrogenic Aluminum Exposure

Initiation of widespread hemodialysis in the 1970s led to the development of syndromes in end-stage renal failure patients that have been attributed to Al, including the dialysis encephalopathy syndrome (DES, aka:dialysis dementia; Alfrey *et al.*, 1976). Aluminum sources include the dialysate and Al $(OH)_3$, which is used as a phosphate binder. DES has been seen in nondialyzed adult uremics and premature infants exposed to these Al sources. The neurobehavioral toxicity of DES includes speech difficulty, motor incoordination, memory loss, and tremor progressing to convulsions. NFTs, senile plaques, and cerebrovascular amyloid are not seen in DES.

The contribution of Al-containing antiperspirants/deodorants and antacids to AD was assessed by Graves *et al.* (1990). There was a trend toward a higher risk of AD with increased frequency of use of Al-containing antiperspirants. A positive association was found between antacid use and AD, but not between Al-containing antacid use and AD. These results do not clarify the Al–AD controversy.

1.4. Aluminum Exposure and Brain Aluminum in Alzheimer's Disease

The onset of AD is generally in the 5th decade or later, leading to a progressive decline in cognitive function and increasing emotional disturbance. Seizure activity is sometimes seen prior to death, which occurs 6–8 years after AD onset. Hallmark neuropathological signs include NFTs and senile plaques, as shown in Figure 1, and cerebrovascular amyloid. Numerous studies have been conducted utilizing various analytical instrumental techniques to assess the observation of Crapper *et al.* (1973) that Al is higher in AD than age-matched non-AD brains at autopsy. Many studies found higher brain Al in AD, but many did not (see Landsberg *et al.*, 1992, and references therein for some of these studies). These inconsistent findings contribute to the controversies concerning the role of Al in the etiology of AD and the possible ability of AD neuropathology to cause neuronal Al accumulation. Two-year treatment with desferrioxamine—a chelator of iron, Al, and other metals—was reported to reduce the progression of some AD symptoms (Crapper McLachlan *et al.*, 1991b). Desferrioxamine has been reported to reverse Al-induced behavioral effects in rats (Connor *et al.*, 1989; discussed below) and Al-induced NFTs in rabbits (Savory, personal communication).

It is well accepted that Al is a neurotoxin and that it is a contributory factor in the etiology of DES. The contribution of Al to the etiology of neurobehavioral symptoms in persons occupationally exposed to Al, or in persons who develop the ALS-PD complex or AD, is highly controversial. The signs and symptoms of these disorders are varied, contributing to the argument that Al cannot be the primary etiological agent for all of these disorders. The disparity in symptomatology and replicability of studies prevents synthesis of a concise description of Al induced neurobehavioral toxicity in humans. A single animal model of Al intoxication cannot address the disparate syndromes attributed to Al that have been reported in humans. Therefore, in this chapter, the results of studies of Al in animals will be related to AD.

2. Experimental Studies of Aluminum-Induced Neurobehavioral Toxicity

2.1. Historical Review

The initial observation of Al-induced neurotoxicity was of Al tartrate-induced neuronal degeneration in rabbits (Siem, quoted by Döllken, 1898). Aluminum phosphate injection intracerebrally or into the cisterna magna of

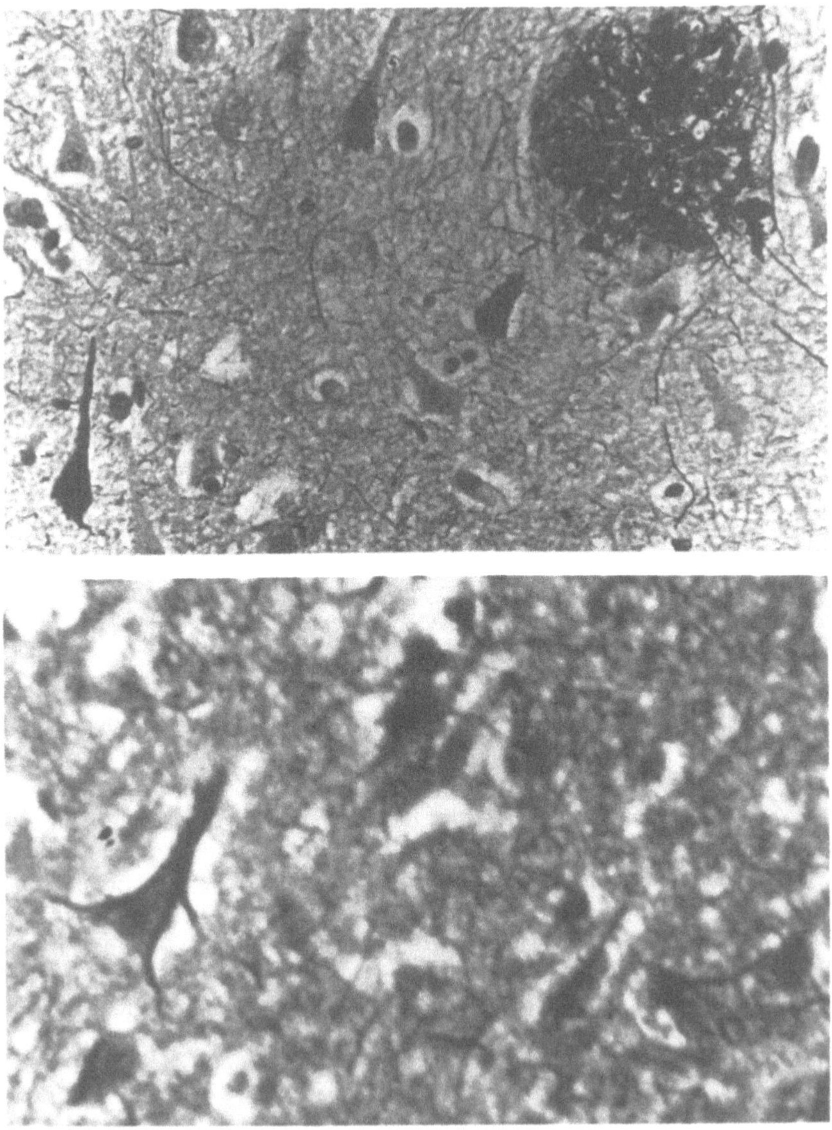

Figure 1 Photomicrographs of silver-stained hippocampus from Alzheimer's disease (top) and hippocampal cortex from rabbit (bottom) postmortem brain. The rabbit received 20 subcutaneous injections of 0.4 mmole/kg Al (lactate). Flame-shaped structures in lower left corner of top panel and left side of bottom panel are neurofibrillary tangles. The large circular structure in the upper right of the top panel is a senile plaque. Photomicrograph in top panel prepared by Dr. Daniel Perl and published in 1985 in "Alzheimer's disease," by R.J. Wurtman, *Sci. Am.* 252:62. Reprinted with permission.

rabbits produced ataxia, convulsions, and NFTs approximately 9–14 days later (Klatzo *et al.*, 1965). The route of exposure, abrupt onset, and profound convulsions do not mimic AD or its progression. Repeated subcutaneous (sc) injection of Al to rabbits also produced NFTs (DeBoni *et al.*, 1976). Symptom onset was delayed, compared to intracerebroventricular (icv) Al, and was Al dose dependent, rendering these animals more amenable to prolonged study. A prolonged Al-induced neurotoxicity was also obtained with Al powder injection into the cisternal medullary space (Wisniewski *et al.*, 1982; Bugiani and Ghetti, 1982). When scaled to life expectancy, rabbit survival for 8–10 months after Al powder administration parallels human survival for 6–8 years after AD onset. In contrast to AD, Al-treated rabbits did not uniformly develop neurobehavioral pathology and die, even within 1 year.

2.2. Aluminum-Induced Changes in Neuronal Structure and Function in the Rabbit

Evaluation of the ability of Al-induced NFTs to model AD NFTs revealed that the rabbit NFTs were less twisted and more branched than human NFTs and that they contained Al, which was not detected in human NFTs (Terry and Peña, 1965). The Al-induced NFTs were shown to be orderly bundles of 10-nm filaments, not modeling the paired, 10-nm-wide, helix-wound filaments found in AD, ALS-PD, and other neurodegenerative disorders (Wisniewski *et al.*, 1970). The topography of NFTs induced by icv and cisternal medullary-injected Al was reported to be similar and to include limbic cortex, striatum, thalamus, brainstem, medulla, and spinal cord (Wisniewski *et al.*, 1980). This differs from AD, which demonstrates a preponderance of NFTs in the entorhinal and perirhinal cortices and hippocampus (Arriagada *et al.*, 1992). In contrast, Pendlebury *et al.* (1988) reported that icv Al produces NFTs in cortex and hippocampus. Senile plaques are not produced by Al; therefore, neuropathologically, the Al-intoxicated rabbit more closely models ALS-PD complex than AD.

The rabbit has been utilized in most studies of Al-induced neurobehavioral toxicity. The cat and ferret are the only other species shown to develop NFTs rapidly and reliably (Wisniewski *et al.*, 1980). Al-induced NFTs have been extensively studied as one of the few methods to produce persistent NFTs that model, to some extent, AD NFTs. Over 50 research reports have clearly demonstrated the ability of Al to produce NFTs in rabbits. Although ultrastructurally different from AD NFTs, both types of NFTs are space-occupying perikaryal inclusions that extend into the dendrites and, to a lesser extent, the axon, potentially altering neuronal function. Both Al-induced and AD NFTs are found in cortical projection neurons, but not interneurons (Kowall *et al.*, 1989). Shigematsu and McGeer (1992) concluded that Al induces an axoplasmic flow disturbance.

Aluminum-induced NFTs and AD NFTs positively respond to silver stain and SMI 31, an antibody to phosphorylated neurofilaments. However, antibodies against tau and microtubule-associated protein 2 (MAP2) strongly stained AD, but not Al-induced NFTs in the rabbit (Kowall *et al.*, 1989). Tau may be the principal protein subunit of AD NFTs (Yankner and Mesulam, 1991). Shigematsu and McGeer (1992, and personal communication) injected Al (0.075–7.5 μmole as the Cl) icv or into the striatum of rats and found positive immunohistochemical staining with Alz50 (an antibody raised against an AD-specific protein), modeling AD. However, they found no staining to ubiquitin (a protein associated with NFTs), tau, or paired-helical filaments (PHF), in contrast to AD. Oral Al to rats, which do not develop NFTs, reduced MAP2, but not tau or tubulin (Johnson *et al.*, 1992). The similar response to silver and SMA 31 stains in AD and Al-intoxicated rabbit brain probably reflects the presence of NFTs and some similarities in NFT composition between these two types of brains. The differential response to PHF, tau, MAP2, and ubiquitin probably reflects significant differences in their structure and origin, as reviewed by Bertholf (1987).

Considerable recent research on the etiology of AD has focused on the hypothesis that abnormal amyloid precursor protein (APP) metabolism leads to production of excess or abnormal cerebrovascular beta-amyloid protein. Shigematsu and McGeer (1992) found that immunoreactivity to APP and beta-amyloid protein residues increased in striatal axonal cytoplasm, cortical pyramidal axons and cell bodies, and microglia of Al-exposed rats, modeling AD. Clauberg and Joshi (1993) suggest a proposed mechanism of Al-induced beta-amyloid protein production.

The primary neurotransmitter disruption in AD is a reduction of cholinergic function, although numerous other neurotransmitter systems are affected. High-affinity choline uptake was inhibited by Al (Lai *et al.*, 1980), as appears to occur in some AD victims (Pascual *et al.*, 1991). The ability of Al to model the neurochemical changes in AD was assessed in postmortem tissue from Al-intoxicated rabbits (Beal *et al.*, 1989). Decreases in cortical choline acetyltransferase activity, serotonin, norepinephrine, and to a lesser extent glutamate, were seen in the Al model and have been reported in AD brain. The lack of an Al-induced decrease of somatostatin or GABA did not model AD.

A primary advantage of the Al model is the persistence of the structural neuropathological changes, in contrast to those produced by colchicine and the vinca alkaloids.

2.3. Aluminum-Induced Blood-Brain Barrier Changes

Disruption of blood-brain barrier (BBB) integrity can produce neurobehavioral toxicity. Animal studies have shown that peripherally administered Al increases the transmembrane diffusion of a variety of compounds across the BBB, including sucrose and peptides (Kim *et al.*, 1986; Banks and Kastin,

1989). The integrity of the BBB, assessed by permeation of albumin (M_r approx. 70,000) into the brain, was not disrupted by Al (Banks and Kastin, 1989). However, BBB integrity assessment by permeation of a quaternary compound (M_r 246) revealed an Al-induced compromise in integrity (unpublished observations). Aluminum inhibited the transport from brain to blood of some, but not all, tested substances, suggesting a selective noncompetitive inhibition by Al (Banks and Kastin, 1989). There is evidence for disruption of BBB permeability in AD (Kalaria *et al.*, 1991), which might contribute to AD pathology.

3. Neurobehavioral Studies in Species Other Than the Rabbit

As the signs of AD are primarily manifest as a progressive dementia, assessment of the Al-intoxicated animal as a model of AD has focused on neurobehavioral toxicity. Concurrent with their report of elevated Al in AD brain, the laboratory of Donald Crapper McLachlan investigated the effect of bilateral hippocampal or internal capsular Al injection on behavior in the cat (Crapper and Dalton, 1973). Cats were trained in a delayed response task before they were Al injected and their performance tested daily. To perform the task correctly, the cat had to select the well containing food 5–40 sec after viewing placement of the food in one of 2 or 3 wells. Performance declined, suggesting an Al-induced short-term memory deficit. Similar Al treatment attenuated acquisition of a one-way active conditioned avoidance response (CAR), which required walking from the preferred to the nonpreferred side of a two-chamber shuttle-box. This study suggested an Al-induced learning deficit. Electrical self-stimulation of the hypothalamus was not altered by Al, suggesting that the behavioral decrements were not due to an Al-induced alteration in motivation or reinforcement mechanisms. There have been no other reported studies of Al neurobehavioral toxicity in the cat.

Neurobehavioral studies of Al in rodents have generally not produced robust or consistent results. Bowdler *et al.* (1979) studied 60-day-old Al-treated rats for general activity in an open field, coordinated motor activity on a rotorod, sensory response measured as the critical flicker frequency (CFF), and acquisition of a one-way CAR. The rats received daily oral Al gavage, under ether anesthesia, for 21 days. Behavioral results were correlated to brain Al concentration. The validity of their results has been questioned owing to the repeated ether exposure. CAR acquisition did not correlate with brain Al, although increased brain Al correlated with reduced rotorod time, increased distance traveled in the open field, and increased CFF, suggesting Al-induced excitement. King *et al.* (1975) also failed to find an Al-induced attenuation of learning in 230–300-g rats after bilateral hippocampal Al injections, using a CAR. Although brain Al was greatly elevated, NFTs were not seen in these rats, nor have NFTs been reported in any rats exposed to Al. The results suggested that the Al-induced impairment of learning and memory is dependent on the presence of NFTs.

Addition of Al (0.2% as the chloride) to the diet of rats for 12 weeks failed to affect acquisition or extinction of a two-way CAR, although it significantly increased the number of trials to retention, suggesting an Al-induced memory deficit (Commissaris *et al.*, 1982). Locomotor activity was attenuated, in contrast to the results of Bowdler *et al.* (1979), whereas rotorod performance was not, suggesting Al-induced decreased activity but no disturbance of skeletal muscle tone or coordination. Dietary Al exposure for 11 months (0.1% Al) increased trials to CAR acquisition, suggesting an Al-induced behavioral deficit in the absence of NFTs. Exposure of adult rats to four dietary Al concentrations for 1 month failed to demonstrate Al exposure-related changes in mouse-killing behavior, open field activity, latency to a passive CAR, or acquisition of a visual discrimination reversal task (Thorne *et al.*, 1986). However, when total or hippocampal brain Al was compared to activity and adverse affects on the passive CAR and discrimination reversal tasks, positive correlations were obtained. The reported hippocampal Al concentrations are considerably higher than other reports, decreasing confidence in some of the brain Al-behavior correlations. Consumption of drinking water containing 12.8 mg $Al(OH)_3$/ml for 60 days beginning at weaning did not produce significant increases in brain Al or alteration of cognitive function in rats (Thorne *et al.*, 1987). The authors conclude that developing animals are more resistant to the neurotoxic effects of Al. Addition of 0.3% Al to the diet of 120–130-g rats for 1 month did not affect open-field activity but decreased the correct response (long latency to enter a darkened chamber) in a passive CAR task (Connor *et al.*, 1988), suggesting that Al exposure reduced retention of the response.

However, the Al-exposed rats extinguished faster. If extinction is viewed as the acquisition of a new response, these results suggest an Al-induced facilitation of learning. The acquisition and retention of a one-way active CAR and the reacquisition of 8-arm radial maze performance, to assess working memory, were not affected by Al exposure. When Connor *et al.* (1989) repeated their previous study they found no Al effect on passive CAR acquisition but were able to replicate the Al-induced facilitation of extinction. Administration of the Al chelator desferrioxamine after Al exposure attenuated the Al-induced facilitated extinction of the passive avoidance response. The authors note that chelation treatment reversed the behavioral deficit produced by Al. The significance of this result in relation to the apparent ability of Al to enhance learning (facilitation of extinction) is difficult to interpret. Addition of Al chlorohydrate (1%) to the drinking water of mice attenuated acquisition of a two-way CAR at 2 months of age when exposure began at 1 day of age, but not at 1 month of age (Yen-Koo, 1992). Two Al injections by the intraperitoneal or sc route to 1-month-old mice also attenuated CAR acquisition, but not in a dose-dependent manner (Yen-Koo, 1992). Older mice were not tested.

In several of the above studies, Al exposure did not alter open-field activity, suggesting that the effects on behavioral tests attributed to Al were not due to changes in motor activity. However, using motor activity as a measure of neurotoxicity, Golub *et al.* (1989) found decreased total activity and vertical movement in 8–12-week-old mice after consuming a diet for 6 weeks that contained sufficient Al (1000 μg/g) to increase tissue Al two to four times. Consumption by mice of an Al-containing diet throughout gestation and lactation produced no effects on body weight during gestation but some decrease during lactation. Neurobehavioral test battery performance was not affected (Donald *et al.*, 1989; Golub *et al.*, 1992a). However, 18-day-old offspring showed retarded maturation of vertical screen climbing and 21-, 25-, or 39-day-old-offspring showed increased forelimb and hindlimb grip strength, decreased sensitivity to a thermal stimulus, and increased foot splay. Although the neurobehavioral test battery failed to show effects in the maternal Al-exposed mouse, mice consuming Al throughout gestation and lactation developed ataxia, limb splaying, and hindlimb dragging (Golub *et al.*, 1987, 1992a). Similar neuromotor deficits developed in rabbits after repeated Al injections (Yokel, 1984). Increasing dietary Al exposure to 90 days decreased hindlimb grip strength and auditory startle response and increased negative geotaxis latencies (Golub *et al.*, 1992b). Gestational Al exposure in mice produced maze-learning deficits in offspring reared in a nonenriched environment (Tsujii and Hoshishima, 1979). Behavior of the adults was not studied. The relationship between Al-induced developmental toxicity and AD is unclear. Similar studies in adult or aged mice might indicate whether the Al-intoxicated mouse models AD to any extent.

4. Neurobehavioral Studies of Aluminum in the Rabbit

As the rabbit readily develops NFTs, as shown in Figure 1, and a clinical syndrome of Al neurotoxicity, it is a reasonable species in which to expect Al-induced learning and memory deficits. Using a step-down active CAR task, Petit *et al.* (1980) demonstrated a deficit in both acquisition, 10 days after, and reacquisition, 13 days after, icv injection of 5 μmole of Al. Dendritic changes in the Al-treated rabbits consistent with dying back provided a possible cause for the behavioral deficits. A similar study in rabbits Al-treated at 2 days of age with 2 μmole Al revealed less pronounced Al-induced behavioral toxicity in the presence of NFTs and dendritic dying back, suggesting that the immature mammal was less susceptible than the adult to Al (Petit *et al.*, 1985). Similar results were obtained by Wisniewski *et al.* (1980), who observed that icv injection of

0.5 mg Al to infant rabbits produced a lower incidence of mortality and neurological signs and symptoms than seen after injection of 1 mg to adult rabbits. His group (Rabe *et al.*, 1982) demonstrated Al-induced deficits in water-maze acquisition in rabbits that received cisternal medullary Al at age 15 days. Older rabbits were not studied, preventing the comparison of age-dependent effects.

It was recognized that Al-induced motor deficits might confound interpretation of behavioral studies (Crapper and Dalton, 1973). Many of the studies cited above included a measure of general activity to assess Al-induced motor deficits. This concern was an incentive to select a behavioral procedure in the rabbit with minimal dependence on motor function. The rabbit is not typically used in behavioral studies. However, the classically conditioned-defensive eyeblink reflex has been extensively studied in the rabbit for three decades, particularly in the past decade by physiological psychologists. Their studies have revealed the neural pathways and structures subserving this simple form of learning and memory. The hippocampus and cerebellum are two of the structures intimately involved. The site of the engram appears to be the cerebellum (Lavond *et al.*, 1993). This behavior was selected for the study of Al toxicity in the rabbit owing to the low probability that behavioral changes would result from Al-induced motor effects.

We studied acquisition, retention, and extinction of a differential classically conditioned eyeblink response in the rabbit by recording nictitating membrane (third eyelid) extensions as the conditioned response (CR). Differential conditioning was conducted using two auditory conditioning stimuli (CS), 1 and 8 kHz, only one of which was paired with the unconditioned stimulus (UCS), namely shock, in a trace paradigm. Rabbits received 100 conditioning trials/daily session for 8 consecutive days. In 50 trials, one tone was paired with the UCS; in the other 50 trials the other tone was presented alone. Ten days after completion of acquisition training, rabbits received 100 trials of each tone alone, in a retention/extinction session. (See Yokel *et al.* [1988] for a timetable of this experimental procedure.) Rabbits were Al-intoxicated by repeated treatment with various Al doses. Intoxication was evidenced by the production of NFTs, elevated tissue Al, and an inverse relationship between weight gain and Al dose. Response acquisition was attenuated in adult (approximately 6-month-old) rabbits that received 20 sc injections of Al (200 or 400 μmole/kg/injection as the lactate) over 1 month.

Behavioral testing began 2.5 weeks after completion of Al injections. Rabbits that received 50 or 100 μmole/kg injections did not differ from controls (Yokel, 1983). These results suggest an Al-induced learning deficit after the higher Al exposures. All groups of rabbits demonstrated differential conditioning. Presentation of explicitly unpaired CSs and UCSs failed to elicit CRs, further suggesting that the CRs seen during conditioning were learned, not due to nonassociative processes such as sensitization or pseudoconditioning. The percentage of unconditioned responses (UCRs) as a function of UCS intensity was not decreased in the rabbits that received 400 μmole Al/kg compared to controls, nor were the latencies or magnitudes of the UCRs, suggesting that the attenuation of response acquisition was not due to Al-induced sensory or motor impairment of the CR. Comparison of the percentage of CRs during the initial trials of the retention/extinction session, as a measure of memory (Yokel *et al.*, 1988), revealed greater response savings in the control than 400 μmole Al/kg-loaded rabbits, suggesting an Al-induced memory deficit. The repeated sc injection of 400 μmole Al/kg produced NFTs in rabbit cortex, but not the hippocampus or cerebellum (Forrester and Yokel, 1985), suggesting that the Al-induced behavioral toxicity occurred in the absence of NFTs in the primary structures mediating this behavior.

To assess the developmental stage of the mammal during which Al is most behaviorally toxic, rabbits were exposed to Al *in utero,* during the first month postpartum, during the second month postpartum, at approximately 6 months of age (above), and as aged rabbits (Yokel, 1984, 1985, 1987, 1989; Yokel *et al.*, 1988). Two–3.4-year-old retired breeders, which were used as aged rabbits, demonstrated cerebellular Purkinje cell changes that model the aged human brain (Woodruff-Pak, 1988). Rabbits were Al exposed *in utero* by sc Al injection to the pregnant doe, whereas all other rabbits received sc Al injections. Conditioned response acquisition began at 7 or 11 weeks of age in rabbits exposed to Al *in utero* and during the first postpartum month and at 11 and 15 weeks of age in rabbits exposed to Al during the second postpartum month. Aluminum-induced learning and memory deficits were only seen in adult and retired breeder rabbits (for a summary figure see Yokel *et al.*, 1988), suggesting that the mature and aged mammalian brain is more susceptible to Al-induced toxicity than the immature brain. This is consistent with the observations of others, discussed above. This is also consistent with the occurrence of AD in late adult and aged humans. It is not known, however, if high-level Al exposure early in the life of the rabbit would result in behavioral deficits in adult or later life, because this study was not conducted. When adult rabbits were given an icv Al or control injection and tested 6 to 10 days later for acquisition or retention of the eyeblink reflex, Al-induced learning and memory deficits were seen (Pendlebury *et al.*, 1988; Solomon *et al.*, 1988). The acquisition/retention deficit correlated with an NFT index based on the percentage of cells with NFTs in the frontal and parietal cortex, hippocampus, pons, and medulla. The deficit was independent of CS modality; tone, light, and vibration were compared—further indicating that the behavioral deficit was not due to sensory changes.

In a further attempt to rule out sensory disruption as the cause of classical conditioning deficits in Al-intoxicated rabbits, Solomon *et al.* (1990) used stimulation of the medial geniculate N., in the auditory pathway, as the CS. They found an Al-induced retention deficit, as seen with other CS modalities. An Al-induced conditioned reflex deficit was also demonstrated in rats after oral dosing of 2.5 mg/kg Al (as alum) for 6 months (Vasiukovich *et al.*, 1978). Classical conditioning studies in humans are discussed in Section 5.

The Relationship between Aluminum-Induced Structural and Functional Changes

Considerable early work showed a correlation between NFTs and neurobehavioral toxicity. To more directly test the hypothesis that Al-induced neurobehavioral toxicity can occur in the absence of NFTs, Farnell *et al.* (1982) studied hippocampal slices from Al-exposed rabbits prior to NFT development. Stimulus-evoked potentials and long-term potentiation were inhibited, supporting the hypothesis. Numerous studies have shown cellular effects (reviewed by Crapper McLachlan *et al.*, 1991a) and neurobehavioral toxicity in the absence of NFTs, including many of the studies in rodents, which do not form NFTs.

5. Neurobehavioral Investigations in Humans, Including Alzheimer Disease Subjects

Bowdler *et al.* (1979) used several neuropsychiatric tests in a study of ninety-three 56–90-year-old ambulatory healthy humans. Serum Al was determined in all subjects. When the 25 subjects with the highest serum Al were compared to the 25 with the lowest serum Al, there were significant differences in some neuropsychiatric test responses. However, mean serum Al concentrations in this study were approximately 100-fold greater than normal serum Al, shedding doubt on these results. Bolla *et al.* (1992) found a decline in visual memory in those hemodialysis patients who exhibited higher serum Al. Serum Al was measured after injection of desferrioxamine, given to indicate Al body burden. Patients with lower vocabulary skills showed further Al-neurocognitive declines, suggesting subtle Al-induced neurotoxicity in this population.

Comparisons of the methods used to diagnose and assess AD and its severity (McKhann *et al.*, 1984) to the methods utilized in studies of the Al-intoxicated animal reveal few similarities in the assessment mechanisms. The clinical diagnosis of AD typically includes neuropsychiatric evaluation, often the MMSE (Folstein *et al.*, 1975) and the Alzheimer's Disease Assessment Scale (ADAS; Rosen *et al.*, 1984). Of the components of those evaluations, the recall of objects, test instructions, or previously presented words were closely modeled by recall of food position, in the studies of Crapper and Dalton (1973). Although tremor is evaluated in the ADAS, and frequently reported in Al-intoxicated rabbits, it has not been systematically evaluated in Al-intoxicated animals. Of the other laboratory assessments used to diagnose AD, including the EEG and evoked potentials, CT, regional blood flow, PET, and MRI, the effects of Al have been described only on EEG and evoked potentials. Slow-wave EEG activity increased in both AD and Al-intoxicated subjects (McKhann *et al.*, 1984; Crapper, 1973). Evoked potential latencies decreased in AD (McKhann *et al.*, 1984) whereas amplitude decreased in Al-intoxicated rabbits (Crapper, 1973).

A further test of the ability of the Al-intoxicated rabbit to model AD is the comparison of the ability of AD subjects and Al-intoxicated rabbits to learn the classically conditioned eyeblink reflex. The eyeblink reflex has been extensively utilized in human studies. Attenuated conditioning was observed in institutionalized demented patients diagnosed with either senile dementia or cerebral arteriosclerosis (Solyom and Barik, 1965). Three recent studies have demonstrated a deficit in eyeblink reflex acquisition in AD subjects. Woodruff-Pak *et al.* (1990) obtained 35%–40% CRs in controls after 80 conditioning trials, whereas AD subjects emitted only about 10% CRs. The 20 AD subjects, 11 of whom were living in institutional housing, had an average Blessed Memory test score of 17.55. Solomon *et al.* (1991) obtained 40% CR in controls and 10% in AD patients after 7 blocks of 10 trials/block. In this study the CS was presented alone in the first trial of each block and paired with the UCS in the other 9 trials, generating an intermittent CS–UCS pairing procedure. The AD patients had average MMSE and ADAS scores of 12.8 and 32, respectively. A study of AD patients with an average MMSE score of 20 compared to age/gender matched controls revealed 75% CR in the controls and 50% in the AD subjects after 160 trials (unpublished results). Neither group had reached asymptotic acquisition. Extinction rate was comparable in the two groups. These studies show an AD-induced attenuation of eyeblink reflex acquisition that appears to correlate with the degree of dementia. Therefore, attenuation of eyeblink acquisition was obtained in both Al-intoxicated adult and aged rabbits and AD patients, using very similar experimental paradigms and equipment.

The ability of the Al-intoxicated rabbit to model the AD patient, assessed by eyeblink conditioning, suggested the possibility that the rabbit model would be useful in the development of drugs to treat AD. This hypothesis was tested using the calcium channel entry enhancer 4-aminopyridine (4-AP), which was shown to reverse AD symptoms (Wesseling *et al.*, 1984). Intravenous injection of 1 μmole/kg 4-AP to Al-intoxicated rabbits 15 min before each eyeblink conditioning session resulted in facilitated eyeblink acquisition and retention, compared to vehicle-treated Al-intoxicated rabbits (unpublished results). Replication of this study resulted in a less robust effect of 4-AP. A second study in AD subjects showed less robust effects of 4-AP than the original study (Davidson *et al.*, 1988). The ability of the rabbit Al model to predict therapeutic utility for AD has not been further tested.

6. Summary

Aluminum has been implicated as an etiological agent in several disorders that manifest in different patterns of neurobehavioral toxicity. It is clear that Al has neurotoxic potential. It is not clear whether Al is an etiological agent in AD. Aluminum is able to produce NFTs that only partially model AD NFTs, providing the most practical method to produce persistent AD-like NFTs. Neurobehavioral effects of Al in rodents are not robust or consistent. The rabbit readily develops NFTs in response to brain Al and demonstrates neurobehavioral toxicity. The classically conditioned eyeblink reflex has been used to assess Al-induced behavioral toxicity in the rabbit and as an assay for potential anti-Alzheimer disease drugs. Alzheimer subjects show decrements in acquisition of the classically conditioned eyeblink reflex that parallel those seen in the rabbit. The Al-intoxicated cat and rabbit provide an analogous model for some of the neuropathological, neurochemical, and behavioral changes seen in Alzheimer's disease.

7. References

Alfrey, A.C., LeGendre, G.R., and Kaehny, W.D., 1976, The dialysis encephalopathy syndrome: Possible aluminum intoxication, *N. Engl. J. Med.* 294:184–188.

Anonymous, 1992, Is aluminium a dementing ion?, *Lancet* 339:713–714.

Arriagada, P.V., Marzloff, K., and Hyman, B.T., 1992, Distribution of Alzheimer-type pathologic changes in nondemented elderly individuals matches the pattern in Alzheimer's disease, *Neurology* 42:1681–1688.

Banks, W.A., and Kastin, A.J., 1989, Aluminum-induced neurotoxicity: Alterations in membrane function at the blood-brain barrier, *Neurosci. Biobehav. Rev.* 13:47–53.

Beal, M.F., Mazurek, M.F., Ellison, D.W., Kowall, N.W., Solomon, P.R., and Pendlebury, W.W., 1989, Neurochemical characteristics of aluminum-induced neurofibrillary degeneration in rabbits, *Neuroscience* 29:339–346.

Bertholf, R.L., 1987, Aluminum and Alzheimer's disease: Perspectives for a cytoskeletal mechanism, *CRC Crit. Rev. Clin. Lab. Sci.* 25:195–210.

Bolla, K.I., Briefel, G., Spector, D., Schwartz, B.S., Wieler, L., Herron, J., and Gimenez, L., 1992, Neurocognitive effects of aluminum, *Arch. Neurol.* 49:1021–1026.

Bowdler, N.C., Beasley, D.S., Fritze, E.C., Goulette, A.M., Hatton, J.D., Hession, J., Ostman, D.L., Rugg, D.J., and Schmittdiel, C.J., 1979, Behavioral effects of aluminum ingestion on animal and human subjects, *Pharmacol. Biochem. Behav.* 10:505–512.

Bugiani, O., and Ghetti, B., 1982, Progressing encephalomyelopathy with muscular atrophy, induced by aluminum powder, *Neurobiol. Aging* 3:209–222.

Clauberg, M., and Joshi, J.G., 1993, Regulation of serine protease activity by aluminum: Implications for Alzheimer disease, *Proc. Natl. Acad. Sci. USA* 90:1009–1012.

Commissaris, R.L., Cordon, J.J., Sprague, S., Keiser, J., Mayor, G.H., and Rech, R.H., 1982, Behavioral changes in rats after chronic aluminum and parathyroid hormone administration, *Neurotoxicol. Teratol.* 4:403–410.

Connor, D.J., Jope, R.S., and Harrell, L.E., 1988, Chronic, oral aluminum administration to rats: Cognition and cholinergic parameters, *Pharmacol. Biochem. Behav.* 31:467–474.

Connor, D.J., Harrell, L.E., and Jope, R.S., 1989, Reversal of an aluminum-induced behavioral deficit by administration of deferoxamine, *Behav. Neurosci.* 103:779–783.

Crapper, D.R., 1973, Experimental neurofibrillary degeneration and altered electrical activity, *Electroencephalogr. Clin. Neurophysiol.* 35:575–588.

Crapper, D.R., and Dalton, A.J., 1973, Alterations in short-term retention, conditioned avoidance response acquisition and motivation following aluminum-induced neurofibrillary degeneration, *Physiol. Behav.* 10:925–933.

Crapper, D.R., Krishnan, S.S., and Dalton, A.J., 1973, Brain aluminum distribution in Alzheimer's disease and experimental neurofibrillary degeneration, *Science* 180:511–513.

Crapper McLachlan, D.R., Kruck, T.P., Lukiw, W.J., and Krishnan, S.S., 1991a, Would decreased aluminum ingestion reduce the incidence of Alzheimer's disease?, *Can. Med. Assoc. J.* 145:793–804.

Crapper McLachlan, D.R., Dalton, A.J., Kruck, T.P.A., Bell, M.Y., Smith, W.L., Kalow, W., and Andrews, D.F., 1991b, Intramuscular desferrioxamine in patients with Alzheimer's disease, *Lancet* 337:1304–1308.

Davidson, M., Zemishlany, Z., Mohs, R.C., Horvath, T.B., Powchik, P., Blass, J.P., and Davis, K.L., 1988, 4-Aminopyridine in the treatment of Alzheimer's disease, *Biol. Psychiatry* 23:485–490.

DeBoni, U., Otvos, A., Scott, J.W., and Crapper. D.R., 1976, Neurofibrillary degeneration induced by systemic aluminum, *Acta Neuropathol.* 35:285–294.

Döllken, V., 1898, Ueber die wirkung des aluminiums mit besonderer berücksichtigung der durch das aluminium verursachten läsionen im centralnervensystem, *Arch. für Exper. Pathol. Pharmacol.* 40:98–120.

Donald, J.M., Golub, M.S., Gershwin, M.E., and Keen, C.L., 1989, Neurobehavioral effects in offspring of mice given excess aluminum in diet during gestation and lactation, *Neurotoxicol. Teratol.* 11:345–351.

Farnell, B.J., DeBoni, U., and Crapper McLachlan, D.R., 1982, Aluminum neurotoxicity in the absence of neurofibrillary degeneration in CA1 hippocampal pyramidal neurons *in vitro*, *Exp. Neurol.* 78:241–258.

Folstein, M.F., Folstein, S.E., and McHugh, P.R., 1975, "Mini-Mental State" A practical method for grading the cognitive state of patients for the clinician, *J. Psychiatr. Res.* 12:189–198.

Forrester, T.M., and Yokel, R.A., 1985, Comparative toxicity of intracerebroventricular and subcutaneous aluminum in the rabbit, *Neurotoxicology* 6(3):71–80.

Golub, M.S., Gershwin, M.E., Donald, J.M., Negri, S., and Keen, C.L., 1987, Maternal and developmental toxicity of chronic aluminum exposure in mice, *Fundam. Appl. Toxicol.* 8:346–357.

Golub, M.S., Donald, J.M., Gershwin, M.E., and Keen, C.L., 1989, Effects of aluminum ingestion on spontaneous motor activity of mice, *Neurotoxicol. Teratol.* 11:231–235.

Golub, M.S., Keen, C.L., and Gershwin, M.E., 1992a, Neurodevelopmental effect of aluminum in mice: Fostering studies, *Neurotoxicol. Teratol.* 14:177–182.

Golub, M.S., Han, B., Keen, C.L., and Gershwin, M.E., 1992b, Effects of dietary aluminum excess and manganese deficiency on neurobehavioral endpoints in adult mice, *Toxicol. Appl. Pharmacol.* 112:154–160.

Graves, A.B., White, E., Koepsell, T.D., Reifler, B.V., van Belle, G., and Larson, E.B., 1990, The association between aluminum-containing products and Alzheimer's disease, *J. Clin. Epidemiol.* 43:35–44.

Johnson, G.V.W., Watson, A.L., Lartius, R., Uemura, E., and Jope, R.S., 1992, Dietary aluminum selectively decreases MAP-2 in brains of developing and adult rats, *Neurotoxicology* 13:463–474.

Kalaria, R.N., Golde, T.E., Cohen, M.L., and Younkin, S.G., 1991, Serum amyloid P in Alzheimer's disease: Implications for dysfunction of the blood-brain barrier, *Ann. NY Acad. Sci.* 640:145–148.

Kim, Y.S., Lee, M.H., and Wisniewski, H.M., 1986, Aluminum-induced reversible change in permeability of the blood-brain barrier to [^{14}C]sucrose, *Brain Res.* 377:286–291.

King, G.A., De Boni, U., and Crapper, D.R., 1975, Effect of aluminum upon conditioned avoidance response acquisition in the absence of neurofibrillary degeneration, *Pharmacol. Biochem. Behav.* 3:1003–1009.

Klatzo, I., Wiśniewski, H., and Streicher, E., 1965, Experimental production of neurofibrillary degeneration. 1: Light microscopic observations, *J. Neuropathol. Exp. Neurol.* 24:187–199.

Kowall, N.W., Pendlebury, W.W., Kessler, J.B., Perl, D.P., and Beal, M.F., 1989, Aluminum-induced neurofibrillary degeneration affects a subset of neurons in rabbit cerebral cortex, basal forebrain and upper brainstem, *Neuroscience* 29:329–337.

Lai, J.C.K., Guest, J.F., Leung, T.K.C., Lim, L., and Davison, A.N., 1980, The effects of cadmium, manganese and aluminium on sodium-potassium-activated and magnesium-activated adenosine triphosphatase activity and choline uptake in rat brain synaptosomes, *Biochem. Pharmacol.* 29:141–146.

Landsberg, J.P., McDonald, B., and Watt, F., 1992, Absence of aluminium in neuritic plaque cores in Alzheimer's disease, *Nature* 360:65–68.

Lavond, D.G., Kim, J.J., and Thompson, R.F., 1993, Mammalian brain substrates of aversive classical conditioning, *Annu. Rev. Psychol.* 44:317–342.

Longstreth, W.T., Rosenstock, L., and Heyer, N.J., 1985, Potroom palsy? Neurologic disorder in three aluminum smelter workers, *Arch. Intern. Med.* 145:1972–1975.

Marlowe, M., Stellern, J., Errera, J., and Moon, C., 1985, Main and interaction effects of metal pollutants on visual-motor performance, *Arch. Environ. Health* 40:221–225.

McKhann, G., Drachman, D., Folstein, M., Katzman, R., Price, D., and Stadlan, E.M., 1984, Clinical diagnosis of Alzheimer's disease: Report of the NINCDS-ADRDA Work Group under the auspices of Department of Health and Human Services Task Force on Alzheimer's disease, *Neurology* 34:939–944.

McLaughlin, A.I.G., Kazantzis, G., King, E., Teare, D., Porter, R.J., and Owen, R., 1962, Pulmonary fibrosis and encephalopathy associated with the inhalation of aluminium dust, *Br. J. Ind. Med.* 19:253–263.

Mussi, I., Calzeferri, G., Buratti, M., and Alessio, L., 1984, Behaviour of plasma and urinary aluminium levels in occupationally exposed subjects, *Int. Arch. Occup. Environ. Health* 54:155–161.

Pascual, J., Fontán, A., Zarranz, J.J., Berciano, J., Flórez, J., and Pazos, A., 1991, High-affinity choline uptake carrier in Alzheimer's disease: Implications for the cholinergic hypothesis of dementia, *Brain Res.* 552:170–174.

Pendlebury, W.W., Perl, D.P., Schwentker, A., Pingree, T.M., and Solomon, P.R., 1988, Aluminum-induced neurofibrillary degeneration disrupts acquisition of the rabbit's classically conditioned nictitating membrane response, *Behav. Neurosci.* 102:615–620.

Perl, D.P., and Good, P.F., 1991, Aluminum, Alzheimer's disease, and the olfactory system, *Ann. NY Acad. Sci.* 640:8–13.

Petit, T.L., Biederman, G.B., and McMullen, P.A., 1980, Neurofibrillary degeneration, dendritic dying back, and learning-memory deficits after aluminum administration: Implications for brain aging, *Exp. Neurol.* 67:152–162.

Petit, T.L., Biederman, G.B., Jonas, P., and LeBoutillier, J.C., 1985, Neurobehavioral development following aluminum administration in infant rabbits, *Exp. Neurol.* 88:640–651.

Rabe, A., Lee, M.H., Shek, J., and Wisniewski, H.M., 1982, Learning deficit in immature rabbits with aluminum-induced neurofibrillary changes, *Exp. Neurol.* 76:441–446.

Rifat, S.L., Eastwood, M.R., Crapper McLachlan, D.R., and Corey, P.N., 1990, Effect of exposure of miners to aluminium powder, *Lancet* 336:1162–1165.

Rosen, W.G., Mohs, R.C., and Davis, K.L., 1984, A new rating scale for Alzheimer's disease, *Am. J. Psychiatry* 141:1356–1364.

Shigematsu, K., and McGeer, P.L., 1992, Accumulation of amyloid precursor protein in damaged neuronal processes and microglia following intracerebral administration of aluminum salts, *Brain Res.* 593:117–123.

Sjögren, B., Gustavsson, P., and Hogstedt, C., 1990, Neuropsychiatric symptoms among welders exposed to neurotoxic metals, *Br. J. Ind. Med.* 47:704–707.

Solomon, P.R., Pingree, T.M., Baldwin, D., Koota, D., Perl, D.P., and Pendlebury, W.W., 1988, Disrupted retention of the classically conditioned nictatating membrane response in rabbits with aluminum-induced neurofibrillary degeneration, *Neurotoxicology* 9:209–222.

Solomon, P.R., Koota, D., and Pendlebury, W.W., 1990, Disrupted retention of the classically conditioned nictitating membrane response in the aluminum-intoxicated rabbit using electrical stimulation of the brain as a conditioned stimulus, *Neurobiol. Aging* 11:523–528.

Solomon, P.R., Levine, E., Bein, T., and Pendlebury, W.W., 1991, Disruption of classical conditioning in patients with Alzheimer's disease, *Neurobiol. Aging* 12:283–287.

Solyom, L., and Barik, H.C., 1965, Conditioning in senescence and senility, *J. Gerontol.* 20:483–488.

Spofforth, J., 1921, Case of aluminium poisoning, *Lancet* 200:1301.

Terry, R.D., and Peña, C., 1965, Experimental production of neurofibrillary degeneration. 2: Electron microscopy, phosphatase histochemistry and electron probe analysis, *J. Neuropathol. Exp. Neurol.* 24:200–210.

Thorne, B.M., Donohoe, T., Lin, K.-N., Lyon, S., Medeiros, D.M., and Weaver, M.L., 1986, Aluminum ingestion and behavior in the Long-Evans rat, *Physiol. Behav.* 36:63–67.

Thorne, B.M., Cook, A., Donohoe, T., Lyon, S., Medeiros, D.M., and Moutzoukis, C., 1987, Aluminum toxicity and behavior in the weanling Long-Evans rat, *Bull. Psychonom. Soc.* 25:129–132.

Tsujii, H., and Hoshishima, K., 1979, The effect of the administration of trace amounts of metals to pregnant mice upon the behavior and learning of their offspring, *Shinshu Daigaku Nogakubu Kiyo (J. Fac. Agric. Shinshu U.)* 16:13–27.

Vasiukovich, Lla., Lamentova, T.G., Astakhova, L.F., and Konstantinova, I.N., 1978, [Reexamination of the maximum permissable concentration of aluminum in water], *Gig. Sanit.* 101–103.

Wesseling, H., Agoston, S., VanDam, G.B.P., Pasma, J., DeWit, D.J., and Havinga, H., 1984, Effects of 4-aminopyridine in elderly patients with Alzheimer's disease, *N. Engl. J. Med.* 310:988–989.

White, D.M., Longstreth, W.T., Rosenstock, L., Claypoole, K.H.J., Brodkin, C.A., and Townes, B.D., 1992, Neurologic syndrome in 25 workers from an aluminum smelting plant, *Arch. Intern. Med.* 152:1443–1448.

Wisniewski, H., Terry, R.D., and Hirano, A., 1970, Neurofibrillary pathology, *J. Neuropathol. Exp. Neurol.* 29:163–176.

Wisniewski, H.M., Sturman, J.A., and Shek, J.W., 1980, Aluminum chloride induced neurofibrillary changes in the developing rabbit: A chronic animal model, *Ann. Neurol.* 8:479–490.

Wisniewski, H.M., Sturman, J.A., and Shek, J.W., 1982, Chronic model of neurofibrillary changes induced in mature rabbits by metallic aluminum, *Neurobiol. Aging* 3:11–22.

Woodruff-Pak, D.S., 1988, Aging and classical conditioning: Parallel studies in rabbits and humans, *Neurobiol. Aging* 9:511–522.

Woodruff-Pak, D.S., Finkbiner, R.G., and Sasse, D.K., 1990, Eyeblink conditioning discriminates Alzheimer's patients from non-demented aged, *Neuroreport* 1:45–48.

Yankner, B.A., and Mesulam, N.-M., 1991, β-Amyloid and the pathogenesis of Alzheimer's disease, *N. Engl. J. Med.* 325:1849–1857.

Yasui, M., Yase, Y., Ota, K., and Garruto, R.M., 1991, Aluminum deposition in the central nervous system of patients with amyotrophic lateral sclerosis from the Kii peninsula of Japan, *Neurotoxicology* 12:615–620.

Yen-Koo, H.C., 1992, The effect of aluminum on conditioned avoidance response (CAR) in mice, *Toxicol. Ind. Health* 8:1–7.

Yokel, R.A., 1983, Repeated systemic aluminum exposure effects on classical conditioning of the rabbit, *Neurotoxicol. Teratol.* 5:41–46.

Yokel, R.A., 1984, Toxicity of aluminum exposure during lactation to the maternal and suckling rabbit, *Toxicol. Appl. Pharmacol.* 75:35–43.

Yokel, R.A., 1985, Toxicity of gestational aluminum exposure to the maternal rabbit and offspring, *Toxicol. Appl. Pharmacol.* 79:121–133.

Yokel, R.A., 1987, Toxicity of aluminum exposure to the neonatal and immature rabbit, *Fundam. Appl. Toxicol.* 9:795–806.

Yokel, R.A., 1989, Aluminum produces age-related behavioral toxicity in the rabbit, *Neurotoxicol. Teratol.* 11:237–242.

Yokel, R.A., Provan, S.D., Meyer, J.J., and Campbell, S.R., 1988, Aluminum intoxication and the victim of Alzheimer's disease: Similarities and differences, *Neurotoxicology* 9:429–442.

12

Trimethyltin Neurotoxicity in the Rat as an Analogous Model of Alzheimer's Disease

Michael L. Woodruff and Ronald H. Baisden

1. Usefulness of Animal Models of Neurological Disorders

1.1. Criteria for Assessing the Validity of an Animal Model

Animal models of diseases have been used in experiments to study the cause of a disease, the mechanisms underlying the cellular, organ, or systemic pathology of a disease, and to compare the efficacy of proposed therapeutic interventions to existing clinical treatments. A model may be useful for one of these purposes but not for the others. For example, a particular behavior, such as activity in an open field, may be changed in a dose-related fashion by drugs that have a recognized clinical effect in psychiatric patients, but the rat behavior may not have any clear relationship to the human psychopathology. This model is not necessarily useful for study of the cause and progression of the disease or of its pathophysiology. Rather, its validity relates to a consistent parametric relationship between the effect of a drug on this behavioral measure and the clinical efficacy of the same drug. Newly designed drugs that produce greater effects on the animal model without significant detrimental side effects would then be likely candidates for clinical trials.

This example may hardly appear to warrant the designation "model" and, indeed, seems circular. The rat behavior becomes a model of the human disease only because drugs that are used with at least some degree of success clinically also change the behavior in the rat. The psychopharmacologist who uses this behavior must either become a strict behaviorist, meticulously describing the

MICHAEL L. WOODRUFF and RONALD H. BAISDEN • Department of Anatomy and Cell Biology, J.H. Quillen College of Medicine, East Tennessee State University, Johnson City, Tennessee 37614.

Toxin-Induced Models of Neurological Disorders, edited by Michael L. Woodruff and Arthur J. Nonneman. Plenum Press, New York, 1994.

effects of drugs on the rat's behavior, or be forced to make assumptions about the cognitive structure of the rat and to construct intervening variables so as to explain the behavior in human clinical terms. The latter choice is not necessary. It adds nothing to the validity of the animal model for the purpose of preclinical testing because all that is required of the model is an accurate description of the rat's behavior and a consistent relationship between animal experimental results and clinical efficacy of the drug.

Indeed, the motoric measure of clinical effectiveness might not even be a behavior as one customarily defines the term. An example of this strategy drawn from preclinical pharmacology (Treit, 1985) is the ability of antianxiety agents to suppress seizures in rats and mice. Because of the known correlation between the clinical effectiveness of commonly used anxiolytics and their ability to inhibit seizures in rats, the dose-response curve for inhibition of pentylenetetrazol seizures in rodents is a fairly good predictor of the ability of a newly proposed anxiolytic compound, and this test is widely used as a preclinical screen by pharmaceutical companies.

This model meets the fourth criterion proposed by McKinney and Bunney (1969) in their presentation of criteria for validating animal models of human psychopathology. That is, if a putative clinical therapeutic agent exists, the dependent variable(s) measured in the model should react to therapeutic intervention in a way that is predictable from the effects of the same intervention when applied to humans. The remaining three criteria proposed by McKinney and Bunney are similarity of inducing conditions, similarity of behavioral states, and common underlying mechanisms. Presumably the greater number of these criteria that a particular model meets, the more valid the model. However, as suggested in the opening paragraphs of this chapter, a model of a human disease process may be very useful without meeting all, or even most, of these criteria. Therefore, we propose that as a measure of validity the purpose of the model is as important, or maybe even more important, than the number of points of contact between the model and the naturally occurring disease state. Further, this position may be especially true for development of models of neurological diseases such as Alzheimer's disease (AD) where large gaps exist in our knowledge of the linkage among clinical symptoms, the histological and cellular pathologies, and the fundamental pathological mechanism(s).

1.2. Correlative, Analogous, and Homologous Animal Models

However, we do not deny that a given model will be more useful in studying a disease the more similarities that obtain between the two entities. In the example provided in the opening paragraphs of this chapter, only one similarity was given between the disease and the model—that is, both responded in a predictable way to drug treatment. This type of model may be viewed as the type that

Gage *et al.* (1985) designated as *correlative*. The symptoms of interest presented in the model do not have to be analogous to those of the human disease. Therefore, this type of model may be created without much knowledge concerning the pathology of the endogenous disease being modeled. As suggested above, and stated explicitly by Gage *et al.* (1985), the most frequent use of correlative models is as screening devices for new therapeutic treatments. The model of AD proposed below is of at least this type.

Gage *et al.* (1985) define a second type of model as the *analogous* model. Referring specifically to neurological disorders, an analogous model of a disease would be one in which both the anatomical regions of the central nervous system known to be damaged in the human disease and appropriate behavioral abnormalities are present. The actual causes of the damage in the experimental model and in the human disease do not have to be the same. Essentially all of the models described in this book, with the possible exceptions of the 1-methyl-4-phenyl-1,2,3,6-tetrahydropyridine (MPTP) and ferric models of Parkinson's disease, and the models of motor neuron disease described by Strong and Garutto, represent analogous models. The model of AD proposed below can also be viewed as an analogous model.

The *homologous* model is the final type of model described by Gage *et al.* (1985). As the name suggests, this model assumes that the same disease state is found in animals and humans. Clearly, different behavioral repertoires must be considered when a model of a neurological disease is evaluated as a candidate for categorization as homologous. However, even with this consideration, homologous models are difficult to find in searching for models of human neurodegenerative diseases, and the model proposed below is not homologous. Indeed, of the models discussed in this text, only the MPTP model of Parkinson's disease in monkeys is incontestably homologous to at least some cases of the human disorder.

2. Features of Alzheimer's Disease and Animal Models

2.1. Existing Toxicant-Induced Models of Alzheimer's Disease

The two best-known toxicant-induced animal models of AD are described in this text. These are the consequences of exposure to aluminum in the rabbit and destruction of the basal forebrain cholinergic neurons, usually by microinjection of a toxin (Smith, 1988), although Juárez de Ku and Meserve (this volume) have presented data indicating that early exposure to PCBs can substantially decrease forebrain cholinergic activity. Neither experimentally induced disruption of cholinergic activity nor aluminum-induced cortical neuropathology is a verifiable homologous model. However, some (e.g., Roberts, 1986) support can be mustered for the hypothesis that exposure to aluminum salts is at least a part of the etiology of AD, and animal models designed to study this possibility have been devised (Yokel, this volume; Solomon *et al.*, this volume).

The position that aluminum is the sole etiological factor in AD is not held even by most proponents of the weaker position that aluminum or aluminum silicates are one factor in the complex etiology leading to AD. Indeed, many authorities doubt that aluminum has any causative role in AD (e.g., Terry, 1986; Ward, 1986). Therefore, the aluminum model may best be regarded as an analogous model that includes neuropathological (neurofibrillary tangles and senile plaques) and behavioral (impaired memory formation) elements that also occur in AD.

No single model of AD can be considered comprehensive. As exemplified by the aluminum model and the cholinergic depletion model, each model of AD reproduces only some of the neuropathological hallmarks of the disease, and all the models tend to use memory dysfunction as the validating behavioral variable. The model presented below is a continuation of this general strategy. Indeed, until more extensive knowledge concerning the etiology of AD is achieved, it will be difficult to move beyond the analogous model. Parenthetically, it could be argued that without a more homologous model of AD its cause will remain elusive.

Despite this last caveat, the goal of this chapter is to develop an argument for exposure to trimethyltin (TMT) as an analogous model of AD, which generally mimics some of the changes seen in the neocortex and hippocampus of AD victims as well as certain of the behavior deficits that are caused by the disease (i.e., memory dysfunction, disarray of personal grooming, wandering). To provide a basis for comparison in our attempt to accomplish this goal a brief overview of some of the changes that accompany AD will be presented. Neuropathological changes in the brains of AD victims will be presented in Sections 2.2 through 2.4 of this chapter. The "classic" neuropathological hallmarks of AD will be described first, followed by a description of discrete patterns of neuropathology. Reference to reactive synaptogenesis (axonal sprouting) within the hippocampus in AD victims will conclude the section. A brief synopsis of the behavioral consequences of AD will be presented in Section 3 and the proposed model will be presented in Section 4.

2.2. Traditional Neuropathological Markers of Alzheimer's Disease

At the gross level of analysis the classic neuropathological markers for AD include atrophy of the association areas of the frontal, parietal, and occipital lobes, while microscopically neurofibrillary tangles (NFTs) and neuritic or "senile" plaques are also observed (Figure 1A–C). In addition to these microscopic changes, Hirano bodies and granulovacuolar degeneration occur within hippo-

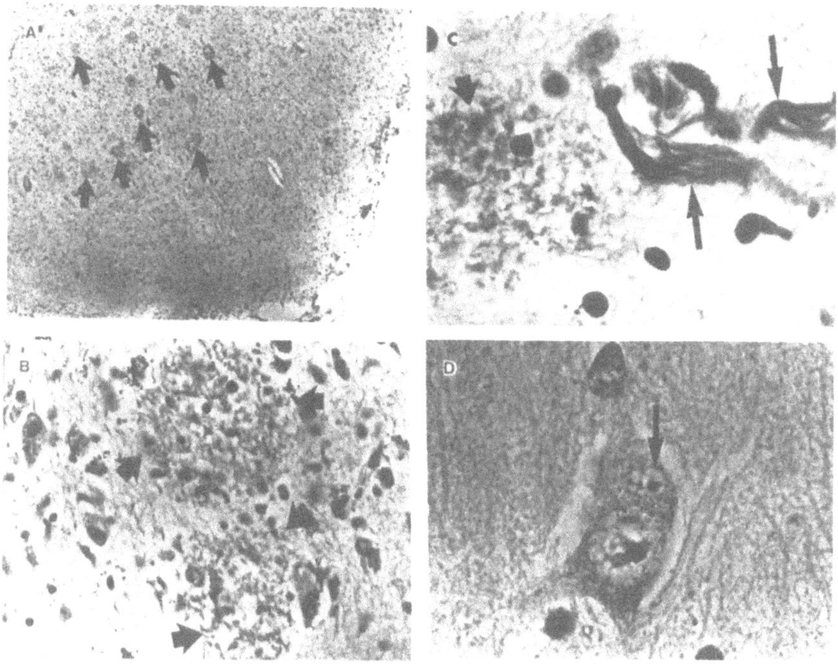

Figure 1 **A**, **B**, and **C** present photomicrographs of silver-stained sections of brain tissue from an Alzheimer's victim. The arrowheads in all three panels point to plaques. Note that the plaques are located below the first two layers of cortex (**A**) and are not homogeneous in appearance (**B** and **C**), but contain particles and strands of differing density, shape, and size. The longer arrows in (**C**) point to neurofibrillary tangles. The arrow in (**D**) points to an example of granulovacuolar degeneration in a hippocampal pyramidal neuron stained with hematoxylin and eosin.

campal neurons. Hirano bodies (Hirano, 1965) are eosinophilic cytoplasmic inclusions that appear in several neurodegenerative diseases. They are presumed to be made of thin actin filaments (Goldman, 1983). Granulovacuolar degeneration (Figure 1D) appears as a granule contained in a membrane-bound vacuole within the cytoplasm of hippocampal pyramidal cells. As with neurofibrillary tangles, granulovacuolar degeneration (Ball and Lo, 1977) is a type of cytoskeletal abnormality and appears to be a form of tubulin (Price *et al.*, 1985).

2.3. Regional and Laminar Differences in Neuropathological Markers of Alzheimer's Disease

As indicated above, cortical atrophy, when present, in the brains of AD victims is substantially more prevalent in the association areas of the cortex and the hippocampal formation than in the primary motor and sensory areas. This observation of apparent gross differences in loss of neocortical tissue corresponds to the loss of memory and higher cognitive functions with relative preservation of basic sensory and motor function seen in AD victims.

Microscopic analyses suggest that even within areas that are prone to gross atrophy not all neuronal layers and subfields demonstrate equal pathology. For example, some data suggest that laminar differences in neuropathology may exist in association areas of the neocortex. For example, morphometric studies by Mann *et al.* (1985) and Lewis *et al.* (1987) indicate that NFTs are more often found in the pyramidal neurons of layers III and V than in neurons in other lamina. However, although NFT formation is a valid predictor of neuron death it is not the only predictor. Thus, the prevalence of NFTs in layer III and layer V neurons does not necessarily entail that AD is a disease of large projection neurons. Indeed, Braak and Braak (1986) have found that at least in the medial parts of the orbitofrontal cortex small interneurons are more likely to be affected in AD than are pyramidal cells. Further, Vogt *et al.* (1990) have observed that the pattern of neuron loss in the posterior cingulate gyrus demonstrates five variations that may correlate to different clinical manifestations of the disease. Therefore, it cannot be safely stated that neuron loss in one or two particular layers of association neocortex is always more prevalent, or more prevalent earlier in the course of the disease, than loss in other lamina in the same neocortical field. There appear to be regional variations as well as possible variations that correspond to the clinical presentation of the disorder (Vogt *et al.*, 1990).

However, more compelling evidence exists for laminar and subfield-specific neuropathology within the entorhinal cortex and hippocampal formation. For example, Hyman and Van Hoesen (1989) report that in the brains of AD victims both NFTs and neuronal necrosis are found far more frequently in neurons within layers II and IV of the entorhinal cortex than the other layers of this cortical area. Layer II gives rise to the glutaminergic perforant pathway that provides the major neocortical input to the hippocampus, whereas layer IV is the target of efferents from the subiculum and subfield CA1 of the hippocampus proper. Hyman and Van Hoesen also report that NFTs are found with high frequency in the projection neurons CA1 and the subiculum, whereas CA3 pyramids are relatively free of NFTs. These investigators report that plaque formation is higher in the terminal field of the perforant pathway in the dentate gyrus than in other regions of the hippocampus, thus suggesting significant axonal degeneration in this area. Support for this contention was provided by Hyman *et al.* (1987b), who found that glutamate was significantly reduced in the outer molecular layer of the dentate gyrus in the brains of AD victims. These observations are interpreted to indicate that memory impairments observed in AD are not due to total degeneration of the hippocampus, but to its isolation by loss of its connections to the neocortex.

2.4. Reactive Synaptogenesis (Axonal Sprouting) of Cholinergic Input to the Dentate Gyrus of Alzheimer's Disease Victims

In 1972 Lynch *et al.* reported that the density of histochemically visualized acetylcholinesterase (AChE) in the outer molecular layer of the rat dentate gyrus increased significantly after destruction of the ipsilateral entorhinal cortex. Destruction of the ipsilateral entorhinal cortex destroys the perforant pathway that provides cortical input to the dentate gyrus. Many of the terminals of cholinergic inputs from the medial septal nucleus to the dentate gyrus overlap the terminal zone of the perforant pathway. When the perforant pathway degenerates, the intact septal afferents "sprout" and occupy area previously occupied by the perforant pathway terminals (Lynch *et al.*, 1972; Stanfield and Cowan, 1982).

From its first presentation by Alzheimer (1907) AD has been viewed as a degenerative disorder—that is, a disease of decreased neuronal plasticity. More recently degeneration of the cholinergic forebrain has been specifically linked to AD. Therefore, the possibility that some form of reactive synaptogenesis might occur in the brains of AD victims was not considered until the middle of the 1980s when Geddes *et al.* (1985) and Hyman *et al.* (1987a) presented evidence that the cholinergic input to the dentate gyrus expanded in AD victims.

Although surprising when contrasted to the view that AD is simply a disease of cell loss, the observation of reactive synaptogensis in the septodentate cholinergic pathway is not surprising in light of the data presented by Hyman and Van Hoesen (1989) and their collaborators. The early and primary sites of degeneration in AD are the cells that give rise to the perforant pathway. The consequent denervation of the dentate gyrus causes unknown cellular events that stimulate reactive synaptogenesis of the axon terminals of the cholinergic neurons that remain in the medial septal nucleus. Thus, not only is the hippocampus isolated from the cortex in AD, but reactive synaptogenesis within the cholinergic inputs to the dentate gyrus results in formation of an aberrant subcortical input to the structure. Both the loss of neocortical input to hippocampus and the subsequent reactive synaptogenesis in the septohippocampal cholinergic input have been neglected in previous animal models of AD. The model proposed below includes both of these variables.

3. Summary of the Clinical Abnormalities Accompanying Alzheimer's Disease

An analogous model of a neurological disease must include not only some of the neuropathological markers of the disease, but also at least some of the behavioral and/or cognitive dysfunctions. The model of AD proposed below presents three behavioral deficits in the rat that parallel deficits seen in human AD victims. However, just as all of the neuropathological hallmarks of AD are not present in the model of the disease proposed below, several of the clinical aspects of the disease are also not present. Indeed, as may be determined from the following brief description of the disease, some of the clinical characteristics of AD can be considered to be uniquely human (e.g., aphasia and anomia) and not reproducible in an animal model.

The behavioral and cognitive abnormalities that are typically recognized as clinically diagnostic of AD may be divided into three stages (Hamdy, 1990). These stages are not discretely demarcated from one another, and the progression of the disease varies significantly from patient to patient.

The first stage of AD is characterized most markedly by working memory impairment that produces difficulties in the acquisition of new information. There may be some difficulty in finding the correct word to name any object in speech or writing (anomia). The visuospatial impairments that are noticeable in the later stages of AD are absent or minimal. There may be personality changes, but because the patient retains some insight and may react to his or her condition, personality changes that occur in Stage 1 may not be caused by the disease itself. Stage 1 of AD generally lasts from 1 to 3 years. The neurological exam, the results of brain-imaging procedures, and the electroencephalogram (EEG) are generally within normal limits.

The second stage of AD lasts from 2 to 10 years. During this period memory loss becomes increasingly severe and involves long-term (remote) memories, as well as recent memory. Insight into the condition is lost, and the ability of patients to groom themselves deteriorates. This latter difficulty may be secondary to the development of apraxia. Indeed, the development of two or more of the conditions of apraxia, agnosia, aphasia, and a worsening of anomia if this condition was present in Stage 1, or its development if absent from Stage 1, are defining characteristics of Stage 2 of AD. If apraxia and aphasia develop they usually appear in the later stages of Stage 2.

In addition to these disorders of higher cognitive function, the patient will demonstrate loss of judgment. Patients in Stage 2 will often become very restless and exhibit frequent pacing if confined in a room, or walk about continuously if permitted outside (so-called wandering). Other than some loss of acuity of smell, the basic sensory exam will be normal, as will the motor exam. However, brain-imaging techniques will generally reveal some brain atrophy and the EEG may be abnormal.

Finally, severe impairment of all cognitive functions, abnormal sensory and motor results in the neurological exam, and complete inability to care for oneself characterize Stage 3 of AD.

The model proposed below includes three of the behaviors that characterize the clinical manifestation of AD. These are impairment in putative working memory tasks, increased locomotor activity in home and novel environments, and disruption in normal grooming patterns. Additionally, as described in the next section, this model becomes analogous to AD because it also includes three of the neuropathological signs of AD, namely degeneration of part, but not all of the hippocampal circuit, degeneration within the association areas of the neocortex, and reactive synaptogenesis in the cholinergic septodentate pathway.

4. Effects of Trimethyltin on Brain and Behavior as an Analogous Model of Alzheimer's Disease

4.1. Caveats Concerning Predicting the Effects of Trimethyltin

At least two factors influence the consistency of the effects on brain and behavior of trimethyltin (TMT). These must be considered if the model of AD proposed in this chapter is used. First, the effects of TMT vary significantly with species. For example, Brown *et al.* (1984) report that the LD_{50} for TMT is approximately 3 mg/kg for hamsters, gerbils, and marmosets, but 12.6 mg/kg for the rat. These authors suggest that the fact that TMT exhibits significant binding to hemoglobin in rat, but not in the other species, may account for the observed differences in LD_{50} among the species.

However, there are also species differences in the pattern and severity of neuropathology associated with TMT, and these are not readily explained. For example, neurons in the retina, lateral geniculate nucleus, and visual cortex show marked degeneration in marmosets exposed to TMT, but these areas show little if any change in TMT-treated rats (Brown *et al.*, 1984). Further, the ventral (temporal) end of the dentate gyrus shows significant loss of granule cells in the rat, with relative sparing of the dorsal (septal) end of this part of the hippocampal formation, whereas fields CA1 and CA3c show greater loss of pyramidal cells in the dorsal, rather than the ventral, hippocampus in this species (Chang and Dyer, 1985). The mouse shows a pattern of damage to the hippocampus opposite to that of the rat (Chang *et al.*, 1983), whereas the dentate granule cells of the hamster appear to be unaffected by TMT (Brown *et al.*, 1984). Therefore, the model offered in this chapter is limited to the rat. The pattern of neuronal necrosis for the rat will be described in more detail below.

The second caveat concerning this model has to do with variation in potency of batches of TMT. O'Callaghan *et al.* (1989) have reported that different batches of TMT, even when obtained from the same supplier, exhibit vast differences in neurotoxicity. We have also observed this in our lab. Therefore, it may be necessary to order several batches of TMT and test each one for potency before beginning an experiment. However, once a potent batch of TMT has been identified, the shelf life appears to be almost indefinite if the material is stored over desiccant (O'Callaghan *et al.*, 1989, and personal observation).

4.2. Behavioral Consequences of Trimethyltin Exposure in Rats

The effects of TMT on the nervous system have been extensively studied. Case studies of the consequences of human exposure to TMT have been reported (e.g., Fortemps *et al.*, 1978; Ross *et al.*, 1981). Memory impairment, confusion, and agitation are the most prominent behavioral effects of TMT poisoning in humans. The behavioral consequences of TMT exposure in rats may be divided into acute and chronic effects. Beginning within 3 days after exposure to 6 or 7 mg/kg (calculated as free base weight) by intragastric gavage and continuing for between 20 and 30 days, rats given TMT demonstrate some or all of the following behaviors: tremor, seizures, tail mutilation, irritability, hypothermia, and weight loss (Dyer *et al.*, 1982, and personal observations). The severity of these effects depends on the age of the rat. Younger and therefore lighter-weight rats demonstrate less severe acute effects.

The acute effects typically subside during the second postexposure month, but the rats continue to show hyperactivity (Ruppert *et al.*, 1982; Cannon *et al.*, 1992), disrupted patterns of self-grooming (Cannon *et al.*, 1992), deficits in radial-arm and water-maze acquisition (Walsh *et al.*, 1982b; Woodruff *et al.*, 1991), impaired performance of the differential reinforcement of low response rates (DRL) operant schedule (Mastropaolo *et al.*, 1984; Woodruff *et al.*, 1991), and impaired acquisition of passive avoidance (Walsh *et al.*, 1982a). Both the acute and chronic behavioral effects of TMT resemble those described after damage to several of the structures of the limbic system (Isaacson, 1982), and the chronic effects have usually been interpreted (Dyer *et al.*, 1982; McMillan and Wenger, 1985) to most resemble those seen after damage to the hippocampus or its connections.

In summary, the long-term behavioral effects of TMT in rats include deficits in working memory, increased locomotor activity, and disruption of grooming. These are all behaviors that are also seen during early to mid-stage AD. Further, TMT-induced damage to the hippocampus, entorhinal cortex, and neocortical association areas appears to account for the behavioral deficits seen in TMT-exposed rats.

4.3. Patterns of Trimethyltin-Induced Neural Toxicity in Rats

The effects of TMT on the hippocampus have been the most often noted neuropathological consequences of this toxicant. Detailed observations by Chang and Dyer (1983) are representative of the literature in this area. These investigators followed the consequences of exposure to 6 mg/kg of TMT on the hippocampus of rats 8 hr, 3 days, 15 days, 30 days, and 60 days following exposure. Necrotic changes were found in the granule cells of the dentate gyrus at 3 days postexposure. No changes were observed in the pyramids of CA3a,b at this time, but CA3c and CA1 showed early signs of neuropathology. Evidence of necrosis

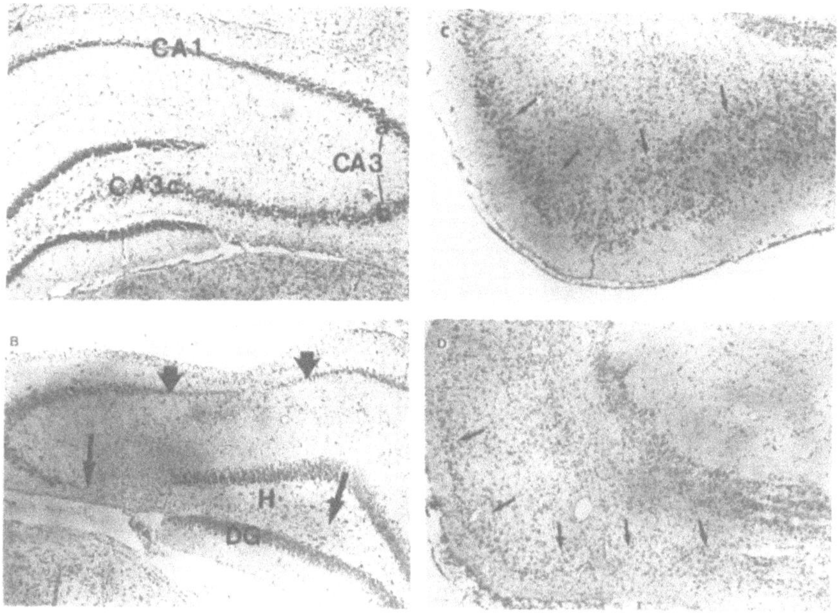

Figure 2 A thionin-stained section through the dorsal hippocampus of a normal rat is presented in (**A**). The subfields of Ammon's horn are labeled and the pattern of destruction produced by TMT is easily seen when the section in (**A**) is compared to a similar section from a rat given 6 mg/kg TMT (**B**). That part of CA3 between the arrows that normally occupies the hilus (H) of the dentate gyrus has been destroyed. CA1 is also substantially damaged (between arrowheads), but CA3a,b and the dentate granule cells (DG) are comparatively spared. TMT also damages layers II and IV of the entorhinal cortex (arrows in **C** and **D**) as can be seen when the thionin-stained section through the entorhinal cortex of a normal rat (**C**) is compared to a similar section from a TMT-treated rat (**D**).

in the granule cells subsided by 15 days and were essentially absent thereafter, but significant loss of the pyramidal cells in CA1 and CA3c was evident at the later time-points. At day 60, active signs of neuropathology (e.g., swollen neurons, chromatolysis) were virtually absent, but cell loss from CA1 and CA3c was extensive, especially in the septal half of the hippocampus (Figure 2A,B). The pattern of damage to the hippocampus just described has been confirmed by Balaban *et al.* (1988) and Whittington *et al.* (1989) using silver stains. Additionally, and of importance for development of this model, Naalsund *et al.* (1985) and Whittington *et al.* (1989) have noted degenerating neurons within the subiculum using the Fink-Heimer stain.

In addition to the hippocampus, areas of the cortex related to the limbic system also are extensively damaged by TMT. For example, both the entorhinal and pyriform cortex suffer cell loss (Balaban *et al.*, 1988; Brown *et al.*, 1979;

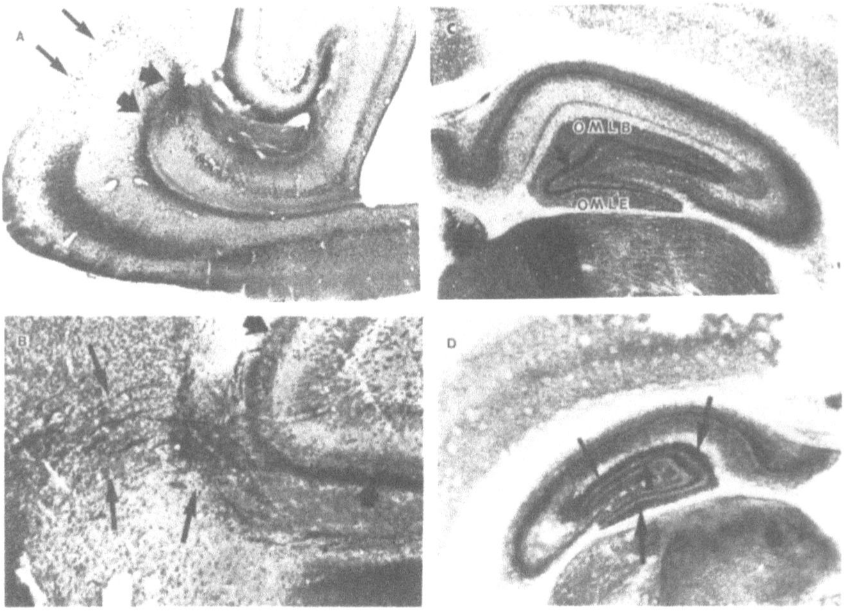

Figure 3 As can be seen in the photomicrographs of Fink-Heimer–stained sections presented in (**A**) and (**B**), the destruction produced in the entorhinal cortex by TMT causes necrosis of layer II neurons of the entorhinal cortex (longer arrows in **A** point to argyrophilic neurons) and degeneration of the perforant pathway (arrowheads in **A** and longer arrows in **B**). The degeneration of the terminals of the perforant pathway (arrowheads in **B**) in the outer molecular layer of the dentate gyrus causes an increase in the staining density of AChE (arrows in **D**), a putative indication of reactive synaptogenesis of the cholinergic input from the medial septal nucleus/diagonal band to this area. As can be seen from comparing the photomicrograph of an AChE-stained section from a normal rat to a similar section from a TMT-treated rat, AChE staining does not increase in the infragranular region of the inner molecular layer (arrowheads in **C** and **D**), but increases in the outer molecular layer of both the buried (OMLB) and exposed (OMLE) blades of the dentate gyrus.

Bouldin *et al.*, 1981; Chang and Dyer, 1983; Whittington *et al.*, 1989). Neuronal loss within the entorhinal cortex is especially prevalent in the cells of layer II (Figure 2C,D), and as a consequence the perforant pathway demonstrates substantial degeneration (Figure 3A,B). Neurons of the olfactory tubercle also degenerate after exposure to TMT (Balaban *et al.*, 1988; Chang and Dyer, 1983).

The neuron loss in the areas just described is sufficient evidence that more subtle losses produced by TMT in neuron populations elsewhere in the neocortex may be overlooked. However, when counts of neurons are made in parietal and frontal association areas, statistically significantly fewer than normal numbers of neurons are found in these regions (Cannon *et al.*, 1992). Thus, while the ability of TMT to destroy parts of the hippocampus is most noticeable, this toxicant also causes loss of neurons in other cortical areas.

4.4. Reactive Synaptogenesis within the Cholinergic Septodentate Projection after Exposure to Trimethyltin

As indicated above, the cholinergic innervation of the dentate gyrus increases in AD victims, at least in those that come to autopsy in the early stage of the disease. Loss of neurons that give rise to the perforant path, the terminals of which overlap the distribution of the terminal zone of the cholinergic afferents, presumably sets the occasion for reactive synaptogenesis in the cholinergic projection from the medial septal nucleus/diagonal band to the outer molecular layer of the dentate gyrus. The occurrence of reactive synaptogenesis in this system also implies that degeneration of projection neurons of the entorhinal cortex precede degeneration of cholinergic septal/diagonal-band neurons.

Reactive synaptogenesis also occurs within the cholinergic innervation of the outer molecular layer (OML) of the dentate gyrus of the rat after exposure to TMT. Hagan *et al.* (1988) first noted that with visual inspection AChE staining in the OML appeared to be intensified in rats given an oral dose of 7 mg/kg TMT. Woodruff and Baisden (1990) used densitometric analysis to quantify the amount of histochemical staining for AChE in control rats and rats given 6 mg/kg TMT, and they found a significant increase staining density in the OML of the TMT-treated rats (Figure 3C,D). The proposal that this increase in AChE staining was attributable to increased cholinergic innervation within the OML was strengthened by observations that choline acetyltransferase (ChAT) activity in the OML increased significantly in TMT-treated rats (Cannon *et al.*, 1991; Naalsund *et al.*, 1985). Therefore, although TMT does not produce the decrease in cholinergic markers generally found in the basal forebrain and cortex of AD victims, it does produce expansion of the cholinergic innervation of the OML of the dentate gyrus similar to that reported to occur in at least early-stage AD victims.

5. Conclusions and Restrictions of the Model

5.1. Summary of the Analogy between the Effects of Trimethyltin on the Rat and Alzheimer's Disease

The effects of a neurotoxic dose of TMT on the rat brain and of the degenerative processes of AD share the following characteristics. (1) Both are progressive. Unlike an acute lesion of the basal forebrain that destroys the nucleus basalis, the neuropathological changes produced by TMT continue for as long as 60 days after one exposure (Whittington *et al.*, 1989). Therefore, the effectiveness of nootropic drugs intended to slow progressive behavioral impairment could be tested at varying times after exposure to TMT. (2) The hippocampus is largely isolated from neocortex in both cases as both the principal neurons of the entorhinal cortex, subfield CA1 of the hippocampus, and the subiculum are targets of both forms of neuropathology. However, the dentate gyrus and much of CA3 is spared. (3) Other limbic areas, both neocortical (e.g., orbital and pyriform cortices) and subcortical (the basolateral amygdala), suffer neuron loss. (4) Motor and sensory cortical areas are far less affected than the association cortex. (5) The cholinergic septodentate projection exhibits reactive synaptogenesis.

At least three behavioral changes seen after TMT exposure may be construed as being analogous to clinical signs of AD that appear during the first two stages of the disease. Working memory is impaired in both AD and the rat exposed to TMT. Although no motor abnormalities are seen in the first two stages of AD, locomotor activity often increases significantly. Rats given TMT also exhibit increased locomotor activity. Finally, the ability to self-groom effectively is disrupted.

5.2. Limitations of the Model

Because several of the neuropathological and behavioral consequences of TMT exposure in the rat parallel changes in these measures in human AD victims, we believe that TMT toxicity in the rat is an analogous model to Alzheimer's disease. However, there is no doubt that the model has substantial limitations. First, amyloid-containing plaques are not only one of the traditional neuropathological markers for AD, but amyloid deposition as the cause of plaque formation is one of the best-supported hypotheses offered to explain the etiology of neuronal death seen in AD (e.g., Selkoe, 1991), and exposure of rats to tin compounds does not cause amyloid-containing plaques to appear in the brain. Nor are NFTs found in the brains of rats exposed to TMT. Thus, the two primary cellular correlates of AD are not produced by TMT. This indicates that study of this model is not likely to increase understanding of the etiology of AD.

Second, the mechanism by which TMT causes neuron death is not known. This makes it difficult to draw parallels to factors that might cause neuron death in AD. Third, although markers for some transmitters decrease after TMT exposure (McMillan and Wenger, 1985), the cholinergic neurons of the basal forebrain are not appreciably damaged by this toxicant, whereas the cholinergic deficit in AD is profound and related to the memory impairment observed in its victims. Fourth, the match between the distribution of neuron loss in hippocampus and neocortex produced in the rat by TMT and that which occurs in humans suffering from AD is far from perfect. Finally, unlike the conditioned eyeblink response used by Solomon *et al.* (this volume), the behaviors disrupted by TMT in the rat are not the exact same behaviors disrupted in humans by Alzheimer's disease.

Despite these limitations we conclude this chapter by suggesting that the TMT model in the rat might make a relatively inexpensive (as compared to aged rats) model for preclinical testing of proposed nootropic drugs or other potential treatments for AD such as neurotransplantation.

ACKNOWLEDGMENTS. Preparation of this chapter was supported by a grant to M.L.W. from the National Institute of Environmental Health Sciences.

6. References

Alzheimer, A., 1907, Uber eine eigenartige Erkankung der Hirnude Algemeine, *Z. Psychiatr.* 64:146–148.

Balaban, C.D., O'Callaghan, J.P., and Billingsley, M.L., 1988, Trimethyltin-induced neuronal damage in the rat brain: Comparative studies using silver degeneration stains, immunocytochemistry and immunoassay for neuronotypic and gliotypic proteins, *Neuroscience* 26:337–361.

Ball, M.J., and Lo, P., 1977, Granulovacuolar degeneration in the ageing brain and in dementia, *J. Neuropathol. Exp. Neurol.* 36:474–487.

Bouldin, T.W., Goines, N.D., Bagnell, C.R., and Krigman, M.R., 1981, Pathogenesis of trimethyltin neuronal toxicity: Ultrastructural and cytochemical observations, *Am. J. Pathol.* 104:237–249.

Braak, H., and Braak, E., 1986, Ratio of pyramidal cells versus non-pyramidal cells in the human frontal neocortex and changes in ratio with ageing and Alzheimer's disease, *Prog. Brain Res.* 40:185–212.

Brown, A.W., Aldridge, W.N., Street, B.W., and Verschoyle, R.D., 1979, The behavioral and neuropathologic sequelae of intoxication by trimethyltin compounds in the rat, *Am. J. Pathol.* 97:59–82.

Brown, A.W., Verschoyle, R.D., Street, B.W., Aldridge, W.N., and Grindley, H., 1984, The neurotoxicity of trimethyltin chloride in hamsters, gerbils and marmosets, *J. Appl. Toxicol.* 4:12–21.

Cannon, R.L., Hoover, D.B., and Woodruff, M.L., 1991, Trimethyltin increases choline acetyltransferase in rat hippocampus, *Neurotoxicol. Teratol.* 13:241–244.

Cannon, R.L., Paul, D.J., Baisden, R.H., and Woodruff, M.L., 1992, Alterations in self-grooming sequences in the rat as a consequence of hippocampal damage, *Psychobiology* 20:205–218.

Chang, L.W., and Dyer, R.S., 1983, A time-course study of trimethyltin-induced neuropathology in rats, *Neurobehav. Toxicol. Teratol.* 5:443–459.

Chang, L.W., and Dyer, R.S., 1985, Septotemporal gradients of trimethyltin-induced hippocampal lesions, *Neurobehav. Toxicol. Teratol.* 7:43–49.

Chang, L.W., Wenger, G.R., McMillan, D.E., and Dyer, R.S., 1983, Species and strain comparison of acute neurotoxic effects of trimethyltin in mice and rats, *Neurobehav. Toxicol. Teratol.* 5:337–350.

Dyer, R.S., Walsh, T.J., Wonderlin, W.F., and Bercegeay, M., 1982, The trimethyltin syndrome in rats, *Neurobehav. Toxicol. Teratol.* 4:127–133.

Fortemps, E., Amand, G., Bombir, A., Lauwerys, R., and Laterre, E.C., 1978, Trimethyltin poisoning: Report of two cases, *Int. Arch. Occup. Environ. Health* 41:1–6.

Gage, F.H., Björklund, A., Isacson, O., and Brundin, P., 1985, Uses of neurotransplantation in animal models of neurodegenerative diseases, *in* "Neural Transplantation and Regeneration" (G.D. Das and B. Wallace, eds.), pp. 103–124, Springer-Verlag, New York.

Geddes, J.W., Monaghan, D.T., Cotman, C.W., Lott, I.T., Kim, R.C., and Chui, H.C., 1985, Plasticity of hippocampal circuitry in Alzheimer's disease, *Science* 230:1179–1181.
Goldman, J.E., 1983, The association of actin with Hirano bodies, *J. Neuropathol. Exp. Neurol.* 42:146–152.
Hagan, J.J., Jansen, J.H.M., and Broekkamp, C.L.E., 1988, Selective behavioral impairment after acute intoxication with trimethyltin (TMT) in rats, *Neurotoxicology* 9:53–74.
Hamdy, R., 1990, Clinical presentation, *in* "Alzheimer's Disease: A Handbook for Caregivers" (R.C. Hamdy, J.M. Turnbull, L.D. Norman, and M.M. Lancaster, eds.), pp. 3–15, C.V. Mosby, St. Louis, MO.
Hirano, A., 1965, Pathology of amyotropic lateral sclerosis, *in* "Slow, Latent and Temperate Virus Infections" (D.C. Gajdusek, C.J. Gibbs, and M. Alpers, eds.), pp. 23–27, National Institute of Neurological Diseases and Blindness, Bethesda, MD.
Hyman, B.T., and Van Hoesen, G.W., 1989, Hippocampal and entorhinal cortex cellular pathology in Alzheimer's disease, *in* "The Hippocampus: New Vistas" (V. Chan-Palay and C. Köhler, eds.), pp. 449–512, Alan R. Liss, New York.
Hyman, B.T., Kromer, L.J., and Van Hoesen, G.W., 1987a, Reinnervation of the hippocampal perforant pathway zone in Alzheimer's disease, *Ann. Neurol.* 21:259–267.
Hyman, B.T., Van Hoesen, G.W., and Damasio, A.R., 1987b, Alzheimer's disease: Glutamate depletion in perforant pathway terminals, *Ann. Neurol.* 22:37–40.
Isaacson, R.L., 1982, "The Limbic System" (2nd ed.), Plenum Press, New York.
Lewis, D.A., Campbell, M.J., Terry, R.D., and Morrison, J.H., 1987, Laminar and regional distributions of neurofibrillary tangles and neuritic plaques in Alzheimer's disease: A quantitative study of visual and auditory cortices, *J. Neurosci.* 7:1799–1808.
Lynch, G., Matthews, G.B., Mosko, S., Parks, T., and Cotman, C., 1972, Induced acetylcholinesterase-rich layer in rat dentate gyrus following entorhinal lesions, *Brain Res.* 42:311–318.
Mann, D.M.A., Yates, P.O., and Marcyniuk, B., 1985, Some morphometric observations on the cerebral cortex and hippocampus in presenile Alzheimer's disease, senile dementia of the Alzheimer type and Down's syndrome in middle age, *J. Neurol. Sci.* 69:139–159.
Mastropaolo, J.P., Dacanay, R.J., Luna, B.H., Tuck, D.L., and Riley, A.L., 1984, Effects of trimethyltin chloride on differential-reinforcement-of-low-rate responding, *Neurobehav. Toxicol. Teratol.* 6:193–199.
McKinney, W.T., and Bunney, W.E., 1969, Animal model of depression: Review of evidence and implications for research, *Arch. Gen. Psychiatry* 21:240–248.
McMillan, D.E., and Wenger, G.R., 1985, Neurobehavioral toxicology of trialkyltins, *Pharmacol. Rev.* 37:365–379.
Naalsund, J.V., Allen, C.N., and Fonnum, F., 1985, Changes in neurobiological parameters in the hippocampus after exposure to trimethyltin, *Neurotoxicology* 6:145–158.
O'Callaghan, J.P., Niedzwiecki, D.M., and Means, J.C., 1989, Variations in the neurotoxic potency of trimethyltin, *Brain Res. Bull.* 22:637–642.
Price, D.L., Struble, R.G., Altschuler, R.J., Casanova, M.F., Cork, L., and Murphy, D.B., 1985, Aggregation of tubulin in neurons in Alzheimer's disease, *J. Neuropathol. Exp. Neurol.* 44:366.
Roberts, E., 1986, Alzheimer's disease may begin in the nose and may be caused by aluminosilicates, *Neurobiol. Aging* 7:561–567.
Ross, W.D., Emmett, E.A., Steiner, J., and Tureen, R., 1981, Neurotoxic effects of occupational exposure to organotins, *Am. J. Psychiatry* 138:1092–1095.
Ruppert, P.H., Walsh, T.J., Reiter, L.W., and Dyer, R.S., 1982, Trimethyltin-induced hyperactivity: Time course and pattern, *Neurobehav. Toxicol. Teratol.* 4:135–139.
Selkoe, D.J., 1991, The molecular pathology of Alzheimer's disease, *Neuron* 6:487–498.
Smith, G., 1988, Animal models of Alzheimer's disease: Experimental cholinergic denervation, *Brain Res. Rev.* 13:103–118.

Stanfield, B.B., and Cowan, W.M., 1982, The sprouting of septal afferents to the dentate gyrus after lesions of the entorhinal cortex of adult rats, *Brain Res.* 232:162–170.

Terry, R.D., 1986, Does Alzheimer's disease spread, and is it causally related to aluminum?, *Neurobiol. Aging* 7:570.

Treit, D., 1985, Animal models for the study of anti-anxiety agents: A review, *Neurosci. Biobehav. Rev.* 9:203–222.

Vogt, B.A., VanHoesen, G.W., and Vogt, L.J.,.1990, Laminar distribution of neuron degeneration in posterior cingulate cortex in Alzheimer's disease, *Acta Neuropathol.* 80:581–58.

Walsh, T.J., Gallagher, M., Bostock, E., and Dyer, R.S., 1982a, Trimethyltin impairs retention of a passive avoidance task, *Neurobehav. Toxicol. Teratol.* 4:163–167.

Walsh, T.J., Miller, D.B., and Dyer, R.S., 1982b, Trimethyltin, a selective limbic system neurotoxicant, impairs radial-arm maze performance, *Neurobehav. Toxicol. Teratol.* 4:177–183.

Ward, C.D., 1986, Commentary on "Alzheimer's disease may begin in the nose and may be caused by aluminosilicates," *Neurobiol. Aging* 7:574–575.

Whittington, D.L., Woodruff, M.L., and Baisden, R.H., 1989, The time course of trimethyltin-induced fiber and terminal degeneration in hippocampus, *Neurotoxicol. Teratol.* 11:21–33.

Woodruff, M.L., and Baisden, R.H., 1990, Exposure to trimethyltin significantly enhances acetylcholinesterase staining in the rat dentate gyrus, *Neurotoxicol. Teratol.* 12:33–39.

Woodruff, M.L., Baisden, R.H., and Nonneman, A.J., 1991, Anatomical and behavioral sequelae of fetal brain transplants in rats with trimethyltin-induced neurodegeneration, *Neurotoxicology* 12:427–444.

Index

FSC
www.fsc.org
MIX
Papier | Fördert gute Waldnutzung
FSC® C083411

Kolibri 360 GmbH
Ferdinand-Jühlke-Straße 7
99095 Erfurt, Deutschland
produktsicherheit@kolibri360.de